W0246196

IUTAM SYMPOSIUM ON OPTIMIZATION OF MECHANICAL SYSTEMS

SOLID MECHANICS AND ITS APPLICATIONS
Volume 43

Series Editor: **G.M.L. GLADWELL**
Solid Mechanics Division, Faculty of Engineering
University of Waterloo
Waterloo, Ontario, Canada N2L 3G1

Aims and Scope of the Series

The fundamental questions arising in mechanics are: *Why?*, *How?*, and *How much?* The aim of this series is to provide lucid accounts written by authoritative researchers giving vision and insight in answering these questions on the subject of mechanics as it relates to solids.

The scope of the series covers the entire spectrum of solid mechanics. Thus it includes the foundation of mechanics; variational formulations; computational mechanics; statics, kinematics and dynamics of rigid and elastic bodies; vibrations of solids and structures; dynamical systems and chaos; the theories of elasticity, plasticity and viscoelasticity; composite materials; rods, beams, shells and membranes; structural control and stability; soils, rocks and geomechanics; fracture; tribology; experimental mechanics; biomechanics and machine design.

The median level of presentation is the first year graduate student. Some texts are monographs defining the current state of the field; others are accessible to final year undergraduates; but essentially the emphasis is on readability and clarity.

For a list of related mechanics titles, see final pages.

IUTAM Symposium on
Optimization of Mechanical Systems

Proceedings of the IUTAM Symposium
held in Stuttgart, Germany,
26–31 March 1995

Edited by

D. BESTLE

Department of Machine Dynamics,
Technical University of Cottbus,
Germany

and

W. SCHIEHLEN

Institute B of Mechanics,
University of Stuttgart,
Germany

KLUWER ACADEMIC PUBLISHERS
DORDRECHT / BOSTON / LONDON

A C.I.P. Catalogue record for this book is available from the Library of Congress.

ISBN-13: 978-94-010-6555-9   e-ISBN-13: 978-94-009-0153-7
DOI: 10.1007/978-94-009-0153-7

Published by Kluwer Academic Publishers,
P.O. Box 17, 3300 AA Dordrecht, The Netherlands.

Kluwer Academic Publishers incorporates
the publishing programmes of
D. Reidel, Martinus Nijhoff, Dr W. Junk and MTP Press.

Sold and distributed in the U.S.A. and Canada
by Kluwer Academic Publishers,
101 Philip Drive, Norwell, MA 02061, U.S.A.

In all other countries, sold and distributed
by Kluwer Academic Publishers Group,
P.O. Box 322, 3300 AH Dordrecht, The Netherlands.

All Rights Reserved
© 1996 Kluwer Academic Publishers
Softcover reprint of the hardcover in 1st edition 1996

No part of the material protected by this copyright notice may be reproduced or
utilized in any form or by any means, electronic or mechanical,
including photocopying, recording or by any information storage and
retrieval system, without written permission from the copyright owner.

CONTENTS

PREFACE

The International Union of Theoretical and Applied Mechanics (IUTAM) initiated and sponsored an International Symposium on Optimization of Mechanical Systems held in 1995 in Stuttgart, Germany. The Symposium was intended to bring together scientists working in different fields of optimization to exchange ideas and to discuss new trends with special emphasis on multibody systems.

A Scientific Committee was appointed by the Bureau of IUTAM with the following members:

> S. Arimoto (Japan)
> F.L. Chernousko (Russia)
> M. Géradin (Belgium)
> E.J. Haug (U.S.A.)
> C.A.M. Soares (Portugal)
> N. Olhoff (Denmark)
> W.O. Schiehlen (Germany, Chairman)
> K. Schittkowski (Germany)
> R.S. Sharp (U.K.)
> W. Stadler (U.S.A.)
> H.–B. Zhao (China)

This committee selected the participants to be invited and the papers to be presented at the Symposium. As a result of this procedure, 90 active scientific participants from 20 countries followed the invitation, and 49 papers were presented in lecture and poster sessions.

The scientific lectures were devoted to the following topics:

- o Modeling and Optimization of Multibody Systems
- o Design of Mechanisms
- o Sensitivity Analysis of Multibody Systems
- o Optimization of Mechanical Systems
- o Optimal Motion Control
- o Optimization in Structural Mechanics
- o Optimal Design of Materials
- o Applications in Engineering

○ Selected Problems in Optimization
○ Optimization Methods and Algorithms
○ Multicriteria Optimization

Since many of the presentations are related to more than one of these topics, the papers in this volume are arranged in alphabetical order with respect to the first author. The papers indicate the wide scope of engineering applications of optimization methods and show some first steps of using sensitivity analysis and optimization in the design of dynamic systems. The presentations and discussions during the Symposium will certainly stimulate further theoretical and applied investigations in this challenging field of optimization. The publication of the proceedings may promote this development.

Generous financial and material support contributed to the success of the Symposium. The help of the following sponsors is gratefully acknowledged:

International Union of Theoretical and Applied Mechanics
Deutsche Forschungsgemeinschaft, Bonn
International Science Foundation, Washington
Kluwer Academic Publishers, Dordrecht
Dr.–Ing. h.c. F. Porsche AG, Stuttgart
Ford–Werke AG, Köln
Ingenieurgesellschaft für Angewandte Technologie mbH, Pöcking
Krupp Maschinentechnik GmbH, Essen
Paurat GmbH, Voerde
Robert Bosch GmbH, Stuttgart
City of Stuttgart
University of Stuttgart, Stuttgart

The success of the Symposium would not have been possible without the excellent work of the Local Organizing Committee. Members of the Committee were:

W.O. Schiehlen (Chairman), R. Prommersberger (Secretary),
D. Bestle, P. Eberhard, A. Eiber, L. Gaul, E. Ramm, K.H. Well.

In addition, sincere thanks are due to all the members of the Institute B of Mechanics of the University of Stuttgart, all of them contributed to the success of the Symposium. Furthermore, thanks are due to Kluwer Academic Publishers for the efficient cooperation.

Stuttgart, July 1995   Dieter Bestle
                                                                      Werner Schiehlen

LIST OF PARTICIPANTS

(Chairmen are identified by an asterisk)

Ambrosio, J., IDMEC/IST, Av. Rovisco Pais, 1096 Lisboa Codex, PORTUGAL

Avello, A. , Dipartimento Mecanica Aplicada, CEIT., P. Manuel Lardizabal, 15, 20009 San Sebastian, SPAIN

Balandin, D.V., Res. Institute for Math. and Cyb., Nizhny Novgorod State University, 10 Uljanova Str., Nizhny Novgorod, 603 005, RUSSIA

Banerjee, A., Institut für Mechatronik, Bergwerkstraße, 47445 Moers, GERMANY

Banichuk, N., Dept. Solid Mech. & Struct. Design, Inst. for Problems in Mech., Vernadskogo, 101, Moscow 117526, RUSSIA

Barthold, F.-J., Inst. f. Baumechanik u. Num. Mech., Universität Hannover, Appelstr. 9A, 30167 Hannover, GERMANY

Beiner, L., Center for Techn. Education, P.O. Box 305, 58102 Holon, ISRAEL

Bendsoe, M.P., Mathematical Institute, Building 303, The Technical University of Denmark, DK – 2800 Lyngby, DENMARK

Berbyuk, V., Inst. of Appl. Probl. in Mech. and Math., Ukrainien Academy of Sciences, Naukova Str. 3–B, Lviv, 290601, UKRAINE

Bestle, D., Institut B für Mechanik, Universität Stuttgart, Pfaffenwaldring 9, 70550 Stuttgart, GERMANY

Bischof, C., Math. and Comp. Science Division, Argonne National Laboratory, 9700 S. Cass. Ave., Argonne, IL 60439, U.S.A.

Bletzinger, K.-U., Institut für Baustatik, Universität Stuttgart, Pfaffenwaldring 7, 70550 Stuttgart, GERMANY

Bolotnik, N.N., Institute for Problems in Mechanics, The Russian Academy of Sciences, pr. Vernadskogo 101, Moscow 117526, RUSSIA

Brauchli, H., Institut für Mechanik, HG F 41.4, ETH Zentrum, 8092 Zürich, SWITZERLAND

* **Campen, D.H. van**, Mechanical Engineering Dept., Technical University Eindhoven, P.O. Box 513, 5600 MB Eindhoven, THE NETHERLANDS

* **Chernousko, F.L.**, Institute for Problems in Mechanics, The Russian Academy of Sciences, pr. Vernadskogo 101, Moscow 117526, RUSSIA

Choi, D.H., Dept. of Mech. Design & Prod. Engineering, Hanyang University, Haengdang–Dong 17, Sungdong–Ku, Seoul, SOUTH KOREA 133–791

Dobrynina, I.S., Institute for Problems in Mechanics, Russian Acad. of Science, pr. Vernadskogo 101, Moscow 117526, RUSSIA

Eberhard, P., Institut B für Mechanik, Universität Stuttgart, Pfaffenwaldring 9, 70550 Stuttgart, GERMANY

Eiber, A., Institut B für Mechanik, Universität Stuttgart, Pfaffenwaldring 9, 70550 Stuttgart, GERMANY

* **Eschenauer, H.**, FOMAAS, Universität Siegen, Paul Bonatzstr. 9–11, 57076 Siegen, GERMANY

Formal'sky, A.M., Institute of Mechanics, Moscow State University, Michurinsky pr. 1, Moscow 119899, RUSSIA

Gaul, L., Institut A für Mechanik, Universität Stuttgart, Pfaffenwaldring 9, 70550 Stuttgart, GERMANY

Géradin, M., LTAS – Dynamique des Constructions Mécaniques, Institut de Mécanique, 21, Rue E. Solvay, 4000 Liège, BELGIUM

Gordon, T.J., Dept. of Aeronautical and Automotive Engineering, University of Technology, Loughborough, Leics, LE11 3TU, U.K.

* **Griewank, A.**, TU Dresden, Institut für wiss. Rechnen, Mommsenstraße 13, 01062 Dresden, GERMANY

Guest, S.D., Dept. of Engineering, University of Cambridge, Trumpington Street, Cambridge, CB2 1PZ, U.K.

Gumpert, W., Franz–Mehring Str. 6, 09112 Chemnitz, GERMANY

Hammer, V.B., Dept. of Solid Mechanics, Technical University of Denmark, Building 404, 2800 Lyngby, DENMARK

Hansen, J.M., Dept. of Solid Mechanics, Building 404, The Technical University of Denmark, 2800 Lyngby, DENMARK

Hansen, M.R., Institute of Mech. Eng., University of Aalborg, Pontoppidanstraede 101, 9220 Aalborg East, DENMARK

Hilf, K.–D., IWR, Im Neuenheimer Feld 368, 69120 Heidelberg, GERMANY

Horak, J., Dept. of Machine Design, Fac. of Mech. Eng.,West Bohemia University, Americka 42, 31600 Plzen, CZECH REPUBLIC

Johnson, A., Sundial Engineering, 3073 Bateman St., Berkeley, CA 94705, U.S.A.

Kim, M.–S., Dept. of Mech. Design & Prod. Engineering, Hanyang University, Haengdang–Dong 17, Sungdong–Ku, Seoul, SOUTH KOREA 133–791

Kiriazov, P., Institute of Mechanics, Bulg. Acad. of Sciences, Acad. G. Bonchev Str., bl. 4, 1113 Sofia, BULGARIA

Koski, J., Tampere University of Technology, Department of Mechanical Engineering, P.O. Box 589, 33101 Tampere, FINLAND

Kraft, D., Labor für Regelungs– und Steuerungstechnik, Fachhochschule München, Dachauer Straße 98b, 80335 München, GERMANY

Krause, R., Daimler Benz AG, Goldsteinstr. 235, 60528 Frankfurt, GERMANY

Krog, L.A., Dept. of Mechanical Engineering, Aalborg University, Pontoppidanstraede 101, 9220 Aalborg East, DENMARK

Kust, O., Meerestechnik II, Technische Universität Hamburg–Harburg, Eissendorfer Straße 42, 21071 Hamburg, GERMANY

Lauritzen, S., Institute of Mech. Eng., University of Aalborg, Pontoppidanstraede 101, 9220 Aalborg East, DENMARK

Lennartsson, A., Dept. of Mechanics, Royal Institute of Technology, Lindstedtsvagen 25, 10044 Stockholm, SWEDEN

Lund, E., Institute of Mech. Eng., University of Aalborg, Pontoppidanstraede 101, 9220 Aalborg, DENMARK

Markine, V.L., Lab. of Eng. Mech. & Marine Tech., Delft University of Technology, P.O. Box 5033, 2600 GA Delft, THE NETHERLANDS

Marti, K., Inst. für Math. und Rechneranw., Univ. der Bundeswehr München, Werner–Heisenberg Weg 39, 85577 Neubiberg, GERMANY

McPhee, J., Systems Design Eng., University of Waterloo, Waterloo, Ontario N2L 3G1, CANADA

* **Meijers, P.**, Lab. of Mech. Eng. & Marine Tech., Delft University of Technology, P.O. Box 5033, 2600 GA Delft, THE NETHERLANDS

Moder, T., Lehrstuhl f. Höhere und Num. Math., TU München, 80290 München, GERMANY

Müller, P.C., Sicherheitstechnische Meß– und Regelungstechnik, Universität Wuppertal, Gaußstraße 20, 42097 Wuppertal, GERMANY

Mullineux, G., Dept. of Manuf. and Eng. Sys., Brunel University, Uxbridge, Middlesex UB8 3PH, U.K.

Olhoff, N., Dept. of Mechanical Engineering, Aalborg University, Pontoppidanstraede 101, 9220 Aalborg East, DENMARK

Pagalday, J.M., Dipartimento Mecanica Aplicada, CEIT., P. Manuel Lardizabal, 15, 20009 San Sebastian, SPAIN

* **Pedersen, P.**, Dept. of Solid Mechanics, Building 404, The Technical University of Denmark, 2800 Lyngby, DENMARK

Pellegrino, S., Dept. of Engineering, University of Cambridge, Trumpington Street, Cambridge, CB2 1PZ, U.K.

Pesch, V., Ford Motor Co., 21175 Oakwood Blvd., Dearborn, MI 48124, U.S.A.

Petrov, E.P., Shilovskij v'ezd 5/12a, Kharkov, 310013, UKRAINE

* **Pfeiffer, F.**, Lehrstuhl B für Mechanik, TU München, Arcisstraße 21, 80333 München, GERMANY

* **Popp, K.**, Institut für Mechanik, Universität Hannover, Appelstraße 11, 30167 Hannover, GERMANY

Ptchelintsev, A.V., Building Research Institute, 1 Tatehara, Tsukuba–City, Ibaraki, Prefecture 305, JAPAN

* **Ramm, E.**, Institut für Baustatik, Universität Stuttgart, Pfaffenwaldring 7, 70550 Stuttgart, GERMANY

Rauh, J., Forschung Fahrzeugdynamik, F1M/SD, c/o Mercedes Benz AG, E222, 70322 Stuttgart, GERMANY

Ringertz, U., Dept. of Lightweight Structures, Royal Institute of Technnology, 100 44 Stockholm, SWEDEN

Roytman, A.B., P.O.B. 1018, Zaporozhye 330104, UKRAINE

Rozvany, G., Universität Essen, FB 10, Postfach 10 37 64, 45117 Essen, GERMANY

Schiehlen, W.O., Institut B für Mechanik, Universität Stuttgart, Pfaffenwaldring 9, 70550 Stuttgart, GERMANY

* **Schittkowski, K.**, Mathematisches Institut, Universität Bayreuth, Universitätsstraße 30, 95440 Bayreuth, GERMANY

Schulz, M., Institut für Mechanik, ETH Zentrum, 8092 Zürich, SWITZERLAND

Schwerin, R. von, IWR, Im Neuenheimer Feld 368, 69120 Heidelberg, GERMANY

* **Sharp, R.S.**, Automotive Studies Group, School of Mechanical Engineering, Cranfield University, Bedford MK 43 OAL, U.K.

Sigmund, O., Dept. of Solid Mechanics, Building 404, The Technical University of Denmark, 2800 Lyngby, DENMARK

Slagmaat, M.T.P. van, TNO Road Vehicles Res. Institute, P.O. Box 6033, 2600 JA Delft, THE NETHERLANDS

* **Snyman, J.A.**, Dept. of Mechanical Eng., University of Pretoria, Pretoria 0002, SOUTH AFRICA

* **Soares, C.A.M.**, CEMUL, Instituto Superior Técnico, Universidade Técnico de Lisboa, Av. Rovisco Pais, 1096 Lisboa Codex, PORTUGAL

* **Stadler, W.**, Division of Engineering, San Francisco State University, 1600 Holloway Avenue, San Francisco, CA 94132, U.S.A.

Statnikov, R.B., Profsoyuznaga Str. 43, block, apt 370, Moscow 117420, RUSSIA

Tada, Y., Dept. of Comp. and Sys. Eng., Faculty of Engineering, Kobe University, Rokkodai, Nada, Kobe 657, JAPAN

Taylor, J.E., Dept. of Aerospace Eng., University of Michigan, Ann Arbor, Michigan 48109–2140, U.S.A.

Tortorelli, D., Dept. of Mech. and Ind. Eng., University of Illinois at Urbana–Champaign, 1206 W. Green Street, Urbana, IL 61801, U.S.A.

Twyman, B.A., Dept. of Manuf. and Eng. Sys., Brunel University, Uxbridge, Middlesex UB8 3PH, U.K.

Valasek, M., Dept. of Mechanics, Fac. of Mech. Eng., Czech Technical University of Prague, Karlovo nam. 13, 12135 Praha 2, CZECH REPUBLIC

Wang, S., 306 IATL, University of Iowa, Iowa City, Iowa 52242, U.S.A.

Weber, C., FOMAAS, Universität Siegen, Paul Bonatzstr. 9–11, 57076 Siegen, GERMANY

Weber, C.-T., Institut für Mechanik, Otto–von–Guericke–Univ. Magdeburg, Universitätsplatz 2, 39106 Magdeburg, GERMANY

* **Well, K.H.**, Inst. für Flugmechanik und Flugregelung, Universität Stuttgart, Forststr. 86, 70176 Stuttgart, GERMANY

Wentscher, H., Inst. für Robotik und Systemdynamik, DLR Oberpfaffenhofen, Postfach 1116, 82230 Wessling, GERMANY

Wimmer, J., Forschung Fahrzeugdynamik, F1M/SD, c/o Mercedes Benz AG, E222, 70322 Stuttgart, GERMANY

Winckler, M.J., Inst. für Wiss. Rechnen, Im Neuenheimer Feld 368, 69120 Heidelberg, GERMANY

Yoshimura, M., Dept. of Precision Engineering, Kyoto University, Sakyo–ku, Kyoto 606–01, JAPAN

Zhang, W.H., Institute of Mechanics, University of Liège, 21 rue E. Solvay, 4000 Liège, BELGIUM

WELCOME ADDRESS

HEIDE ZIEGLER
Rector of the University of Stuttgart

Dear Professor Schiehlen,
Dear Professor Olhoff,
Dear colleagues and guests,

I am happy to welcome you to this Symposium on Optimization of Mechanical Systems. Optimization has become a keyword in a variety of research areas today, especially in those disciplines that are related to application–oriented research. The application of optimality concepts to the design of multibody systems has, by comparison, been somewhat neglected – despite the fact that the multibody systems approach itself is already widely accepted for analyzing the dynamic behavior of mechanical systems like robots, vehicle systems, machines, and mechanisms. Therefore, the organization of a conference for scientists from various fields of research, devoted to an exchange of ideas concerning possible developments in multibody dynamics, ought to be highly appreciated not only by those scientists themselves, but also by all those interested in the long–term applicability of multibody systems in general.

As Rector of the University of Stuttgart I naturally welcome such activities, and I am pleased that this university will be your host institution for the next few days. Let me, perhaps, outline the main features of the University of Stuttgart, in order to give you an indication of the place you have come to.

The foundation of the "United High and Vocational School" in Stuttgart in 1829 was one of the events which marked the beginning of the Industrial Age in Württemberg. Apart from training craftsmen, the curriculum of this school also offered a more general, theoretically oriented education. This concept of a general education reaching beyond the mere conveyance of knowledge is still valid today. Thus, on July 4, 1967, the former Technical College was renamed "University of Stuttgart". Since 1988 the university has been organized into 14 schools or faculties. The close interconnection of research and teaching, one of the trademarks of excellent German universities, makes it possible both to provide a very good education for students aiming for positions in the business world, government or science as well as to hold a top position in basic and applied research. The annual budget of the university amounts to 500 million German marks – not counting construction costs. 200 million are provided by public and private trustees for research projects, 36 per cent of those 200 million are provided by industry.

The University of Stuttgart can boast 12 so–called collaborative research centers, and Professor Schiehlen contributes to more than one of them. Collaborative Research Centers, which are funded by the German Research Association (Deutsche Forschungsgemeinschaft) are the pride of every German university, and the University of Stuttgart, together with the Technical University in Munich, holds the national top position in this respect – a result of efficient research and straightforward management.

Disciplines in the engineering and natural science departments range from architecture, civil engineering and surveying, electrical engineering, and mechanical engineering to chemistry, physics, mathematics and computer science. Three specialists should be noted: One, the University of Stuttgart is the only civil university where aerospace engineering can be studied in a basic course. Two, the curriculum of technical biology combines the basics of natural science with its technological applications. Three, the university offers a graduate course called "Infrastructure Planning", which is taught in English, for civil engineers and architects that come from developing countries.

There are many ways in which the university interrelates with the city of Stuttgart and the surrounding area. The university presents an important economic factor, since it employs 5,000 men and women as well as providing many innovative impulses to industry and technology. And there is a certain quality of life here which I hope you will have a chance to experience: the city itself is located in an appealing countryside amongst forests and vineyards; yet, it also boasts a number of cultural attractions ranging from its internationally renowned ballet to art museums such as the National Gallery which was built by James Stirling or the television tower, the first of its kind worldwide and designed by an engineer and ex–rector of the University of Stuttgart, Fritz Leonhardt.

Since you, as you are assembled in this hall, certainly present a multibody system, I am convinced that your dynamic behavior will equal the opportunities which this environment offers you over the next five days. It, therefore, only remains for me to formally open the Symposium on Optimization of Mechanical Systems and to wish you all every kind of success.

WELCOME ADDRESS

N. OLHOFF
Secretary of the Congress Committee of IUTAM

Rector, Prof. Heide Ziegler,
Mr. Chairman, Prof. Werner Schiehlen,
Ladies and Gentlemen,

True science does not recognize state boundaries, nationalities, or political systems. Cooperation between scientists from different countries and parts of the World has a long tradition.

Organized meetings between scientists in the field of mechanics were initiated more than 70 years ago, namely in 1922, when Prof. Theodore von Kármán and Prof. Tullio Levi–Civita organized the world's first conference in hydro– and aero–mechanics. Two years later, in 1924, the First International Congress encompassing all fields of mechanics, i.e., analytical, solid and fluid mechanics, including applications, was held in Delft, The Netherlands. From then on (with exception of the year 1942) International Congresses in Mechanics have been held every four years. Out of this emerged the "International Union of Theoretical and Applied Mechanics", IUTAM, which organizes congresses and symposia all over the world.

The disruption of international scientific cooperation caused by the Second World War was deeper than that caused by the First World War, and the need for reknotting ties seemed stronger than ever before when the mechanics community reassembled in Paris for the Sixth Congress in 1946. Under these circumstances, at the Sixth Congress in Paris, it seemed an obvious step to strengthen bonds by forming an international union, and as a result IUTAM was created and statutes adopted. Then, the next year, in 1947, the Union was admitted to ICSU, the International Council of Scientific Unions. This council coordinates activities among various other scientific unions to form a tie between them and the United Nations Educational, Scientific and Cultural Organization, UNESCO.

Today, IUTAM forms the international umbrella organization of more than 40 national adhering organizations of mechanics from nations all over the world. Furthermore, a large number of international scientific organizations of general or more specialized branches of mechanics are connected with IUTAM as Affiliated Organizations. As a few examples, let me mention: the European Mechanics Society (EUROMECH), the International Association of Computational Mechanics (IACM), the International Association for Vehicle System Dynamics (IAVSD), the International Congress of Mechanical Behaviour of Materials (ICM), and the International Congress on Fracture (ICF).

It goes without saying that IUTAM carries out an exceptionally important task of scientific cooperation on mechanics on the international scene. For those of you that may be not aware, I may inform that the Chairman of this symposium, Professor Werner Schiehlen, earlier, for no less than eight years, has served in the immensely important, and also very laborious job, of being the Secretary–General of IUTAM. I need not say that Prof. Schiehlen's service in this job has been very highly esteemed and appreciated by all the Adhering and Affiliated Organizations of IUTAM.

As I mentioned before, IUTAM organizes international Congresses and Symposia all over the world. Thus, the Nineteenth International Congress of Theoretical and Applied Mechanics will be held in Kyoto, Japan, from 25 – 31 August 1996, i.e., next year. Announcements of this congress have been widely distributed and published in many scientific journals, and in fall 1995, a final announcement of the congress will be distributed to all who have responded to prior announcements.

As always in the past, the Congress will cover the entire field of theoretical and applied mechanics. The scientific program will consist of two General (opening and closing) Lectures, fifteen invited Sectional Lectures, six Mini–symposia on designated topics for special emphasis, and a large number of Contributed Papers to be presented in Lecture Sessions or in Seminar Presentation Sessions. As a novel feature of the Nineteenth Congress relative to the earlier ones, IUTAM's Congress Committee has selected forty "Pre–nominated Sessions" on typical topics of mechanics which are not covered by the Mini–symposia. The topics of these pre–nominated sessions are listed in the announcements of the congress, and the Congress Committee has pre–appointed the chairpersons for these sessions and asked them to stimulate scientists within the respective sub–fields to submit contributed papers for possible presentation at the Congress.

As it may be of particular interest for the participants in the present Symposium on Optimization of Mechanical Systems, I would like to mention that at the Congress in Kyoto in 1996,

–   One of the Mini–symposia is devoted to structural optimization, and will be chaired by Prof. M.P. Bendsøe and co–chaired by Prof. R.T. Haftka. This Mini–symposium will be arranged jointly with the newly founded International Society for Structural and Multidisciplinary Optimization (ISSMO), whose application for membership as an

Affiliated Organization of IUTAM is pending. The Founding President of ISSMO, Prof. G.I.N. Rozvany, also attends the present Symposium.

- A Pre–nominated Session is devoted to multibody dynamics, and will be chaired by Prof. W. Schiehlen and co–chaired by Prof. M. Geradin (who also serves as a member of the Scientific Committee of the current Symposium, and is here today) and Prof. J. Angeles.

- Also, a Pre–nominated Session will be devoted to contact and friction problems. This Mini–symposium will be arranged jointly with an Affiliated Organization of IUTAM, namely the International Association for Vehicle System Dynamics, whose Secretary–General, Prof. R.S. Sharp, also serves as a member of the Scientific Committee of the current Symposium, and is with us today.

The present Symposium is exceptionally interesting because it is interdisciplinary within the framework of mechanics. Thus, the Symposium combines and covers two extremely important, and very rapidly expanding fields of mechanics: optimization on the one hand, and analysis and synthesis of multibody systems on the other. IUTAM found that the proposal from the University of Stuttgart for such a symposium was not only very timely, but also very well founded in the outstanding research carried out in these areas at the University of Stuttgart, and the proposal for the Symposium was readily accepted and granted by the General Assembly of IUTAM.

On behalf of IUTAM, I wish to express my sincere thanks to the University of Stuttgart, in particular to Rector, Professor Heide Ziegler, for the invitation to host this significant scientific event, and I wish to welcome all the invited participants for their readiness to come and to contribute to the success of the Symposium by very active participation in the lectures, the poster session, the scientific discussions, as well as in the social program.

Finally, I would like to mention that to sponsor a scientific meeting is one thing, to organize one is another. A heavy burden is placed on the shoulders of the Chairman who is in charge of the scientific and the practical local arrangements, and of the associates who are assisting him. All who have tried this before know perfectly well how much work that has to be done in organizing a meeting like this one.

Thus, we should all feel obliged, not only to the International Scientific Committee, but also very much to the Local Organizing Committee, and in particular to the Chairman of both Committees, Professor Werner Schiehlen, who has carried the heaviest load and responsibility.

It is up to us now, Ladies and Gentlemen, to harvest the fruits of the Organizers' work. Let us contribute our share to make this IUTAM Symposium a meeting that will be long remembered as a very successful one!

On behalf of IUTAM, I greet you all and wish you great success!

OPENING ADDRESS

WERNER SCHIEHLEN
Chairman of the Symposium

Dear Professor Olhoff,
Highly Honoured Guests,
Dear Colleagues from so many countries,

On behalf of the Scientific Committee I call this meeting to order, and in the name of the Local Organizing Committee, I welcome you to the Symposium on Optimization of Mechanical Systems.

The logo of our Symposium is called the "Optimization Puzzle". This unfinished "Puzzle" is a graphical representation for the interdisciplinary approach of optimal design for mechanical systems like rotating machines, moving vehicles or vibrating structures, respectively. The three fundamentals of our subject are related to multibody system dynamics, to structural mechanics and to optimization theory.

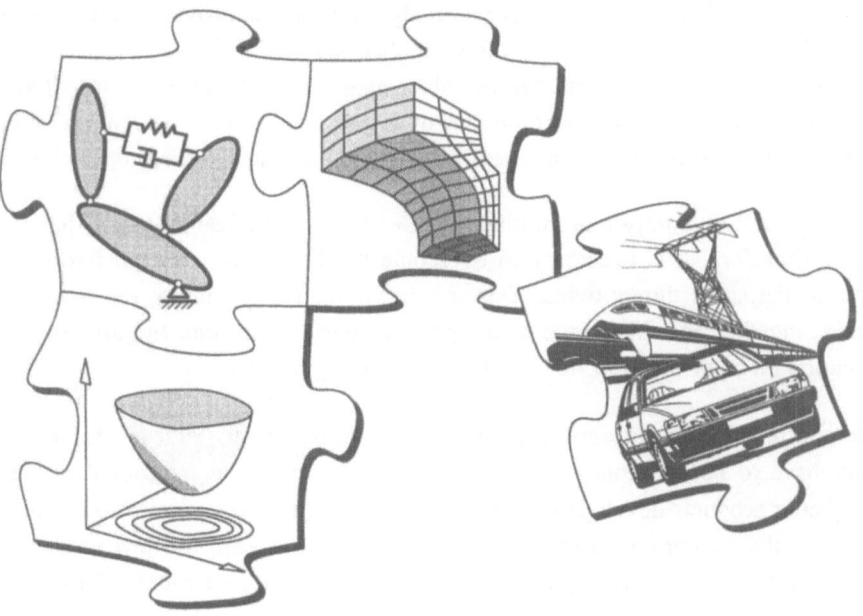

Firstly, the method of multibody systems has been developed during the last three decades starting out from the theory of gyroscopes and the urgent needs of stabilization of satellites in space. Today we can rely on sophisticated formalisms for modeling and simulating multibody systems. Many phenomena like flexibility, friction, impact and active control have been considered resulting often in mechatronic systems. Fast computers, in particular, transputers and power PCs allow real time simulations.

Secondly, the problems of structural mechanics are treated by the method of finite elements and, more recently, by the method of boundary elements, both often devoted as computational mechanics. These methods are very well developed and broadly used in civil engineering, in metal forming and many branches of mechanical engineering. Buckling of shells, contact problems, fracture of structural elements, interaction with fluids, noise generation, reliability and structural optimization are only a few topics originating from the research in structural mechanics. The software tools require often the most advanced computation facilities like workstation clusters, parallel machines or super computers. In contrary to multibody dynamics, optimization of the design is a well established research field in structural mechanics.

Thirdly, the theory of optimization is part of the mathematical sciences with broad application in economics and engineering. There are numerous methods available and even the choice of an efficient optimization algorithm is a nontrivial problem. The gradient methods are characterized by a low number of function evaluations, what is very essential for multibody systems. On the other hand stochastic optimization algorithms have better access to the global optimum of systems with many design variables and constraints. Thus, often a combination of different algorithms is the proper choice.

Each research community like multibody dynamics or optimization theory has its own culture. Therefore, it is necessary to overcome the fences, to work on interfaces and not to do the same things twice. This is what is meant by putting the pieces of the "Puzzle" together and this is one of the goals of our Symposium. In particular, on the application piece of the "Puzzle" much more work is needed.

There is a broad range of optimization problems in dynamics. First of all, vehicle dynamics have to be mentioned. The worldwide competition on the global markets requires better products developed in shorter periods. Therefore, optimization is an essential part of the concurrent or simultaneous engineering process for new products. Further, all kind of mechatronic systems requires optimization, in particular for the evaluation of control gains.

In biomechanics, the optimization approach often helps to understand nature which usually offers optimal designs.

What are the detailed topics of our Symposium? Altogether 50 papers will be presented within one Poster–Discussion Session and twelve Lecture Sessions all of them guided by a Chairman. The general topics are as follows:

○ Modeling and Optimization of Multibody Systems
○ Sensitivity Analysis of Mechanical Systems
○ Design of Mechanisms
○ Optimization in Structural Mechanics
○ Optimal Design of Materials
○ Mathematical Optimization Methods and Algorithms
○ Multicriteria Optimization and
○ Applications in Engineering

After this survey on the scientific aspects of the Symposium let me add a few remarks on the relation between IUTAM and Stuttgart.

In 1967, exactly twenty–eight years ago, the first IUTAM Symposium was held in Stuttgart under the chairmanship of Professor Eckkehart Kröner. This first Symposium was devoted to the "Generalized Cosserat Continuum and the Continuum Theory of Dislocations with Applications". Twelve years later in 1979 the second IUTAM Symposium took place in Stuttgart. Professor Richard Eppler was chairman of the Symposium on "Laminar–Turbulent Transition". Six years ago, in 1989, I had the pleasure to chair the IUTAM Symposium on "Nonlinear Dynamics in Engineering Systems". Thus, in a quarter of a century altogether four IUTAM Symposia were held at the University of Stuttgart in all the major fields which are Solid Mechanics, Fluid Mechanics and Dynamics.

But there are also other relations between the University of Stuttgart and IUTAM. From 1950 to 1963 Professor Richard Grammel of Stuttgart represented his country in IUTAM's General Assembly. From 1984 to 1992 I served myself as the Union's Secretary–General. And 1987 the annual Bureau meeting of IUTAM was held in Stuttgart including a visit of Kepler's birthplace in the nearby city of Weil der Stadt, one of the roots of mechanics as a science.

Last but not least I would like to thank all the other sponsors of this Symposium, namely, the University of Stuttgart and the German Research Council, the City of Stuttgart and the industrial companies listed in the Final Programme.

And very personally, I express my gratitude to all my coworkers at the Institute B of Mechanics for their devoted and engaged work. This work started more than one year ago and culminated last week. Especially, I would like to mention Dr. Dieter Bestle who is going to leave the Institute to be a Professor at the Technical University Cottbus in

W. SCHIEHLEN

one of Germany's new states. If you have any problem, please feel free to contact anybody of the Institute's staff, indicated by the letter i on the name tag.

So, I wish you, that this Symposium may stimulate and satisfy your scientific interest and offer you many opportunities to personal contacts with scientists from more than 20 countries active in Optimization of Mechanical Systems. I thank you.

DECOMPOSITION AND SENSITIVITY ANALYSIS FOR SOME DYNAMIC PROBLEMS OF OPTIMAL DESIGN

N.V.BANICHUK and A.D.LARICHEV
Institute for Problems in Mechanics Russian Academy of Science, 117526, Moscow, Russia

1. Introduction

The problems of finding the best precurved surface of thin-shelled structural elements form a special class of optimal design problems [1-5]. These problems are of both theoretical and applied interest (shallow curvilinear panels and corrugated plates and shells) and the corresponding optimal solutions may lead to improving such mechanical characteristics of structures as for example rigidity, stability, and so on. The principal result for the problems [1,2] is that the optimal distribution of the initial curvature of the plate is determined from the distribution of the deflection of an unbent plate. This enables us to obtain analitical solution for a wide class of two-dimensional problems and to determine the most interesting features of the optimal design.

In this paper, we consider nonstationary dynamic optimization problems arising in structural mechanics of shallow shells and prebent plates and concerning the finding the best precurved middle surface of thin-shelled structural elements.Considered problems admit effective decomposition and corresponding analysis is of both theoretical and methodological interest.

2. Some General Aspects of Optimization Modelling

Let us consider some mechanical system (structure) occupied the domain Ω and described by the following equations

$$Au = f, \qquad f = f(h, q) \tag{1}$$

where

$$h \in H, \qquad q \in Q \tag{2}$$

Function $f = f(h, q)$ and the admissible sets H and Q of all realizable designs $h = h(x)$ and loads $q = q(x, t)$ are given.The symbol A in (1) denotes a linear differential operator that contains differentiation with respect to the time $t \in [0, T]$ and with respect to the coordinates of the spatial vector $x \in \Omega$.The coefficients of the operator A depend

1

D. Bestle and W. Schielen (eds.), IUTAM Symposium on Optimization of Mechanical Systems, 1–8.
© 1996 *Kluwer Academic Publishers.*

on x and the design variable $h = h(x)$, that is $A = A(x, h)$. It is also assumed that the initial and boundary conditions, which determine the state variable $u = u(x, t)$ and it's derivatives at time $t = 0$ and the manner in which the structure is supported on the boundary of the domain Ω , are all included in the definition of the differential operator A.

The problem of minimization of quality functional

$$J = (u, g) \to \min_h \qquad (u, g) \equiv \int_0^T (\int_\Omega ug \, d\Omega) \, dt \qquad (3)$$

is analyzed under the constraints (1),(2). Here $g = g(x, t)$ - given function and scalar products with integration over x and x, t are denoted by parenthesis with and without subscript Ω

$$(u, g) \equiv \int_0^T (u, g)_\Omega \, dt, \quad (u, g)_\Omega \equiv \int_\Omega ug \, d\Omega \qquad (4)$$

Let us introduce the adjoint operator A^* and the adjoint variable $v(x, t)$ and assume that the function $v(x, t)$ satisfies the adjoint problem

$$A^* v = g \qquad (5)$$

Using the following equalities

$$(u, g) = (u, A^* v) = (v, Au) \qquad (6)$$

we transform the expression (3) for the quality functional to the form

$$J = (v, f) = \int_0^T (v, f)_\Omega \, dt \qquad (7)$$

It is seen from (6) and (7) that if we apply to the structure various loads q from Q then the functional J will decrease or increase in value, while the values of the functional v will remain unaltered. Therefore, to evaluate the functional J we are only needed to solve one time the adjoint problem (5) and to perform low cost operations in (7). This property can be effectively exploited in the case of optimal design of structures for multiple loading conditions and also in the case of optimal design under conditions of incomplete information.

For considered problem basic sensitivity analysis relation can be written in the following manner

$$\delta_{q,h} J = (v, f_h \delta h + f_q \delta q) \qquad (8)$$

Formula (8) shows that to evaluate the sensitivity of the optimized functional it is required to perform multiplications and integration operation with respect to $x \in \Omega$ and $t \in [0, T]$.

3. Optimal Design of Prebent Plates on Elastic Foundation

The problem of optimization of dynamical stiffness is considered here for an elastic shallow shell (prebent plate) lying on an elastic Vincler's foundation with the known compliance coefficients c, supported along boundary curve Γ of the domain Ω and loaded by transverse excitation forces $q(x, y, t)$. On the boundary segment of the plate Γ_1, the plate is rigidly built-in, while on the segment Γ_2 it is freely supported ($\Gamma = \Gamma_1 + \Gamma_2$). The cirve Γ is the boundary of a region Ω located in the xy plane. Let $h(x, y)$ denote the shape of the middle surface of this plate, which is initially deformed in the absence of external loads. Let $w(x, y, t)$ be a function describing deflections in the direction of the z axis of the middle surface of the plate caused by external loads. The magnitude of $w(x, y, t)$ is measured from the middle surface of the initially deformed plate. Let us assume that the deflections are "small", that is, the typical deflection is smaller than the thickness of the plate. It is supposed that the reaction force of elastic foundation is proportional to the total departure of the middle surface, described by the function $w(x, y, t) + h(x, y)$. Then the dynamical equation and boundary conditions for w are of the form

$$mw_{tt} + Lw + cw = q - ch \tag{9}$$

$$(w)_\Gamma = 0, \quad \left(\frac{\partial w}{\partial n}\right)_{\Gamma_1} = 0, \quad D[\Delta w - \frac{1-\nu}{R}\frac{\partial w}{\partial n}]_{\Gamma_2} = 0 \tag{10}$$

$$Lw \equiv (Dw_{xx})_{xx} + (Dw_{yy})_{yy} + \nu(Dw_{yy})_{xx} +$$

$$\nu(Dw_{xx})_{yy} + 2(1-\nu)(Dw_{xy})_{xy}$$

Were $m, \partial w/\partial n, R, \Delta$ and ν are, respectively, the mass per unit area, the derivative of $w(x, y, t)$ in the direction normal to the boundary curve, the radius of curvature of the boundary, the Laplace operator and Poisson's ratio. The cylindrical rigidity of the plate $D = D(x, y)$ is a given function of the x, y coordinates. Note that the operator L does not depend on h. We shall use the following inital conditions

$$(w)_{t=0} = 0, \quad (w_t)_{t=0} = 0 \tag{11}$$

The magnitude of the area of the middle surface is given and is equal to S. Using the assumption that the function $h(x, y)$ is "small", we transform this isoperimetric condition into the form

$$(\nabla h, \nabla h)_\Omega = 2(S - S_0) \tag{12}$$

Here S_0 is the area of Ω and ∇ is the gradient operator. The function h in (11) describing the initial deflection of the plate must satisfy the boundary condition

$$h(x,y) = 0, \quad (x,y) \in \Gamma \tag{13}$$

We formulate the following optimization problem. We need to find a function $h(x,y)$ that satisfies the isoperimetric condition (12) and boundary condition (13) such that the displacement function $w(x,y,t)$ derived from the initial boundary-value problem (9)-(11) assigns a minimum to the functional (generalized dynamic compliance)

$$J^* = \min_h J(h) = \min_h (w,g) \tag{14}$$

To derive the necessary optimality condition and the corresponding equation for the adjoint variable, we write the variational equation directly derived from (9): $m(\delta w)_{tt} + L\delta w + c\delta w + c\delta h = 0$. Multiplying the left-hand side of this equation by the adjoint function $v(x,y,t)$ and integrating over the region Ω and interval $[0,T]$, we obtain

$$(v, m(\delta w)_{tt} + L\delta w + c\delta w + c\delta h) = 0 \tag{15}$$

Let us assume that the function $v(x,y,t)$ satisfies the boundary conditions (10) and the terminal zero conditions. Noting that for functions satisfying these conditions, the operator L is self-adjoint and performing the integration by parts, we rewrite eq. (15):

$$(\delta w, mv_{tt} + Lv + cv) + (\delta h, cv) = 0 \tag{16}$$

Equation (16) can be used to represent the variation of the minimized compliance functional in the form

$$\delta J = -(\delta w, mv_{tt} + Lv + cv - g) - (\delta h, cv) \tag{17}$$

We derive the adjoint function $v(x,y,t)$ by solving the following non-stationary boundary-value problem

$$mv_{tt} + Lv + cv = g \tag{18}$$

$$(v)_\Gamma = 0, \quad \left(\frac{\partial v}{\partial n}\right)_{\Gamma_1} = 0, \quad D\left[\Delta v - \frac{1-\nu}{R}\frac{\partial v}{\partial n}\right]_{\Gamma_2} = 0 \tag{19}$$

$$(v)_{t=T} = 0, (v_t)_{t=T} = 0 \tag{20}$$

For function $v(x,y,t)$ determined in this fashion, the first integral on the right-hand side of (17) is equal to 0. It follows from conditions

(12),(13) that for an arbitrary variation of the design variable δh, the following equality must be true:

$$(\delta h, \Delta h)_\Omega = 0 \qquad (21)$$

A necessary condition for a minimum of the functional J is $\delta J = 0$. Therefore, recalling conditions $\delta J = -(\delta h, cv)$ and (21) we obtain the basic design sensitivity analysis relation and the optimality criterion

$$\delta J = (\delta h, \Delta h - \lambda \int_0^T cv\, dt)_\Omega \qquad (22)$$

$$\Delta h = \lambda \int_0^T cv\, dt \qquad (23)$$

were λ is a Lagrangian multiplier. The optimal shape of the prebent plate is determined from the Dirichlet problem (22) and (13). Lagrangian multiplier is found with the help of the isoperimetric condition (12). To determine the minimal value of J one may use the relation (7)

$$J^* = (v, q - Ch) = \int_0^T (v, q - Ch)_\Omega\, dt \qquad (24)$$

4. Optimal Design of Curvilinear Plates Under Stretching and Bending

Consider the problem of dynamic bending of an elastic initially bent plate supported along its boundary curve Γ (which lies in the x, y plane) and is loaded by dynamic transverse forces $q(x, y, t)$ and also by some static loads acting in the x, y plane applied to its edge.

$$mw_{tt} + Lw - K(\varphi)w = q + K(\varphi)h \qquad (25)$$

where K is linear differential operator

$$K(\varphi)w = h(\varphi_{yy}w_{xx} + \varphi_{xx}w_{yy} - 2\varphi_{xy}w_{xy}) \qquad (26)$$

The stress function φ is determined by solving the static problem of the linear two - dimensional theory of elasticity. The function φ does not depend on either w or h. Therefore, this function can be determined early in the solution process.

As in the preceding section, we choose as our design variable the initial shape function $h(x, y)$ for the plate, and as our quality criterion the compliance integral, which is to be minimized under the constraints (12) and (13). Performing the same substitutions, we derive an equation for the adjoint variable $v(x, y, t)$

$$mv_{tt} + Lv - K(\varphi)v = g \tag{27}$$

with boundary and terminal conditions (19),(20), and necessary optimality condition

$$\Delta h = \lambda \int_0^T K(\varphi)v \, dt \tag{28}$$

The minimized functional J is determined from the basic relation (7).Thus we obtain

$$J^* = \int_0^T (v, q + K(\varphi)h)_\Omega \, dt \tag{29}$$

We arrive at an analogous result.The derivation of an optimal distribution of the initial shape for a plate has been reduced to the solution of a static boundary-value problem for function φ, nonstationary boundary-value problem (19),(20),(27) and Dirichlet's problem (13),(28).

5. Optimal Design of Shallow Shells for Certain Classes of Loads

In many cases either adequate information on the applied loads is not available, or different forces from a certain set Q act in succession upon the structure.Because of the fact it is necessary to investigate more general problems of optimal design.

Let an elastic shallow shell (prebent plate), lying on an elastic Vincler's foundation with the known compliance coefficient C, is subjected to lateral loads $q(x, y, t)$ and given inplane forces at the edge.The lateral loading applied to the shell is not fixed beforehand but the set Q containing all possible forces $q \in Q$ is prescribed.We are to find the function $h(x, y)$ which satisfies the isoperimetric condition (12) and boundary condition (13) such that the deflection function $w(x, y, t)$ determined from the solution of the dynamical equation

$$mw_{tt} + Lw - K(\varphi)w + cw = q - ch + K(\varphi)h \tag{30}$$

With boundary and initial conditions (10),(11) minimizes the following cost functional

$$J^* = \min_h J(h) = \min_h \max_q (g, w) \tag{31}$$

where $g(x, y, t)$ - given function.For considered minimax problem the adjoint variable $v(x, y, t)$ determines the deflection distribution for a plane plate without initial bending lying on an elastic foundation under the same inplane forces and the known lateral load $g(x, y, t)$

$$mv_{tt} + Lv - K(\varphi)v + cv = g \qquad (32)$$

Therefore, to determine $v(x, y, t)$ for a given loads, we can employ known analytical and numerical solutions for problems (19),(20),(32) of bending and stretching of plates, lying on an elastic Vincler's foundation. Then, the expression obtained for $v(x, y, t)$ is substituted into the right-hand side of the necessary optimality condition

$$\lambda \Delta h = \int_0^T (cv - K(\varphi)v) \, dt \qquad (33)$$

and the determination of the optimal shape $h(x, y)$ is reduced to a solution of a Poisson equation with zero Dirichlet condition. Taking into account that

$$J = (v, q) + (v, K(\varphi)h - ch) \qquad (34)$$

and that v and h do not depend on q we find the worst load $q^*(x, y, t)$ from the condition

$$q^* : \max_{q \in Q}(v, q) \qquad (35)$$

In the case when Q is defined by the inequalities

$$q(x, y, t) \geq 0, \quad \int_\Omega q(x, y, t) \, d\Omega \leq P$$

we have

$$q(x, y, t) = P\delta(x - x_*(t))\, \delta(y - y_*(t)),$$

$$J = P \int_0^T v(x_*(t), y_*(t), t) \, dt + (v, K(\varphi)h - ch)$$

were P - given constant and $x_*(t), y_*(t)$ - functions determined with the help of maximization of $v(x, y, t)$ with respect to x and y.

6. Some Notes and Conclusions

As demonstrated in this paper, the solution of dynamic problems of optimal design of initially bent plates subjected to lateral loads and in - plane forces is reduced to the solution of the following three boundary value problems. First, we solve a boundary value problem from the plane theory of elasticity and obtain the stress function $\varphi(x, y)$. This problem is fully autonomous and its solution does not depend on lateral loads and initial shape of the prebent plane.

Then, we solve nonstationary boundary-value problem for adjoint equation with corresponding boundary and terminal conditions. Note that (as is well known)the initial boundary-value problem (19),(20),(27) for adjoint variable determines the deflection of an initially straight plate subjected to both the lateral loads and loads acting in the x, y plane.Therefore,to determine adjoint variable for a given lateral dynamic loads and static in-plane forces one may use well-known analytical and numerical techniques for solving dynamic problems in the simultaneous bending and tension of plates.

Then, the expression obtained for adjoint variable is substituted into the right-hand side of the Poisson equation to find the optimal shape for the prebent plate.This (last) problem has been reduced to the classical formulation of Dirichlet's problem for the Poisson equation.Solving the Dirichlet's problem for the Poisson equation presents no difficulties, and for some types of region Ω may be accomplished by purely analytical techniques.For a majority of practically important cases these solutions may be found in popular book or monographs.

7. Acknowledgements

The authors are indepted to doctors V.V.Saurin and A.A.Barsuk for useful discussion. This work was partially supported by grant number 94-01-00808-a from Russian Fund of Fundamental Research and grant number M2A000 from International Science Foundation.

8. References

1. Banichuk,N.V., Larichev,A.D.: Optimization of prebent plates (in Russian), Proc. XII Conf on shells and Plate Theory, 1 (1980), pp.126-130.
2. Banichuk, N.V.,Larichev,A.D.: Optimal design problems for curvilinear shallow elements of structures. Optimal Control Appl. and Meth., 5(1984), pp.197-205.
3. Plaut,R.H., Johnson,L.W. and Parbery,R.: Optimal forms of shallow shells with circular boundary. Parts 1-3, J.Appl. Mech., 51(1984), pp. 526-539.
4. Plaut,R.H., Jonson,L.W. and Olhoff,N.: Optimal forms of shallow cylindrical panels with respect to vibrations and stability. J. Appl. Mech.,53(1986), pp. 135-140.
5. Plaut,R.H., Young,D.T.: A variational principle useful in optimization rectangular-base shallow shells., in Rozvany G. and B.Karihaloo (eds.) Structural Optimization, Kluwer Academic Publisher, Dordrecht, (1988), pp. 241-248.

A MIN-MAX PROBLEM OF OPTIMAL ACTUATOR PLACEMENT FOR LIFTING

L. Beiner
Center for Technological Education Holon
52 Golomb St, Holon 58102, Israel

Abstract

The paper addresses the following problem: for a three-bar lifting linkage, find the locations of the two ends of the actuated link so as to *minimize* the *maximum* value of the force seen by the actuator over a specified range of arm angular positions. A simple analytical solution to this min-max problem is given, showing that the optimal locations are with the actuated link perpendicular either to the fixed link or to the lifting arm, when the arm is in a horizontal position. The method can be used to optimize the actuation of robots, doors, platforms, landing gears, etc., and allows to account for actuator length constraints. It can be extended also to Stewart's platform-type spatial manipulators.

Nomenclature

F - axial force in the actuator
W - load
L - arm length (from arm articulation to load)
a - fixed link length (from arm articulation to lower pinned end of actuated link)
b - moving link length (from arm articulation to upper pinned end of actuated link)
S - length of actuated link, Eq. (2)
ϕ - angular position of the moving link (positive above horizontal)
α - angle between horizontal and the fixed link
r - link length ratio, Eq. (5)
AF - actuation factor, Eq. (6)

1. Introduction and Problem Formulation

A common type of three-bar lifting linkage consisting of a fixed link a, a moving link b, and an actuated (variable length) link S is shown in Fig. 1 in its two possible configurations, namely, with $a < b$ or $a > b$. Assuming a conservative system (no losses) and neglecting inertia forces, power *in* is equal to power *out*, that is,

9

D. Bestle and W. Schielen (eds.), IUTAM Symposium on Optimization of Mechanical Systems, 9–16.
© *1996 Kluwer Academic Publishers.*

$$F\dot{S} = WL \cos\phi\,\dot\phi \tag{1}$$

From Fig. 1, the length of the actuated link can be expressed as

$$S^2 = a^2 + b^2 - 2ab\cos(\alpha+\phi) \tag{2}$$

and by differentiating it with respect to the time

$$\dot{S} = ab\sin(\alpha+\phi)\,\dot\phi/S \tag{3}$$

and replacing into (1), the axial force in the actuator at a given arm position ϕ is obtained as

$$F = \frac{WL\cos\phi}{ab\sin(\alpha+\phi)}\sqrt{a^2+b^2-2ab\cos(\alpha+\phi)} \tag{4}$$

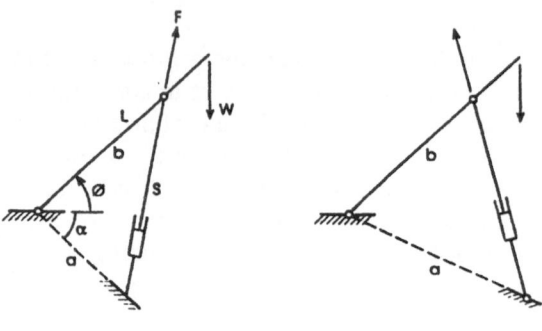

Fig.1: Three-link lifting mechanism

The force equation (4) being symmetrical with respect to a and b, we can define a link ratio r for each of the configurations as follows

$$r = a/b < 1 \quad (a < b) \tag{5a}$$

$$r = b/a < 1 \quad (a > b) \tag{5b}$$

and non-dimensionalize Eq.(4) accordingly in the form of an *actuation factor AF*

$$AF(\phi,\alpha,r=\frac{a}{b}) \equiv \frac{F}{W(L/a)} = \frac{\cos\phi}{\sin(\alpha+\phi)}\sqrt{r^2-2r\cos(\alpha+\phi)+1} \tag{6a}$$

$$AF(\phi,\alpha,r=\frac{b}{a}) \equiv \frac{F}{W(L/b)} = \frac{\cos\phi}{\sin(\alpha+\phi)}\sqrt{r^2-2r\cos(\alpha+\phi)+1} \tag{6b}$$

which is related to the mechanical advantage MA of the lifter (defined by analogy to four-bar linkages [1]) by the following relationships

$$MA \equiv \frac{W}{F} = \frac{1}{AF(L/a)} \quad (a<b) \quad \text{or} \quad \frac{1}{AF(L/b)} \quad (a>b) \tag{7}$$

Now, for a cost-effective actuator choice one must try to locate the actuator ends so that the highest value reached by AF over a specified range of arm angular positions is the lowest possible. We shall call this a *min-max force* problem and state it as follows: find the values of the design variables α and r which *minimize* the *maximum* value reached by AF defined in (6) over a specified range of positions $\phi = \phi_{min} \div \phi_{max}$, that is, solve

$$\max[AF(\phi, \alpha, r)] \rightarrow \min, \qquad \phi = \phi_{min} \div \phi_{max} \qquad (8)$$

This min-max force problem for three-link lifters has been proposed and solved numerically by Shoup [2]. For four-bar linkages which must satisfy certain force transmission requirements, the problem is solved iteratively by using the mechanical advantage and the transmission angle as criteria for improving the successive versions of the design [1]. A general index of the quality of motion and force transmission for spatial mechanisms has been developed by Sutherland and Roth [3-4]. However - as indicated by a recent monograph containing an extensive bibliography [5] - there are only a few known results concerning force transmission *optimization* in linkages, as for instance the position where the mechanical advantage of a four-bar linkage reaches its minimum (Freudenstein's theorem [6]) or its maximum. In this context, the contribution of the present study consists in presenting an analytical solution to the force transmission optimization problem for lifters formulated above.

2. Method of Solution

From the min-max force problem formulation (8), it is intuitively clear that for AF to achieve the lowest maximum value, the maximum of AF with respect to the position variable ϕ must *coincide* with its minima with respect to the design variables α and r. Accordingly, it can be expected to obtain the min-max solution by solving the following set of simultaneous stationarity conditions

$$\partial(AF)/\partial\phi = 0$$
$$\partial(AF)/\partial\alpha = 0 \qquad (9)$$
$$\partial(AF)/\partial r = 0$$

and then determining the nature of each extremum from its second derivative.

- The stationarity condition $\partial(AF)/\partial\phi = 0$

Differentiating (6) twice with respect to ϕ gives after simplification the ϕ-stationarity condition in the following form

$$-\cos\phi\cos^2(\alpha+\phi) + 2\cos\alpha\cos(\alpha+\phi) + \cos\phi - (r+\frac{1}{r})\cos\alpha = 0 \qquad (10)$$

and then the second derivative as

$$\frac{\partial^2(AF)}{\partial\phi^2} = -\sin\phi\sin^2(\alpha+\phi) + 2\sin(\alpha+\phi)[\cos\phi\cos(\alpha+\phi) - \cos\alpha] \qquad (11)$$

Noticing that for $\phi = 0$ the second derivative vanishes while the stationarity condition becomes

$$\cos^2\alpha - (r+\frac{1}{r})\cos\alpha + 1 = 0 \qquad (12)$$

with the roots $\cos\alpha = r$ and $\frac{1}{r}$, we select according to (5) the first root and conclude that for

$$\cos\alpha = r = a/b \quad \text{or} \quad b/a \tag{13}$$

AF has an inflexion point at $\phi = 0$.

- The stationarity condition $\partial(AF)/\partial\alpha = 0$

Differentiating (6) with respect to α, equating to zero and discarding the trivial optimal solution $\phi = 90$ deg (lifting arm in vertical position $\rightarrow F = 0$), the α-stationarity equation is obtained as

$$\cos^2(\alpha+\phi) - (\frac{1}{r}+r)\cos(\alpha+\phi) + 1 = 0 \tag{14}$$

with the roots $\cos(\alpha+\phi) = r$ and $\frac{1}{r}$. Again according to (5), the solution is

$$\cos(\alpha+\phi) = r \tag{15}$$

and by comparing it with (13) we see that the extrema with respect to ϕ and α coincide at $\phi = 0$. Taking the second derivative of (6) with respect to α yields

$$\frac{\partial^2(AF)}{\partial\alpha^2} = \sin(\alpha+\phi)[\frac{1}{r}+r-2\cos(\alpha+\phi)] \tag{16}$$

and by using (13) we get

$$\frac{\partial^2(AF)}{\partial\alpha^2} = \sqrt{1-r^2}\,(\frac{1}{r}-r) > 0 \tag{17}$$

which shows that the extremum with respect to α is a minimum.

- The stationarity condition $\partial(AF)/\partial r = 0$

Differentiating (6) with respect to r, equating to zero and discarding again the trivial optimum $\phi = 90$ deg yields the r-stationarity equation as

$$\frac{r - \cos(\alpha+\phi)}{\sqrt{r^2 - 2r\cos(\alpha+\phi)+1}} = 0 \tag{18}$$

so that an extremum is reached for

$$r = \cos(\alpha+\phi) \tag{19}$$

which is identical to (15) and (13) for $\phi = 0$. Differentiating (18) again with respect to r gives the second derivative as

$$\frac{\partial^2(AF)}{\partial r^2} = \frac{\sin^2(\alpha+\phi)}{[r^2 - 2r\cos(\alpha+\phi)+1]^{3/2}} \tag{20}$$

and by substituting (13) we get

$$\frac{\partial^2(AF)}{\partial r^2} = \frac{1}{\sin\alpha} > 0 \tag{21}$$

indicating that the extremum of AF with respect to r is also a minimum. Therefore, when α and r are related by equation (13), repeated for convenience below

$$\cos\alpha = r = a/b \quad \text{or} \quad b/a \tag{13}$$

the three extrema of AF, namely, the inflexion point with respect to ϕ and the minima with respect to α and r, coincide at $\phi = 0$, which is part of the condition for a solution. From (2) and Fig. 1 we notice that (13) implies either S *perpendicular to* a *at* $\phi = 0$ $(a < b)$ or S *perpendicular to* b *at* $\phi = 0$ $(a > b)$, which are in fact the optimal actuator locations we are looking for: accordingly, condition (13) is called a *min-max design*. The characteristic point of the min-max design is at $\phi = 0$, where it has the following properties:

(a) the actuation factor is unity for both configurations $a < b$ and $a > b$

$$AF(\phi = 0, \cos\alpha = r) = 1 \quad \rightarrow \quad \frac{F}{W(L/a)} = \frac{F}{W(L/b)} = 1 \quad \rightarrow \quad Fa = Fb = WL \quad (22)$$

which expresses in fact the moment equations about O;

(b) the ϕ-stationarity condition (10) is satisfied identically

$$\frac{\partial(AF)}{\partial\phi}(\phi = 0, \cos\alpha = r) = -r^2 + 2r^2 + 1 - r^2 - 1 \equiv 0 \quad (23)$$

indicating an extremum;

(c) the second derivative (11) vanishes, which shows that the extremum is an inflexion point.

To complete the solution, we must prove that the inflexion point represents a maximum of AF. To this end, we first notice that for usual lifting mechanisms the range of variation of ϕ may extend from $\phi_{min} = -20$ deg to $\phi_{max} = 80$ deg, while r can vary between 0 and 1. Using these values, a parametric study of (10) plotted in Fig. 2 shows that for a min-max design, $\partial(AF)/\partial\phi$ has only one root at $\phi = 0$ and is negative everywhere, which means that AF is monotonically decreasing from ϕ_{min} to ϕ_{max}. Therefore, for $\phi \geq 0$ the inflexion point is a maximum and $AF = 1$ represents the *min-max solution*, as illustrated by the AF versus ϕ and r plot given in Fig. 3, where it can be seen that all the curves pass through the inflexion point and are flatter for lower r values. This allows to draw the following final conclusions:

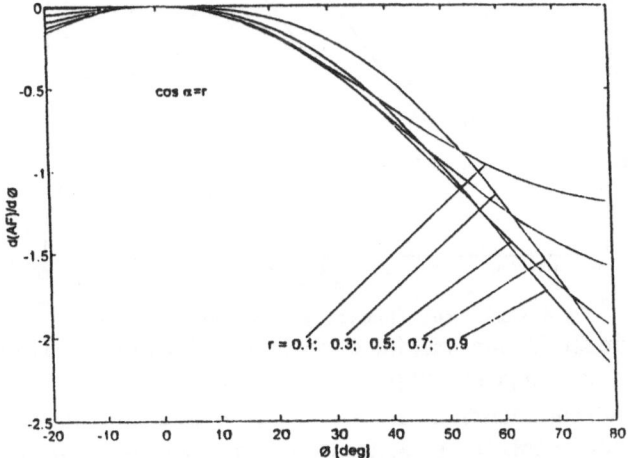

Fig. 2: $\partial(AF)/\partial\phi$ vs. ϕ and r for min - max design

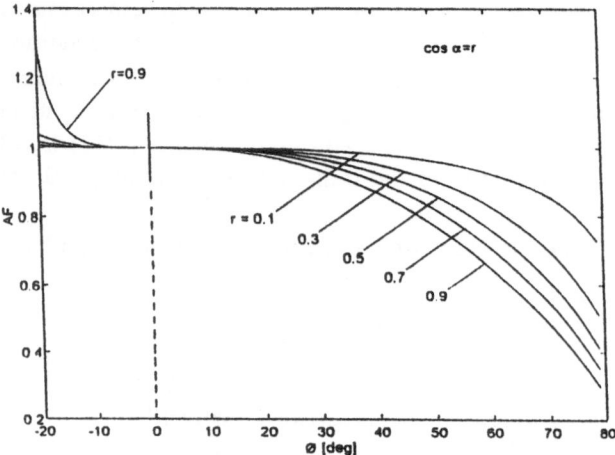

Fig. 3: AF vs. ϕ and r for min-max design

1. For $\phi_{min} = 0$, the min-max solution $AF = 1$ is guaranteed to be the lowest maximum value of AF with respect to any non-min-max design, as illustrated by an example in Fig. 4: this is a global optimal solution.

2. For $\phi_{min} < 0$, that is, when the lifter is required by design to start its motion slightly below the horizontal, the maximum AF value occurs at $\phi = \phi_{min}$ rather than at the inflexion point. In this case, the min-max solution is not strictly valid and the problem must be solved by numerical optimization. However, due to the flatness of the $AF(\phi)$ curves around the inflexion point, $AF = 1$ is still a good suboptimal solution for lower r values, as it will be proven by an example.

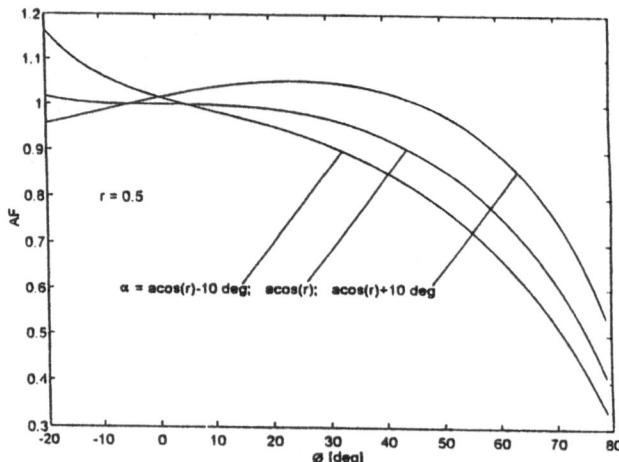

Fig. 4: Comparison of min-max and non-min-max designs

A solution procedure for the min-max force problem can be formulated as follows:

1. Given data: load W, range of arm positions $\phi_{min} \div \phi_{max}$

2. Select from design considerations the arm length L and either the fixed link length a or moving link length b so as to achieve a L/a or L/b ratio as low as possible.

3. From the min-max solution $AF = 1$, compute the maximum actuator force as $F = W(L/a)$ or $F = W(L/b)$ and choose an actuator which will result in a

certain retracted and extended   length of the actuated link, S_{min} and S_{max}, respectively.

4. Choose an r value which together with $\alpha = \cos^{-1} r$ and either b or a computed from Eq.(5) satisfy the geometric constraints

$$S(\phi_{min}) = S_{min} \tag{24}$$
$$S(\phi_{max}) = S_{max} \tag{25}$$

where $S(\phi)$ is defined in (2). If necessary, modify S_{min} and S_{max}.

3. Solution Procedure with Actuator Length Constraints

If the values of S_{min} and S_{max} are dictated by design considerations, the problem must be approached differently. Using for instance (5a) and (13), (2) can be rewritten as

$$S^2 / b^2 = (1 - 2 \cos \phi) r^2 + 2 \sin \phi r \sqrt{1 - r^2} + 1 \tag{26}$$

and by introducing the retracted /extended length ratio

$$k = S_{min} / S_{max} \tag{27}$$

and then dividing (24) by (25) and rearranging, we get the following equation

$$[(1 - 2 \cos \phi_{min}) - k^2 (1 - 2 \cos \phi_{max})] r^2 + 2(\sin \phi_{min} - k^2 \sin \phi_{max}) r \sqrt{1 - r^2} + 1 - k^2 = 0 \tag{28}$$

which must be solved for $0 < r < 1$. An approximate solution can be obtained by assuming $r^2 \ll 1$, in which case (28) reduces to a second order equation. The value of b can be then computed from (26) with S and ϕ assuming either their minimum or maximum values, respectively.

A procedure for solving the min-max problem with actuator length constraints is as follows:

1. Given data: load W, actuated link lengths S_{min} and S_{max}, range of arm positions $\phi_{min} \div \phi_{max}$.
2. Select from design considerations the arm length L.
3. For the given data, compute r, b and a from (28), (26) and (5a), respectively.
4. From the min-max solution $AF = 1$, compute the maximum actuator force required as $F = W(L/a)$ and choose an actuator (note that a similar procedure can be developed using Eq. (5b)).

4. Example of Optimization

As an example illustrating the min-max design method presented above, consider a three-link lifting mechanism with the following specifications :

$W = 14710$ N, $L = 3$ m, $\phi_{min} = -20$ deg, $\phi_{max} = 80$ deg, $S_{min} = 1.0$ m, $S_{max} = 1.8$ m With the above data, we get from (28) $r = 0.4246$ and then from (26) and (5a), $b = 1.3148$ m   and $a = 0.5583$ m, respectively, which together define the min-max design of the lifter. The suboptimal maximum actuator force is obtained from $AF = 1$ as $F = W(L/a) = 14710(3/0.5583) = 79051$ N, which exceeds by only 1.2 % the solution obtained by Shoup [2] using a direct search minimization algorithm with penalty functions.

5. Conclusions

A method for finding the best location for the ends of the actuated link of a three-link lifter is presented. The proposed analytical solution is extremely simple and ensures that the maximum axial force seen by the actuator over the range of arm positions is the lowest obtainable with the specified parameters. The method can be used for optimizing the actuation of lifting mechanisms for robots, doors, platforms, landing gears, etc. , and can accomodate constraints arising from the retracted versus extended length of the actuator. The method can be also modified to optimize the actuation of spatial parallel manipulators of the Stewart's platform type (Fig. 5). Design specifications would include the maximum load, the extreme displacements of the platform, the minimum length of the legs and their actuator stroke. In a first approximation one may consider the resultant of one pair of actuated legs in symmetric operation and optimize the location of the base joint relative to some representative position of the assumed fixed hinge of the platform. The goal is as before to obtain the lowest maximum value of the force seen by the actuators over the entire range of angular positions of the loaded platform. The analysis could be further refined by including other combinations of active and inactive legs so as to reach an overall optimum.

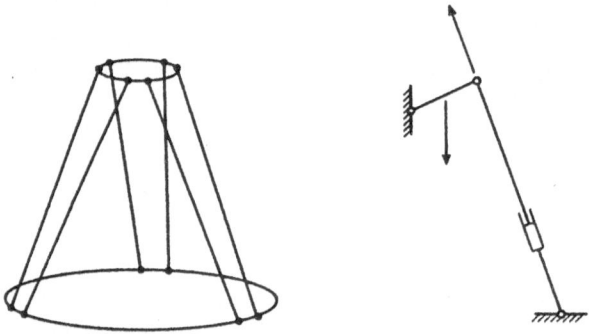

Fig. 5: Stewart's platform actuation

6. References

1. Erdman, A.G. and Sandor, G.N.: *Mechanism Design: Analysis and Synthesis*, Vol.1, pp. 110-126, Prentice Hall (1984).
2. Shoup, T.E.: *A Practical Guide to Computer Methods for Engineers*, Prentice Hall (1979).
3. Sutherland, G. and Roth, B.: "A Transmission Index for Spatial Mechanisms", *Journal of Engineering for Industry*, May 1973, pp. 589-597 .
4. Sutherland, G.H.: "Quality of Motion and Force Transmission", *Mechanism and Machine Theory*, Vol. 16, pp. 221-225, (1981).
5. Erdman, A.G. (ed.): *Modern Kinematics: Developments in the Last Forty Years*, Ch. 11: Optimization in Mechanisms, John Wiley and Sons, pp. 471-519 (1992).
6. Freudenstein, F.: "On the Maximum and Minimum Velocities and Accelerations in Four-Link Mechanisms", *Transactions of the ASME*, Vol. 78, pp. 779-787 (1956).

OPTIMIZATION OF STRUCTURE & MATERIAL PROPERTIES FOR SOLIDS COMPOSED OF SOFTENING MATERIAL

MARTIN P. BENDSØE
Mathematical Institute,
The Technical University of Denmark,
DK-2800 Lyngby, Denmark

JOSÉ M. GUEDES
IDMEC,
Instituto Superior Tecnico,
Av. Rovisco Pais,
P-1096 Lisboa Codex, Portugal.

SHELDON PLAXTON and JOHN E. TAYLOR
Department of Aerospace Engineering,
University of Michigan,
Ann Arbor, MI 48109-2140, USA

Abstract

An existing formulation for the prediction of the optimal material properties tensor for systems made of linear material is extended here to encompass design with nonlinear softening material. The original model for the linear case was stated as a convex, constrained nonlinear programming problem, and this property is preserved in the extension. The development is exemplified for the case of design for minimum global compliance. An extremum problem formulation for elastostatics is incorporated into the design problem, as a convenient way to model the structural analysis of general softening systems. Unlike the case for design with linear material, in the nonlinear problem the form of the solution for optimal material properties, and the associated stress fields as well, evolve with increasing load. Computational results are presented for a two-dimensional example problem, in which a system is designed for different loads while parameters in the material model are held fixed.

1. Introduction

The purpose here is to treat, in analytical form, the design problem for simultaneous prediction of material properties and structural layout. In the present approach, this is accomplished simply by considering design to be characterized in the formulation via a free parametrization of the rigidity tensor of material. Formulations of this kind have been demonstrated recently for structures composed of linearly elastic material, in both a single purpose and multiple purpose design context [4,5]. In these studies the rigidity

17

D. Bestle and W. Schielen (eds.), IUTAM Symposium on Optimization of Mechanical Systems, 17–24.
© 1996 *Kluwer Academic Publishers.*

tensor is allowed to range over all positive, semi-definite tensors, and the design resource (or total cost) is measured through invariants of the tensor. The objective was taken to be 'design for minimum compliance'. Within this formulation a *material optimization problem* can be identified, and thus the optimal local form of the material tensor can be derived. Once the optimal local material properties are determined, the original design problem can be expressed as a simpler equivalent design problem statement involving only the global distribution of resource. For a single loading condition this auxiliary problem takes on a simple form, one similar to that of a variable thickness sheet design problem.

In the developments to follow we will describe an extension of this free material design formulation to the design of a structure composed of a generic form of *nonlinear softening material*. The relevant mechanics is represented in the new formulation in terms of a generalized complementary energy principle developed recently for modelling the equilibrium analysis of such structures [14]. For present purposes the design objective is likewise based on complementary energy. Analytical forms for the optimal material tensors and the global distribution of material can be derived, in much the same way as was indicated above for design with linear materials, and thus the design parameters can be removed from the problem. The reduced problem is then an *equilibrium only* problem, albeit with a nonlinear and non-smooth (optimal) complementary energy functional. Alternatively,by solving analytically only for the optimal *local* properties, the resulting reduced problem is a smooth and convex problem combining equilibrium analysis and the determination of the optimal distribution of bulk resource. This problem is tractable, and a computational example is presented to show the form of results predicted for optimal material distribution.

The work presented in this paper represents a natural extension of the recent developments on simultaneous design of material and structure [see e.g. [4,5]. It constitutes as well a natural progression of developments in modelling for optimal design with advanced materials, and from treatments of topology design using homogenization modelling (see, e.g., the collection of papers in [6,10]. For models that employ the homogenization modelling for design parametrization, the optimal local material parameters can be related directly to a suitable microstructure, as demonstrated in [1,8]. In the context of the present free design parametrization, a different form of 'local structure' is required for the realization of material tensors. Examples of microstructures suitable for this purpose are described in [9,13]. These forms of local structure are not unique, nor are they necessarily of significance here other than to establish the quality of 'realizability'.

2. Problem statement

As indicated in the introduction, availability of an extremum problem formulation for the analysis part of the problem is what makes it possible to treat the design of nonlinear materials conveniently. The type of formulation used in the following development, which amounts to a generalized form of complementary energy principle, is presented in detail in [14]. The portrayal of a general form of nonlinear softening material relies on a feature in the model that has total stress expressed via a superposition of an arbitrary number of independent (constituent) fields. Each such constituent field is represented to be arbitrarily heterogeneous and anisotropic, and constituent stresses may be constrained to lie within a limiting surface. Overall material properties are determined through the model, once the parameters governing each of the the constituent fields are specified as data.

The formulation for equilibrium analysis is stated here in terms of stress fields. With the superposition of P softening components and one strictly linear basis component to make up the total stress, the problem has the form:

$$\min_{\sigma^p, \gamma, \alpha} \left\{ \tfrac{1}{2} \int_{\Omega} (F_{ijkl}^{-1} \gamma_{ij} \gamma_{kl} + \sum_{p=1}^{P} E_{ijkl}^{p}{}^{-1} \sigma_{ij}^{p} \sigma_{kl}^{p}) d\Omega \right\}$$

subject to: [A]

$$\overline{\alpha} - \alpha \leq 0$$

$$div(\gamma_{ij} + \sum_{p=1}^{P} \sigma_{ij}^{p}) + \alpha f = 0$$

$$(\gamma_{ij} + \sum_{p=1}^{P} \sigma_{ij}^{p}) \cdot n = \alpha t \quad \text{on} \quad \Gamma_{T},$$

$$\sigma^p \in \mathbf{K}_p, \ p = 1, ..., P$$

Here E_{ijkl}^{p} are the rigidity tensors for the P softening components and F_{ijkl} is the rigidity tensor for the strictly linear component. Factor α provides for the description of loads in the form of proportional loading. The stresses for the softening components are denoted σ_{ij}^{p}. The structure is subject to body force f and surface traction t on part Γ_T of boundary $\partial\Omega$. Finally, the convex sets of admissible stresses σ_{ij}^{p} for the softening components are denoted by $\mathbf{K}_p$.

Problem (A) is written for a *given material*, and for the analysis problem which it models the combined rigidity tensors, and the information that serves to define sets $\mathbf{K}_p$ altogether comprise the data which govern overall material properties. For the 'design of material properties' problem to be considered below, one or more of these material property tensors are treated as design variables.

Using the *rigidity tensors* as free design variables, the design problem has the form:

$$\inf_{E^p,\,F}\ \min_{\sigma^p,\,\gamma}\left\{\tfrac{1}{2}\int_\Omega (F_{ijkl}^{-1}\,\gamma_{ij}\gamma_{kl}+\sum_{p=1}^{P} E_{ijkl}^{P}{}^{-1}\,\sigma_{ij}^{p}\,\sigma_{kl}^{p})d\Omega\right\} \qquad\qquad \text{[P]}$$

subject to:

$$div(\gamma_{ij}+\sum_{p=1}^{P}\sigma_{ij}^{p})+\overline{\alpha}\,f=0$$

$$(\gamma_{ij}+\sum_{p=1}^{P}\sigma_{ij}^{p})\cdot n=\overline{\alpha}\,t\quad\text{on}\quad\Gamma_T,$$

$$\sigma^p\in\mathbf{K}_p\,,\,p=1,...,P$$

$$E^p>0,\ F>0,$$

$$\int_\Omega\Psi(F)d\Omega\le V_o$$

$$\int_\Omega\Psi(E^p)d\Omega\le V_p,\,p=1,...P$$

Here the design is to be optimal with respect to all positive definite rigidity tensors, and 'material resource' is measured in terms of invariants (symbolized by Ψ in problem statement (P)) of these tensors. In the statement (P) we take the supremum over the rigidity tensors, as we are using a stress based formulation; this is inherent to the analysis case under study. For pure displacement based formulations (see e.g., [5]), the design optimization can be performed over all positive *semi*-definite rigidity tensors.

We choose here to use either the trace or the Frobenius norm to measure resource for all tensors in the formulation, and this means that the invariants $\Psi(F)$; $\Psi(E^p)$ in (P), hereafter represented as 'resource densities ρ ', are given as:

$$\rho_{tr}(F)=F_{ijij}\ ;\ \rho_{tr}(E^p)=E_{ijij}^{p}\,,\ p=1,...,P$$

for the trace measure and

$$\rho_F(F)=\sqrt{F_{ijkl}F_{ijkl}}$$

$$\rho_F(E^p)=\sqrt{E_{ijkl}^{p}E_{ijkl}^{p}}\,,\ p=1,...,P$$

for the Frobenius norm. Note that these measures are homogeneous of degree one. Thus comparing to the conventional 2D problem for the design of material distribution in a sheet (where total cost is proportional to the volume of material), the above 'cost measures' correspond in their role to the sheet thickness.

The solution to problem (P) predicts the optimal distribution of rigidities within the specified softening limits associated with the definition of admissible stresses. (Optimal design with the limits themselves as design variables is treated for arbitrary trussed structures in [15]).

In the case of truss structures modelled as above, design for maximum load carrying capacity using member cross-sectional areas as design variables has been studied in [7] for the case of an elasto-plastic formulation. Truss design for the general softening material is reported in [15,17].

3 Analytical reduction of the problem

Along the lines of the modelling used in [5], parameters that describe the structure are now divided into two groups, namely those parameters that measure the amount of resource assigned to each point of the domain, and a second set that delineates how this resource is used to form the local material tensor. This provides for the following multi-level formulation of the problem:

$$
\inf_{\substack{\rho_p, \rho_o \\ \int_\Omega \rho_o \, d\Omega \leq V_o; \int_\Omega \rho_p \, d\Omega \leq V_p}} \quad \inf_{\substack{E^p > 0, \, F > 0, \\ \Psi(F) = \rho_0, \\ \Psi(E^p) = \rho_p}}
\left\{
\begin{array}{l}
\displaystyle \min_{\sigma^p, \gamma} \left[\frac{1}{2} \int_\Omega (F_{ijkl}^{-1} \gamma_{ij} \gamma_{kl} + \sum_{p=1}^{P} E_{ijkl}^{p}{}^{-1} \sigma_{ij}^{p} \sigma_{kl}^{p}) d\Omega \right] \\[4mm]
\text{subject to:} \\[2mm]
\displaystyle div(\gamma_{ij} + \sum_{p=1}^{P} \sigma_{ij}^{p}) + \overline{\alpha} f = 0 \\[4mm]
\displaystyle (\gamma_{ij} + \sum_{p=1}^{P} \sigma_{ij}^{p}) \cdot n = \overline{\alpha} t \quad \text{on} \quad \Gamma_T, \\[4mm]
\sigma^p \in K_p \,, \, p = 1,...,P
\end{array}
\right\}
\qquad \text{[P1]}
$$

Here the statical admissibility conditions of the inner problem are independent of the design variables. Thus minimization with respect to the pointwise variation of the rigidity tensors can be represented in the form:

$$
\inf_{\substack{E^p > 0, \, F > 0, \\ \Psi(F) = \rho_0, \\ \Psi(E^p) = \rho_p}} \left\{ F_{ijkl}^{-1} \gamma_{ij} \gamma_{kl} + \sum_{p=1}^{P} E_{ijkl}^{p}{}^{-1} \sigma_{ij}^{p} \sigma_{kl}^{p} \right\}
\qquad \text{[P2]}
$$

This characterization is consistent with the assumption of pointwise independent variation of the tensors within fixed values ρ_o, ρ_p of resource. This in turn justifies minimization of the local measure in (P2) at each point of the structure. By direct inspection we can conclude that

$$\inf_{E>0,\, \Psi(E)=\rho} E_{ijkl}^{-1} \sigma_{ij}\sigma_{kl} = \frac{1}{\rho} \sigma_{ij}\sigma_{ij} \qquad \text{[P3]}$$

for any stress field and any rigidity tensor. This result applies for both the trace and Frobenius norm measures of resource. Note that the optimal energy expression in (P3) coincides with the energy of a linearly elastic, zero-Poisson-ratio material of density ρ. The infimum in (P3) is not achieved, as the optimal rigidity tensor is given by:

$$E_{ijkl} = \rho \frac{1}{\sigma_{pq}\sigma_{pq}} \sigma_{ij}\sigma_{kl}$$

This corresponds to a singular orthotropic material, with axes of orthotropy co-aligned with the direction of principal stresses for the field σ_{ij}, and with only one non-zero eigenvalue.

With the introduction of (P3) into (P2), the problem (P1) can now be reduced to the convex problem:

$$\inf_{\rho_p,\, \rho_o} \quad \min_{\sigma^p,\, \gamma} \quad \left[\frac{1}{2} \int_\Omega \left(\frac{1}{\rho_0} \gamma_{ij}\gamma_{ij} + \sum_{p=1}^{P} \frac{1}{\rho_p} \sigma_{ij}^p \sigma_{ij}^p \right) d\Omega \right]$$

subject to:

$$div\left(\gamma_{ij} + \sum_{p=1}^{P} \sigma_{ij}^p\right) + \overline{\alpha} f = 0 \qquad \text{[P4]}$$

$$\left(\gamma_{ij} + \sum_{p=1}^{P} \sigma_{ij}^p\right)\cdot n = \overline{\alpha} t \quad \text{on} \quad \Gamma_T,$$

$$\sigma^p \in K_p, \, p = 1,...,P$$

$$\int_\Omega \rho_o \, d\Omega \leq V_o$$

$$\int_\Omega \rho_p \, d\Omega \leq V_p, p = 1,...P$$

In (P4) the energy measure for each constituent corresponds to the complementary energy of a linear elastic, zero-Poisson-ratio material of density equal to the locally assigned resource value. In this problem we can solve for the resource densities to find

$$\rho_0 = V_o \sqrt{\gamma_{ij}\gamma_{ij}} \Big/ \int_\Omega \sqrt{\gamma_{ij}\gamma_{ij}} d\Omega$$

$$\rho_p = V_p \sqrt{\sigma_{ij}^p \sigma_{ij}^p} \Big/ \int_\Omega \sqrt{\sigma_{ij}^p \sigma_{ij}^p} d\Omega$$

With the insertion of this result in problem statement (P4) , the equivalent but now design independent problem takes the form:

$$\min_{\sigma^p, \gamma} \left\{ \frac{1}{2V_o} \left[\int_\Omega \sqrt{\gamma_{ij}\gamma_{ij}} \, d\Omega \right]^2 + \sum_{p=1}^{P} \frac{1}{2V_p} \left[\int_\Omega \sqrt{\sigma_{ij}^p \sigma_{ij}^p} \, d\Omega \right]^2 \right\}$$

subject to:

$$div(\gamma_{ij} + \sum_{p=1}^{P} \sigma_{ij}^p) + \overline{\alpha} f = 0$$

[P5]

$$(\gamma_{ij} + \sum_{p=1}^{P} \sigma_{ij}^p) \cdot n = \overline{\alpha} t \quad \text{on} \quad \Gamma_T,$$

$$\sigma^p \in \mathbf{K}_p , \ p = 1,...,P$$

This (convex) problem is a generalized minimum complementary energy statement that is applicable for a linear-softening material with a non-smooth energy functional. A derivation of the counterpart to (P5) but for trussed structures, using convex duality arguments, is presented in [2] for the linearly-elastic case expressed via a displacements based minimum compliance formulation.

The computational results presented in this paper are obtained using a code for smooth optimization problems, to solve examples that are interpreted in the form of the (convex and smooth) problem (P4). The smoothness is obtained at the expense of an increased number of variables.

4 Conclusions

It has been shown that an analytical formulation for the design of continuum structures can be extended to cover systems composed of a generic form of elastic/softening materials. A similar development should be available for design with elastic stiffening materials; elastostatics for such systems is represented conveniently using the generalized minimum potential energy formulation [16]. In either case, the optimal material properties can be derived analytically, and this provides for a considerable simplification in the analysis and a commensurate reduction in problem size. The analysis applies as well in two and three-dimensions; the reduction in problem size is especially important in the three-dimensional setting, as a means to render the computational problem into tractable size.

5. Acknowledgments

This work was supported in part by the Danish Technical Research Council, through the Programme of Research on Computer Aided Design (MPB). The support of AGARD (MPB, JMG), JNICT, Portugal (JMG) and the Danish Natural Sciences Research Council (JET) is also gratefully acknowledged.

6. References

1. Allaire, G.; Kohn, R.V. (1993b): "Optimal Design for Minimum Weight and Compliance in Plane Stress using Extremal Microstructures." European J. Mech. A.;1993 (to appear).
2. Bendsøe, M.P.; Ben-Tal, A.; Zowe, J. (1993): "Optimization Methods for Truss Geometry and Topology Design." Structural Optimization (to appear).
3. Bendsøe, M.P.; Diaz, A.; Kikuchi, N. (1993): "Topology and Generalized Layout Optimization of Elastic Structures." loc. cit. Bendsøe and Mota Soares, 1993, pp. 159-206.
4. Bendsøe, M.P.; Diaz, A.; Lipton, R.; Taylor, J.E. (1993a): "Optimal Design of Material Properties and Material Distribution for Multiple Loading Conditions." DCAMM Report no. 469, The Danish Center for Applied Mathematics and Mechanics, The Technical University of Denmark, Lyngby, Denmark, 1993.
5. Bendsøe, M.P.; Guedes, J.M.; Haber, R.B.; Pedersen, P.; Taylor, J.E. (1994): "An Analytical Model to Predict Optimal Material Properties in the Context of Optimal Structural Design." J. Applied Mech. Vol 61, No 4, 930-937 .
6. Bendsøe, M.P.; Mota Soares, C.A. (Eds.) (1993): "Topology Optimization of Structures." Kluwer Academic Press, Dordrecht, The Netherlands, 1993.
7. Bendsøe, M.P.; Olhoff, N.; Taylor, J.E. (1993): "A Unified Approach to the Analysis and Design of Elasto-Plastic Structures with Mechanical Contact." In Rozvany, G.I.N. (Ed.), Optimization of Large Structural Systems, Kluwer Academic Publishers, Dordrecht, The Netherlands, 1993, pp. 697-706.
8. Jog, C.; Haber, R.B. Bendsøe, M.P. (1993): "Topology Design with Optimized, Self-Adaptive Materials." Int. J. Num. Meth. Engng. (to appear).
9. Milton. G.W., Cherkaev, A.V., "Materials with Elastic Tensors that Range Over the Entire Set Compatible with Thermodynamics," In Proc. Joint ASCE-ASME-SES Meet'N' (Herakovich, C.T. & Duva, J.M., Eds.), June 6-9, 1993, University of Virginia, Charlottesville, Virginia, USA, p. 342.
10. Pedersen, P. (Ed.) (1993): "Optimal Design with Advanced Materials." Elsevier, Amsterdam, The Netherlands, 1993.
11. Plaxton, S., Taylor, J.E. (1994): "Applications of a Generalized Complementary Energy Principle for the Equilibrium Analysis of Softening Material." Comp. Meth. Appl. Mech. Engng. 117, 91-103.
12. Schittkowski, K. (1985): "NLPQL: A FORTRAN Subroutine Solving Constrained Nonlinear Programming Problems." Annals Oper. Res., Vol. 5, 1985, pp. 485-500.
13. Sigmund, O. (1993): "Construction of Materials with Prescribed Constitutive Parameters: An Inverse Homogenization Problem." DCAMM Report no. 470, The Danish Center for Applied Mathematics and Mechanics, The Technical University of Denmark, Lyngby, Denmark, 1993. .
14. Taylor, J.E. (1993a): "A Global Extremum Principle for the Analysis of Solids Composed of Softening Material." Int. J. Solids Struct., Vol. 30, 1993, pp. 2057-2069.
15. Taylor, J.E. (1993b): "Truss Topology Design for Elastic/Softening Materials." loc. cit. Bendsøe and Mota Soares, 1993, pp. 451-467.
16. Taylor, J.E. (1994): "A Global Extremum Principle in Mixed Form for Equilibrium Analysis with Elastic/Stiffening Materials (A Generalized Minimum Potential Energy Principle)." J. Appl. Mech. Vol 61-No 4, 914-918.
17. Taylor, J.E.; Logo, J. (1993): "Analysis and Design of Elastic/Softening Truss Structures Based on a Mixed-Form Extremum Principle." In Rozvany, G.I.N. (Ed.), Optimization of Large Structural Systems, Kluwer Academic Publishers, Dordrecht, The Netherlands, 1993, pp. 683-696.

MULTIBODY SYSTEMS MODELING AND OPTIMIZATION PROBLEMS OF LOWER LIMB PROSTHESES

V. BERBYUK
Pidstryhach Institute for Applied Problems of Mechanics and Mathematics of the Ukrainian National Academy of Sciences
3-B, Naukova Str., Lviv, 290601, Ukraine
e-mail:Kalyniak%IPPMM. Lviv.UA@LITech.Lviv.UA

1. Introduction

To study the effect of prosthesis design on the kinematic, dynamic, energetic and other characteristics of an amputee's locomotion and to improve and create new efficient lower limb prostheses it is expendient to use mathematical modeling of a human walk process and dynamic optimization techniques. There are a number of mathematical models of the biped walk, having different degrees of adequacy [1–8]. In addition, different experimental methods have been used for solving these problems [9–11].

In this paper a mathematical model is proposed for investigating the dynamics of a man's skeletal system (MSS) with a below–knee prosthesis. A MSS is simulated by a plane controlled dynamic system of rigid masses. The controlled motions of the system are described by Lagrange's equations of the second kind, and for the expressions for kineto–static balance of the prosthesis under the action of ankle and metatarsal moments and the forces of reactions are derived. An algorithm is construced for solving the problem of human gait dynamics with a below–knee prosthesis. The series of dynamics problems for multibody biothechnical system "Man–Prosthesis" and optimization of structural parameters of the artificial lower extremity of a man has been solved.

2. The Mathematical Model

MSS is simulated by a plane multibody system of rigid masses (Fig. 1). This system comprises an internal body G (trunk) and two legs. Each leg consists of four elements. The two elements with mass and rotatory inertia model the thigh (link OK_i) and shank (link K_iA_i), while the other two massless and inertia–free elements (links $A_iH_iM_i$, M_iT_i) model the foot.

D. Bestle and W. Schielen (eds.), IUTAM Symposium on Optimization of Mechanical Systems, 25–32.
© *1996 Kluwer Academic Publishers.*

In addition to the weights of the trunk, thighs and shanks, the external forces acting on the MSS include the interaction forces between the feet and the ground, which are re-placed by resultant forces R_i, $(i=1,2)$.

It is assumed that the control moments $q_i(t)$, $u_i(t)$, $p_i(t)$, $w_i(t)$ are acting at the hip (point O) , knee (point K_i) , ankle (point A_i) and metatarsal (point M_i) joints, respectively.

There is an important difference between the dynamic of an intact limb, and the prosthetic limbs of a MSS. The mathematical modeling of human gait with a below–knee prosthesis is based on a supposition that the force moments at the prosthetic foot are the passive ones. The values of these moments depend not only from the gait pattern but also on the prosthesis construction.

Henceforth the subscript 1 will refer to the prosthetic leg , 2 to the intact leg.

The set of expressions describing the dynamics of the MSS are:

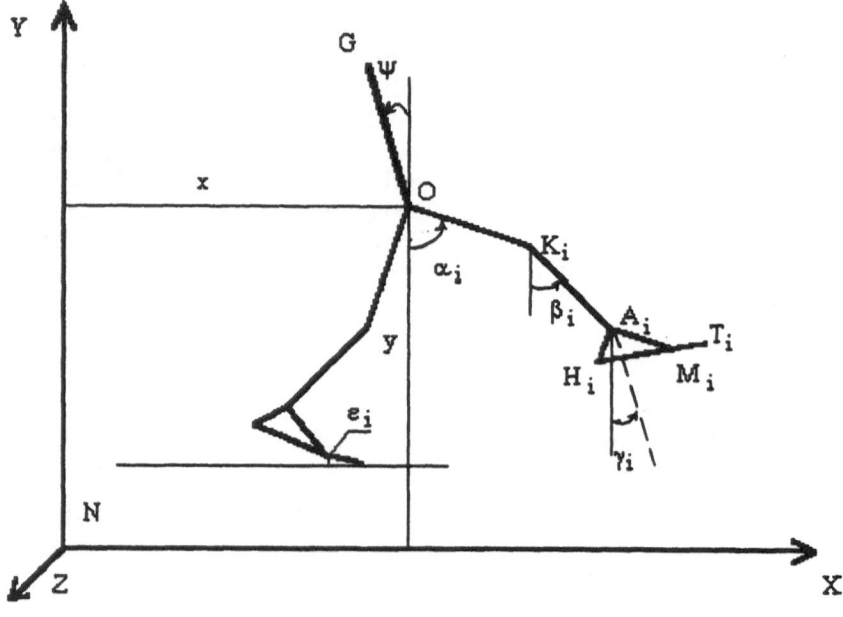

Fig. 1

$$f_1(t) - K_r(\ddot{\psi}\cos\psi - \dot{\psi}^2\sin\psi) = F_{1x}(t) + F_{2x}(t)$$
$$f_2(t) - K_r(\ddot{\psi}\sin\psi + \dot{\psi}^2\cos\psi) = F_{1y}(t) + F_{2y}(t)$$
$$f_{3i}(t) = q_i - u_i + a_i(F_{ix}\cos\alpha_i + F_{iy}\sin\alpha_i) \tag{1}$$
$$f_{4i}(t) = u_i - p_i + b_i(F_{ix}\cos\beta_i + F_{iy}\sin\beta_i)$$
$$f_5(t) = -q_1 - q_2$$

$$f_1(t) = M\ddot{x} + \sum_{i=1}^{2}[K_{ai}(\dot{\alpha}_i\cos\alpha_i) + K_{bi}(\dot{\beta}_i\cos\beta_i)]$$

$$f_2(t) = M(\ddot{y} + g) + \sum_{i=1}^{2}[K_{ai}(\dot{\alpha}_i\sin\alpha_i) + K_{bi}(\dot{\beta}_i\sin\beta_i)]$$

$$f_{3i}(t) = J_i\ddot{\alpha}_i + K_{ai}(\ddot{x}\cos\alpha_i + \ddot{y}\sin\alpha_i) + gK_{ai}\sin\alpha_i + a_iK_{bi}[\ddot{\beta}_i\cos(\alpha_i - \beta_i) + \dot{\beta}_i^2\sin(\alpha_i - \beta_i)]$$

$$f_{4i}(t) = J_{ci}\ddot{\beta}_i + K_{bi}(\ddot{x}\cos\beta_i + \ddot{y}\sin\beta_i) + gK_{bi}\sin\beta_i + a_iK_{bi}[\ddot{\alpha}_i\cos(\alpha_i - \beta_i) - \dot{\alpha}_i^2\sin(\alpha_i - \beta_i)], \quad (i = 1, 2)$$

$$f_5(t) = J\ddot{\psi} - gK_r\sin\psi - K_r(\ddot{x}\cos\psi + \ddot{y}\sin\psi)$$
$$M = m + m_{a1} + m_{a2} + m_{b1} + m_{b2} + m_{f1} + m_{f2}$$

$$J_i = J_{ai} + a_i^2(m_{bi} + m_{fi}), \quad J_{ci} = J_{bi} + b_i^2 m_{fi}$$
$$K_{ai} = m_{ai}r_{ai} + a_i(m_{bi} + m_{fi}),$$
$$K_{bi} = m_{bi}r_{bi} + b_i m_{fi}, \quad K_r = rm$$

$$R_{ix}(t) = F_{ix}(t), \quad R_{iy}(t) = F_{iy}(t)$$
$$p_i - w_i + (y_i - y_{mi})R_{ix} + (x_{mi} - x_i)R_{iy} = 0 \tag{2}$$
$$w_i + (y_{mi} - y_{Ri})R_{ix} + (x_{Ri} - x_{mi})R_{iy} = 0, \quad (i = 1, 2)$$

In equations (1) , (2) : x and y are the Cartesian coordinates of the suspension point O of the legs; ψ, α_i, β_i, γ_i, ε_i are angles that specify the position of the elements of the MSS (Fig. 1); m is the mass of the trunk; r is the distance from the suspension point of the legs to the center of mass of the trunk; J is the moment of inertia of the trunk relative to the Z axis at point O; m_{ai} is the mass of the thigh, a_i is the distance from O to the point K_i; J_{ai} is the moment of inertia of the thigh relative to the Z axis at O; r_{ai} is the distance from O to the center of mass of the thigh ; m_{bi} is the mass of the shank; b_i is the distance from the knee joint to the point A_i ; J_{bi} is the moment of inertia of the shank relative to the Z axis at the point K_i ; r_{bi} is the distance from K_i to the center of mass of shank; m_{fi} is mass of foot located at the ankle joint A_i ; $R_{ix}(t)$, $R_{iy}(t)$ are the horizontal and vertical components of the force R_i ; $F_{ix}(t)$, $F_{iy}(t)$ are the horizontal and vertical components of the principal vector of the reaction forces at the ankle joint A_i; (x_i, y_i), (x_{mi}, y_{mi}), (x_{Ri}, y_{Ri}) are the Cartesian coordinates of the ankle and metatarsal joints, and of the point of application of the vector R_i of the i–th leg, respectively; and g is the acceleration due to gravity.

The expressions (1) are the equations of motion of the mechanical system without feet, written in the form of Lagrange equations of the second kind. The expressions (2) are the conditions for the kineto–static balance of the foot of the i–th leg under the action of the ankle and metatarsal moments and the reactions at the support.

3. Statement of the Problem

The object under consideration is a nonlinear multidimensional controlled dynamical system. All movement of this system is restricted to the sagittal plane NXY of a fixed rectangular Cartesian coordinate system NXYZ (Fig. 1).

Let $Z = \{x, \dot{x}, y, \dot{y}, \psi, \dot{\psi}, \alpha_i, \dot{\alpha}_i, \beta_i, \dot{\beta}_i, \gamma_i, \dot{\gamma}_i, \varepsilon_i, \dot{\varepsilon}_i, \quad (i = 1,2)\}$

be a vector of the phase state, $U=\{R_{ix}, R_{iy}, q_i, u_i, p_i, w_i, i=1,2\}$ be a vector of the controlling stimuli of the MSS, and T be a duration of a double step (stride period).

Consider the optimization problem:

Problem A. It is required to determine the controlled process $\{Z(t), U(t)\}$ $t \in [0,T]$ and parameters $C_p>0$ $C_w>0$ which satisfy the equations of motions (1), the conditions of the kineto–static balance of the feet (2), the boundary conditions

$$Z(0)=Z_0, \quad Z(T)=Z_T \tag{3}$$

given restrictions on the phase coordinates over the time $t \in [0, T]$

$$x(t)=x_H(t), \ y(t)=y_H(t), \tag{4}$$
$$x_i(t)=x_{iA}(t), \ y_i(t)=y_{iA}(t), \ (i=1,2) \tag{5}$$
$$\alpha_i(t)\geq\beta_i(t), \ (i=1,2) \tag{6}$$
$$(x-x_i)^2+(y-y_i)^2\leq(a_i+b_i)^2, \ (i=1,2) \tag{7}$$

given constraints on the controlling stimuli

$$p_1(t)=f_p(t,C_p), \ w_1(t)=f_w(t,C_w) \tag{8}$$

and which minimize the functional [5–8]

$$E =1/L\int_0^T\left\{\sum_{i=1}^2[|q_i(\dot{\psi}-\dot{\alpha}_i)|+|u_i(\dot{\alpha}_i-\dot{\beta}_i)|]+|p_2(\dot{\beta}_2-\dot{\gamma}_2)|+|w_2(\dot{\gamma}_2-\dot{\varepsilon}_2)|\right\}dt \tag{9}$$

In expressions (3)–(9): $x_H(t)$, $y_H(t)$, $x_{iA}(t)$, $y_{iA}(t)$ are the Cartesian coordinates of the hip and ankle joints given from the experimental data of the human locomotion; Z_0, Z_T are the given initial and final phase states of the system; $f_p(t, C_p), f_w(t, C_w)$ are given functions determined the dynamic characteristics of a below–knee prosthesis; C_p, C_w are the vectors of the constructive parameters of a prosthesis structure; L is stride length which is equal to the sum of two step lengths.

The objective functional (9) is the integral over a double step of the sum of the absolute values of the mechanical power of all controlling stimuli acting at intact joints of the MSS.

4. Results and Discussion

Central in the approach proposed for solving problem A is the idea that any optimal control problem can be converted into a standard nonlinear programming problem by parameterizing each of the free variable functions.

It is assumed that there are four phases of the leg action in a double step ($t \varepsilon [0, T]$): the first double support phase ($t \varepsilon [0, \tau_l]$); the prosthetic leg single support ($t \varepsilon [\tau_1, \tau_2]$); the second double support ($t \varepsilon [\tau_2, \tau_3]$), and the intact leg single support phasse ($t \varepsilon [\tau_3, T]$) [12].

There is only one independently variable function in Problem A. If it is assumed that

$$\psi(t) = \begin{cases} A_0 + A_1 t + \dots + A_4 t^4, & t \in [0, \tau_1] \\ B_0 + B_1(t - \tau_2) + \dots + B_8(t - \tau_2)^8, & t \in [\tau_2, T] \end{cases} \qquad (10)$$

then the problem A is converted into the parameter optimization problem: $E=Q(A, B, C_p, C_w) \to$ min, $g(A, B, C_p, C_w)=0$. Here functions Q and g are determined by means of (1)–(9); A, B, C_p, C_w are vectors of the variable parameters.

To solve this parameter optimization problem, the computational algorithm based on Rosenbrock's method [13] has been devised.

Some results of solution problem A for the gaits with slow, natural and fast cadences [14] are shown in the Table (cadence in step/min., the others in SI units) and in Fig. 2–Fig. 6 (in all figures centered curves correspond to prosthetic leg, solid to intact leg of MSS).

In the model a subject height of 1.7 m, mass of 70 kg, and the folowing parameters of the limbs: $m_{ai}=8.12$ kg, $a_i=0.46$ m, $J_{ai}=0.228$ kgm^2, $m_{bi}+m_{fi}=3.36$ kg, $b_i=0.516$ m, $r_{ai}=0.25$ m, $r_{bi}=0.207$ m, $J_{bi}=0.194$ kgm^2 have been considered.

The functions $x_H(t)$ and $y_H(t)$ were specified in the following form [12]: $x_H(t)=0.3+V_t$, $y_H(t)=0.9+0.0234 \sin((4\pi/T)t-1.7)$.

The Cartesian coordinates $x_{iA}(t)$, $y_{iA}(t)$ used in the restrictions (5) are shown in Fig.2-3.

The controlling stimuli of the prosthetic foot were chosen in the form:

$$p_1(t) = C_1(\beta_1 - \gamma_1) + K_1(\dot{\beta}_1 - \dot{\gamma}_1), \quad w_1(t) = C_2(\gamma_1 - \varepsilon_1) + K_2(\dot{\gamma}_1 - \dot{\varepsilon}_1),$$

where C_i, K_i are the parameters of the elastisity and viscoelastisity of the ankle and metatarsal joints, respectively.

The rhythm of the double step was specified by parameters $\tau_1=0.12T$, $\tau_2=0.45T$, $\tau_3=0.67T$.

Energetically optimal laws of motion of a trunk of a MSS are specified by formulas (10) and by the values of the free parameters in the Table.

Figures 4-6 show graphs of control moments in the ankle and knee joints and the vertical component of the support of biotechnical system investigated for gait with natural cadence.

For comparison purposes in Fig. 4–6 the functions obtained from experiments for a human normal gait with natural cadence are shown (dashed curves) [14].

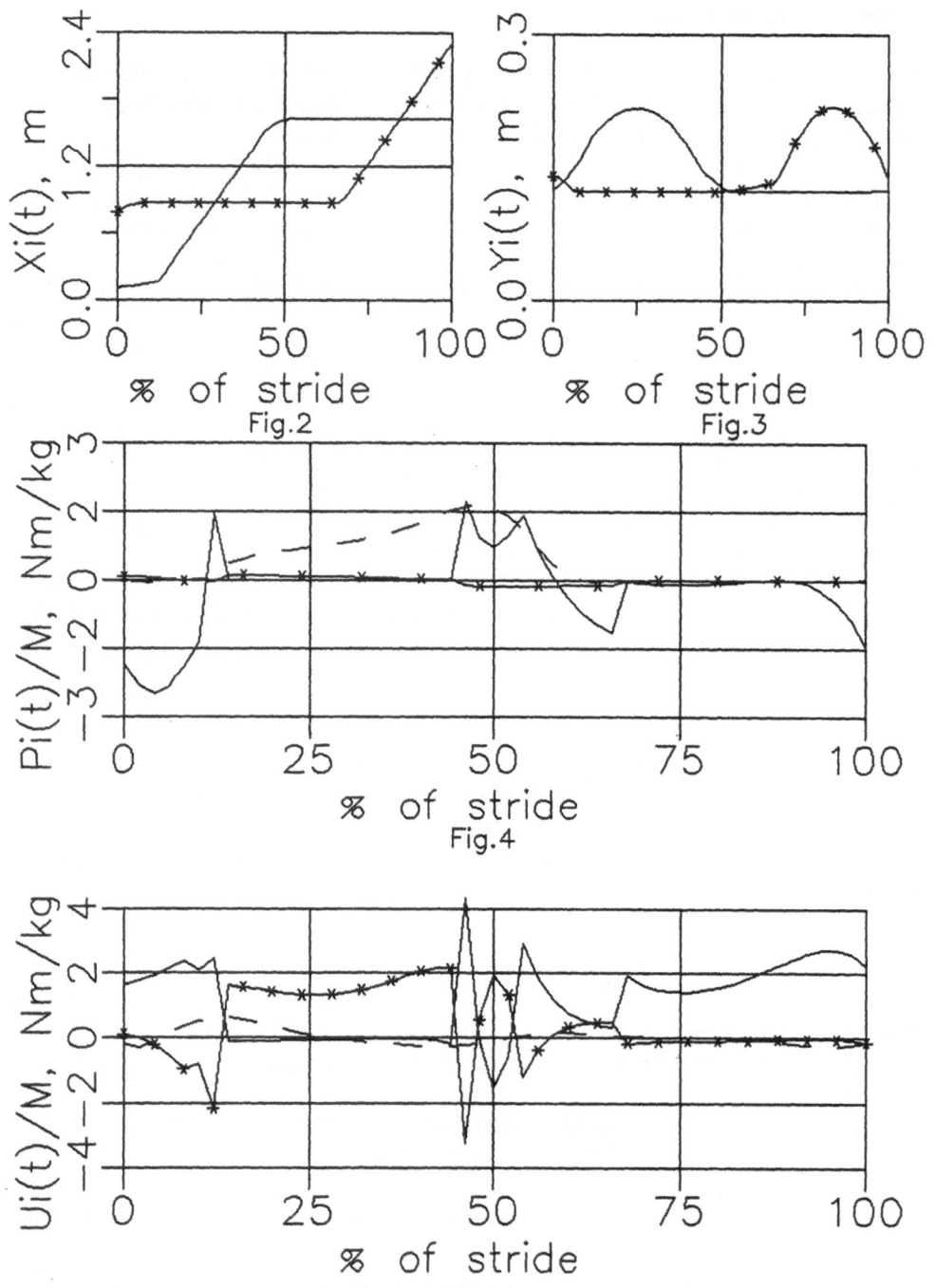

Fig.2

Fig.3

Fig.4

Fig.5

TABLE

Cadence	86.8	105.3	123.3
T	1.383	1.1396	0.9733
V	0.998	1.325	1.685
A_0	-0.314117	-0.154577	0.025723
A_1	-0.348378	-0.350503	-0.355695
A_2	-1.558435	-1.583903	-1.583395
A_3	0.050218	0.098445	0.098497
A_4	-0.000381	0.018242	0.012500
B_4	2.819747	1.285306	4.362961
B_5	0.082244	0.103455	0.148338
B_6	-0.104326	-0.129170	-0.075850
B_7	-0.461138	-0.574321	-0.514498
B_8	3.208211	1.253985	3.032430
C_1	10.22656	9.240574	10.3565
K_1	0.037316	0.634920	1.54595
C_2	17.74291	17.65329	20.1178
K_2	55.51611	57.43469	55.3188
E	297	263	347

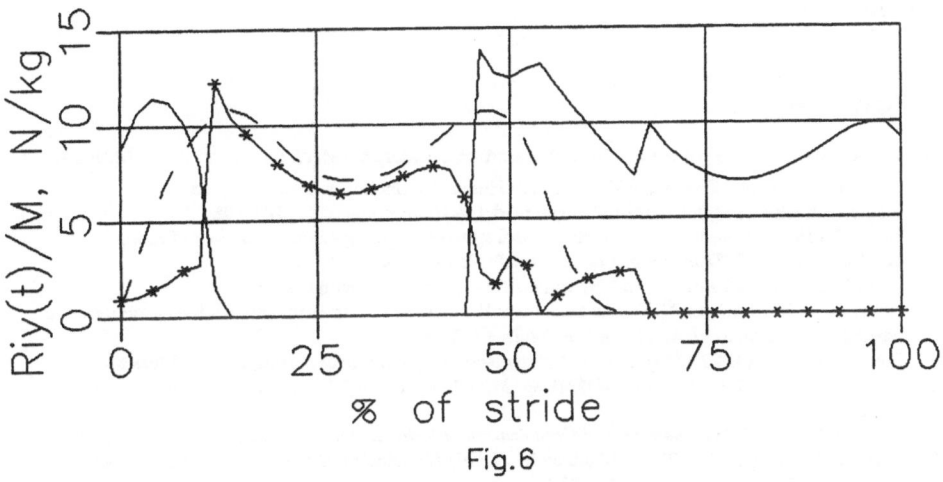

Fig.6

5. Conclusion

In this paper the analysis of a controlled motions of a biotechnical system "Man–Prosthesis" is based on solution of the energy optimal control problem for a plane multibody system. The performance index used is the mechanical work spent to transfer a MSS with below–knee prosthesis from the initial phase state into the final one over the given time.

To solve the nonlinear optimal control problem under given boundary conditions and restrictions on the phase coordinates, a parameter optimization approach has been proposed.

In the framework of proposed mathematical model of biotechnical system "Man–Prosthesis" the folowing conclusions have been drawn.

1. The kinematic, dynamic, and energetic characteristics of controlled motion of a biotechnical system are strongly sensitive to the essential prosthesis parameters, such as spring stiffness and damping properties. For a given gait there are optimal values of the elasticity and viscoelasticity parameters of the ankle and matartasal joints of the prosthetic foot which give minimum energy expended per unit of distance travelled (see Table).

2. The analysis of a number of numerical simulations shows that the natural cadence of the biotechnical system gait gives a minimum to the energy expended per unit of distance travelled (see example 2 in Table).

For normal human locomotion the same result was obtained in [15].

One of the important possible practical applications of the results of the present study may be the optimal design of artificial lower limbs.

6. Acknowlegments

This work was partially supported by the Concern "Ukrprosthesis", Kiev, Ukraine.

7. References

1. Vukobratovich M. (1976) *Walking Robots and Anthropomorphic Mechanism*, Mir Press, Moscow.
2. Aleshinsky S. Y. and Zatsiorsky V. M. (1978) Human locomotion in space analyzed biomechanically through a multi-link chain model, *J. Biomechanics*, **11**, 101-108.
3. Hatze H. (1980) Neuromusculoskeletal control systems modeling - a critical survey of recent developments. *IEEE Transactions on Automatic Control*, **AC-25**. 5, 375-385
4. Larin V. B. (1980)*Control of walking apparatuses* , Naukova Dumka, Kiev.
5. Beletskii V. V. , Berbyuk V. E. and Samsonov V. A. (1982) Parametric optimization of motions of a bipedal walking robot, *J. Mechanics of Solids*, **17**, 24-35.
6. Formalsky A. M. (1982), *Displacement of anthropomorphous mechanisms*, Nauka, Moskow.
7. Beletskii V. V. (1984) *Two-Legged Walking: Model Problems of Dynamics and Control* ,Nauka, Moscow.
8. Berbyuk V. E. (1989)*Dynamics and Optimization of Robototechnical Systems*, Naukova Dumka,Kiev.
9. CapozzoA.,Figure F., Leo T. and Macthett M. (1976) Biomechanical evaluation of above-knee prostheses. *Biomechanics* V-A, 366-372.
10. Winter D.A. and Sienko S.E. (1988) Biomechanics of below-knee amputee gait, *J.Biomechanics*, **21**,3, 361-367.
11. Diandelo D.J., Winter D.A., Ghista D.N., and Newcomber W.R. (1989) Performance assessment of the terry fox jogging prosthesis for above-knee amputees, *J.Biomechanics* ,**22**, 6/7, 543-558.
12. Chow C.K., and Jacobson D.H. (1971) Stadies of human locomotion via optimal programming, *Math. Biosci.*, **10**, 239-306.
13. Rosenbrock H.H.(1960) An automatic method for finding the greatest and least value of a function, *The Computer Jornal*, **3**, 175-184.
14. Winter D. (1991), *The Biomechanics and Motor Control of Human Gait*, University of Vaterloo Press, Canada.
15. Beckett, R., and Chang, K. (1968) An evaluation of kinematics of the gait by minimum energy, *J. Biomechanics*. **1** , 147-159.

MULTI–CRITERIA MULTI–MODEL DESIGN OPTIMIZATION

D. BESTLE AND P. EBERHARD
Institute B of Mechanics, University of Stuttgart
Pfaffenwaldring 9, 70550 Stuttgart, Germany
email: pe@mechb.uni–stuttgart.de

1. Introduction

Due to rapid development of faster and faster computing facilities the multibody system approach which has been studied now for more than three decades [1] is starting to switch from a purely analyzing method to a more synthesizing tool. Optimization methods are applied to optimize multibody systems with respect to their dynamic behavior [2, 3]. The dynamic behavior of multibody systems is determined by parameters like the mass and moments of inertia of the individual bodies, geometrical data, stiffness and damping coefficients, or control parameters of active coupling elements. Each of these parameters may serve as design variable for optimizing the dynamic behavior. For applying optimization algorithms performance criteria have to be defined [4].

Applications to technical problems clearly show that often several conflicting technical specifications and goals have to be taken into consideration. This can only be handled by allowing to define several different performance criteria. Due to the presence of several criteria the design problem has to be considered as a multicriteria optimization problem. The multicriteria optimization approach seems to offer a promising way to handle the situation of conflicting system specifications and requirements and define optimal solutions.

Technical applications also show that analysis of different aspects of a technical system has to be based on different models. For example, vertical vehicle dynamics can be studied with quarter–car or half–car models whereas studies of lateral dynamics require 'bicycle models' or even spatial models. In this paper, therefore, an optimization concept on the basis of simultaneously investigating several different models with shared parameters is demonstrated.

2. Multicriteria Multimodel Design Concept

It is well known that results obtained from numerical analysis of technical systems are only as predictative as the model being used. Phenomena observed from a technical system have to be part of the model if they should be investigated by numerical analysis. The best model, however, need not be the most complex one, but will be the most appropriate one.

D. Bestle and W. Schielen (eds.), IUTAM Symposium on Optimization of Mechanical Systems, 33–40.
© 1996 *Kluwer Academic Publishers*.

If a technical system has to be optimized with respect to a variety of different specifications, it seems to be natural to base system analysis on a variety of different models, each of it appropriate to evaluate the behavior of the system with respect to only some of the specifications. The design concept then has to summarize all these different considerations in order to find optimal solutions with respect to all the specifications. This results in a multidisciplinary multimodel design concept.

According to Fig. 1 the process of design evaluation, i.e. computing the performance values $\psi(p) \in \mathbb{R}^n$ from given design variables $p \in \mathbb{R}^h$, has to be split up into several submodel analyses. Each submodel is specially designed for evaluating the performance of the technical system with respect to a subset of design goals. For explaining the overall design concept, it may be just considered as a black−box function between some input parameters $\bar{p}^i$ of the submodel i and the output criterion values $\bar{\psi}^i = \bar{\psi}^i(\bar{p}^i) \in \mathbb{R}^{n^i}$.

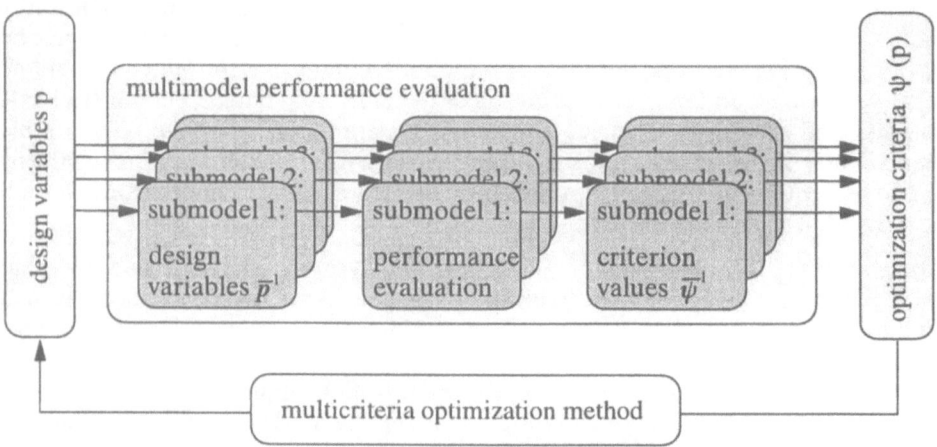

Fig. 1: Multimodel analysis concept

The input parameters $\bar{p}^i$ of the submodels have to be linked to the global set of design variables p. This can be a direct relation

$$\bar{p}^i_j = p_k \quad \text{for some} \quad k \in \{1, ..., h\}, \tag{1}$$

an affine mapping

$$\bar{p}^i = A^i p + \bar{p}^i_0 \tag{2}$$

where $\bar{p}^i_0$ may be used as a bias or for fixing some $\bar{p}^i_j$ to a constant value $\bar{p}^i_{0j}$. If appropriate, also a nonlinear function

$$\bar{p}^i = \bar{p}^i(p) \tag{3}$$

may be defined by the designer. Using this relation, the criterion functions $\bar{\psi}^i$ may also be considered as functions of the global design variables p:

$$\bar{\psi}^i = \bar{\psi}^i(\bar{p}^i(p)) = \bar{\psi}^i(p). \tag{4}$$

The total set of criteria $\psi(p)$ is an union of the subsets of criteria $\overline{\psi}^i(p)$, $i = 1(1)m$:

$$\overline{\psi}(p) = \left[\overline{\psi}^{1^T}, \ \overline{\psi}^{2^T}, \ ..., \ \overline{\psi}^{m^T}\right]^T \in \mathbb{R}^n \quad where \quad n = \sum_{i=1}^{m} \overline{n}^i. \tag{5}$$

In some cases, it is advantageous to re−order the criteria with respect to their types of function which is relevant for sensitivity analysis. This may be performed by a permutation matrix P, i.e.

$$\psi(p) = P\overline{\psi}(p). \tag{6}$$

In the problem formulation phase of the design process, the criteria should be considered to be just an instrument of performance evaluation. It is already part of the multicriteria optimization concept to classify them as objective functions $f_j(p)$, equality constraints $g_j(p) = 0$ or inequality constraints $h_j(p) \leq 0$, Fig. 2. Some of the performance criteria may even be neglected and considered as inactive in a first run in order to simplify the design process. Such a classification may be changed several times within the design process to get a feeling for the technical system and the potentials of optimizing it. Since several criteria may remain as objective functions to be minimized, the problem has to be stated as a multicriteria optimization problem:

$$\underset{p \in \mathcal{P}}{opt} \ \ f(p) \quad where \quad \mathcal{P} := \{p \in \mathbb{R}^h|\ g(p) = 0, \ h(p) \leq 0, \ p^l \leq p \leq p^u\} \tag{7}$$

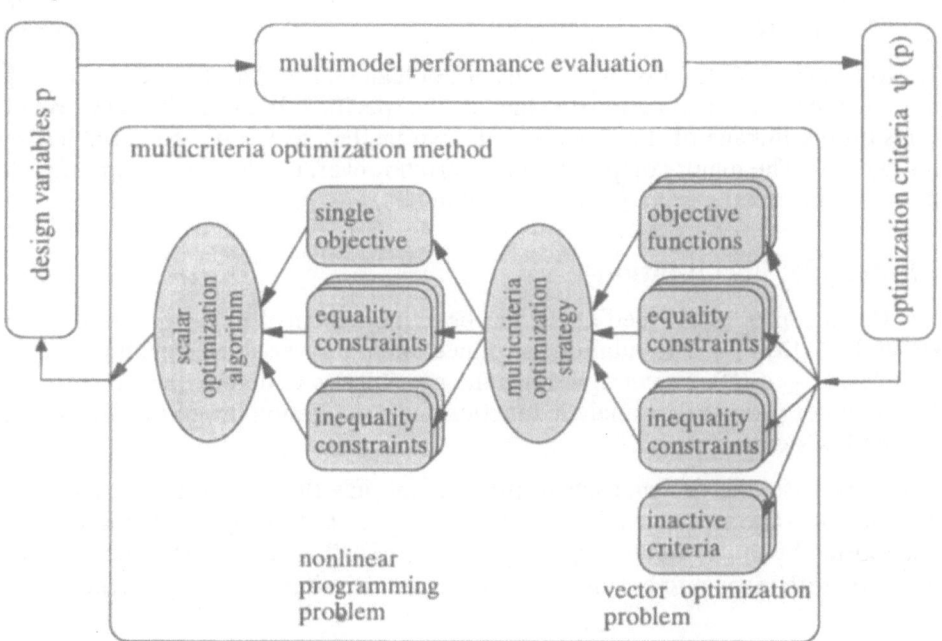

Fig. 2: Multicriteria optimization concept

The operator '*opt*' is defined as simultaneous minimization of the individual objective functions $f_j(p)$. In general, this is not possible due to conflicts arising from different criteria. The multicriteria optimization approach, however, offers a concept of defining optimal solutions also in the situation of conflicting objective functions. Design points $p^P \in \mathcal{P}$ are called Edgeworth−Pareto (EP−) optimal, if there exists no feasible design point p where $f_j(p) \leq f_j(p^P)$ $\forall j \wedge f(p) \neq f(p^P)$, [5]. In general, EP−optimal solutions are not unique and points with different images are not comparable. The designer has to make, therefore, his final decision on an acceptable optimal solution out of the set of EP−optimal points.

EP−optimal solutions of the multicriteria optimization problem have to be found iteratively requiring several performance evaluations. Applied to dynamic system design, performance evaluation involves a time−consuming numerical integration of differential equations of motion. Therefore, not all multicriteria optimization strategies seem to be appropriate for dynamic system design. Strategies which reduce the vector optimization problem to nonlinear programming problems have proven to be very efficient. Several such strategies on the basis of the principles of scalarization, hierarchization or a combination of it have been developed [6]. The resulting nonlinear programming problems can be solved very efficiently with sequential quadratic programming (SQP) algorithms. The drawback, however, is the requirement of gradient information to be computed from the submodels. Taking into consideration the structure given in Fig. 1, we obtain

$$\frac{d\psi}{dp} = \sum_{i=1}^{m} \frac{d\psi}{d\overline{\psi}^i} \frac{d\overline{\psi}^i}{d\overline{p}^i} \frac{d\overline{p}^i}{dp}. \tag{8}$$

Due to equation (6) the first term is just a Boolean matrix. The second term results from the functional relation (4) provided by the specific submodel. The last term depends on the linkage of the modelspecific parameters to the design variables, see Eqs.(1)−(3). The major computational effort, however, results from computing the sensitivity information $d\overline{\psi}^i/d\overline{p}^i$ for each submodel.

3. Submodel Formulation

The design concept described above has been implemented in the program system NEWOPT/AIMS [7]. The submodels connect the input variables $\overline{p}^i$ with the output variables $\overline{\psi}^i = \overline{\psi}^i(\overline{p}^i)$. The designer has the possibilities to define these relations by any computer program, by analytic functions or a simulation program based on the multibody system approach.

In the first case, the designer has to provide not only the criterion values, but also the gradients. The second type of criteria are explicit criteria, gradient computation is supported by the computer algebra program MAPLE. The most important type of criteria with respect to dynamic system design are criteria of integral type:

$$\overline{\psi}_j^i = G_j^{i1}(t^{i1}, y^{i1}, z^{i1}, \overline{p}^i) + \int_{t^{i0}}^{t^{i1}} F_j^i(t, y^i, z^i, \dot{z}^i, \overline{p}^i) \ dt. \tag{9}$$

This type of criteria does not only depend on the system parameters $\overline{p}^i$, but also on the state variables y^i, z^i describing the dynamic behavior of the submodel i. If the multibody system approach is used, the state variables y^i, z^i are given implicitly by differential equations of motion

$$\dot{y}^i = v^i(t, y^i, z^i, \bar{p}^i),$$

$$M^i(t, y^i, \bar{p}^i) \, \dot{z}^i + k^i(t, y^i, z^i, \bar{p}^i) = q^i(t, y^i, z^i, \bar{p}^i) \tag{10}$$

and the initial conditions

$$y^{i0}: \ \Phi^{i0}(t^{i0}, y^{i0}, \bar{p}^i) = 0, \qquad z^{i0}: \ \dot{\Phi}^{i0}(t^{i0}, y^{i0}, z^{i0}, \bar{p}^i) = 0. \tag{11}$$

For each performance evaluation, this initial value problem has to be solved numerically. Simultaneously the performance functions can be computed where the second term evaluates the dynamic behavior within an interesting time interval $[t^{i0}, t^{i1}]$ and the first term accounts for cases where special values for the final state y^{i1}, z^{i1} or a minimal time t^{i1} must be achieved. The final time t^{i1} may be fixed or given implicitly by the final state $H^{i1}(t^{i1}, y^{i1}, z^{i1}, \bar{p}^i) = 0$.

The gradient for this type of criterion function can be computed most reliable and efficient using the adjoint variable approach. This approach results in a set of additional differential equations closely related to the linearized equations of motion. The finite differences approach which is usually used in a context of complicated relations $\bar{\psi}^i(\bar{p}^i)$ has shown to be rather inefficient, inexact and unreliable [3].

4. Application to Vehicle Dynamics

In vehicle dynamics important, but contradicting criteria are riding comfort and riding safety. Even if only comfort is taken into account optimal solutions depend highly sensitive on the kind of excitation of the vehicle and the way of evaluating comfort. In principle, a complex three dimensional model including all effects would be sufficient for investigating the problem. However, experience has shown that three dimensional models with detailed description of the suspension systems will require too much computational time for being included into an iterative, interactive design process. Therefore, the models used have to be simplified to provide just the interesting effects. For example, the spatial vehicle model with simplified suspension systems in Fig. 3 is sufficient to yield information on comfort while driving over a rough road surface.

Such a model may be used for determining optimal values for global geometric data like the wheelbase. If the excitation is the same for the left and the right wheels, e.g. for driving over bumps, such a model may be even reduced to a plain vehicle model, Fig. 4. For detailed analysis of the suspension system, the motion of the car body may be reduced to just a vertical motion. Comfort evaluations for the McPherson suspension and the plain vehicle model are not only performed with respect to a single type of road excitation, but with respect to two and three kinds of excitations, respectively. The McPherson suspension is excited with a single bump and a sinoidal test road, the plain vehicle model is additionally excited by some measured road data. Therefore, these two models have to be considered as two and three submodels in the global design process. Altogether this results in a design model including 6 submodels, 5 design variables chosen as tire stiffness, stiffness and damping coefficients of the front suspension, wheelbase and track and 10 criteria of type (9), i.e.

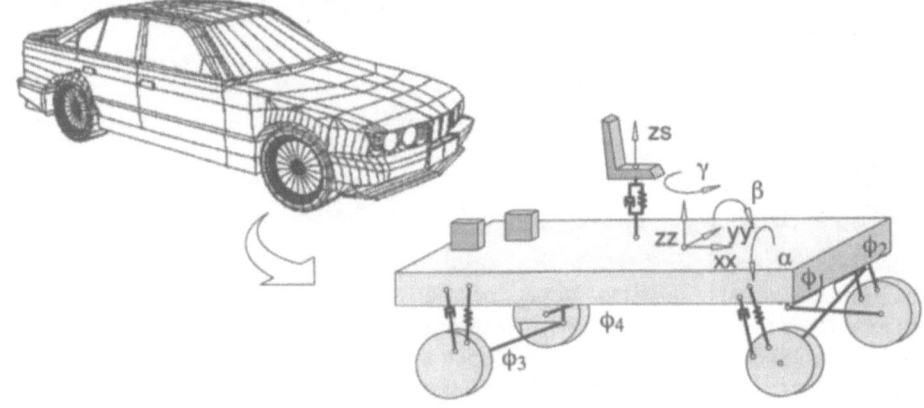

Fig. 3: Spatial vehicle model

$$\psi = \left[\psi^1_{vacc},\ \psi^2_{vacc},\ \psi^3_{vacc},\ \psi^3_{pacc},\ \psi^4_{vacc},\ \psi^4_{pacc},\ \psi^5_{vacc},\ \psi^5_{pacc},\ \psi^6_{vacc},\ \psi^6_{pacc}\right]^T. \qquad (12)$$

The designer has to provide an initial design p^0 resulting in criterion values

$$\psi(p^0) = [2.63, 3.73, 0.43, 0.43, 0.27, 0.16, 0.0033, 0.0039, 0.18, 0.039]^T \qquad (13)$$

which are used to normalize the values of the criteria in the following, i.e. $\hat{\psi}_j = \psi_j/\psi_j(p^0)$. In order to get an idea about what is achievable, the utopian point should be computed. This point is defined as vector consisting of the separately obtainable minima of the criteria yielding

$$\hat{\psi}_{utopia} = [0.002, 0.0009, 0.44, 0.23, 0.41, 0.42, 0.48, 0.45, 0.30, 0.78]^T. \qquad (14)$$

In general, there exists no design point corresponding to this utopian criterion point, and even if such a point exists, it won't be feasible. For achieving feasible EP–optimal design points a multicriteria optimization strategy has to be applied. As an example, results for three different strategies will be shown in the following.

As a first strategy, the weighted objectives method is used. All criteria are considered as objective functions yielding $f = \psi$, and only some bounds on the design variables result in inequality constraints. From comparison of the utopian point (14) with the initial point (13) the values of the 7th, 8th and 10th criterion seem to change less than the others. Therefore, the weighting factors corresponding to these criteria are chosen higher than the others, i.e. $w = [1, 1, 1, 1, 1, 1, 100, 100, 1, 10]^T$. Minimizing the utility function

$$u(p) = \sum_{j=1}^{10} w_j\, f_j(p) \qquad (15)$$

results in optimal criterion values shown in Fig. 5. The results for hierachical optimization are described in [8] and also included in Fig. 5.

A strategy which is closely related to the kind of engineering thinking is goal programming. The method allows to predefine goals which should be achieved. This may be combined with the idea of hierarchization. Here three levels of importance are

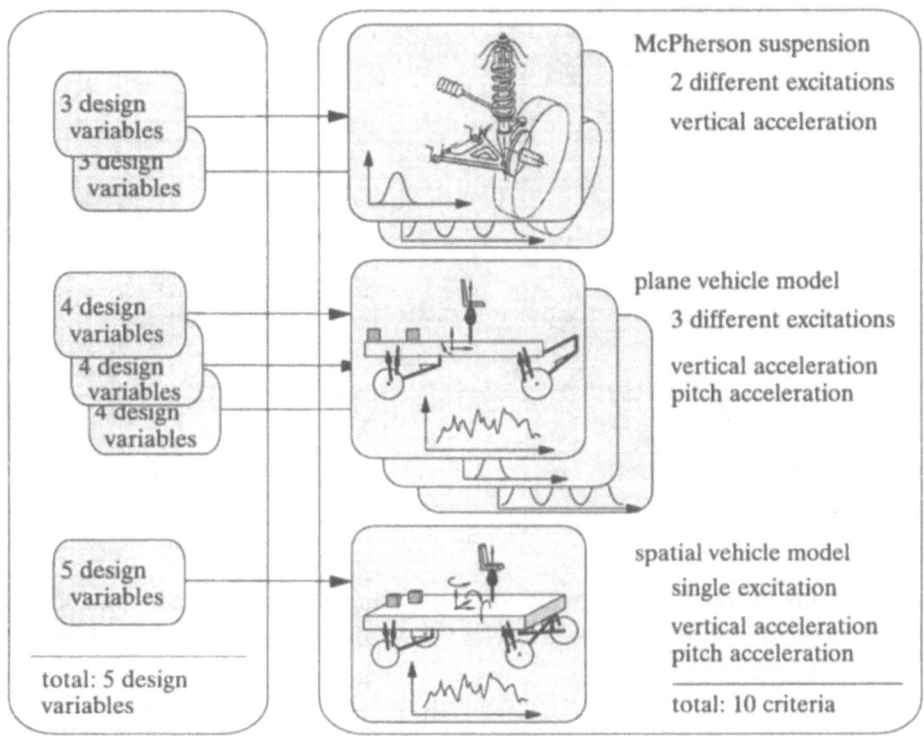

3 design
variables

3 design
variables

McPherson suspension

2 different excitations

vertical acceleration

4 design
variables

4 design
variables

4 design
variables

plane vehicle model

3 different excitations

vertical acceleration
pitch acceleration

5 design
variables

total: 5 design
variables

spatial vehicle model

single excitation

vertical acceleration
pitch acceleration

total: 10 criteria

Fig. 4: Global design model

assigned to the objectives. The group of most important objectives consists of criteria 2 and 5, the second important group of criteria 1, 3, 7 and the least important criteria are 4, 6 and 10. Criteria 8 and 9 are chosen as constraints. The normalized goals are $\hat{\psi}_{goal} = [0.015, 0.016, 0.69, 0.23, 0.48, 0.625, 0.91, -, -, 1.0]^T$.

Not all goals need to be defined at the very beginning of the optimization but can be successively introduced on the basis of the optimization results for the higher levels of importance. The final result and the goals are also given in Fig. 5.

5. Conclusions

Application of optimization methods to dynamic system design is somehow lagging behind the theoretical and algorithmic improvements in optimization theory. This is certainly due to computational effort and the restriction to the classical nonlinear programming problem. Formulation of design problems as such a scalar optimization problem results in solutions which are often even inferior to common sense designs, since only single aspects of the system requirements are taken into consideration. Recent advances in computer technology in combination with multicriteria and multimodel optimization ideas seem to open up new ways for designing dynamic systems. The focus has to change from the solution of the optimization problem to its flexible and convenient formulation. Then, the optimal solution is not a single de-

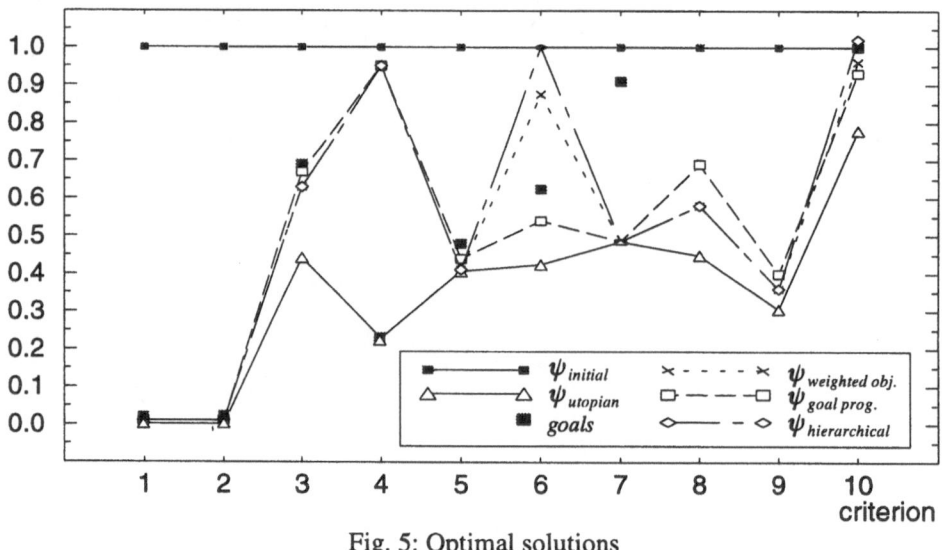

Fig. 5: Optimal solutions

sign point any more, but a set of EP–optimal points from which the designer may chose according to his preferences. All aspects of system requirements can be taken into account simultaneously.

In the paper, a fairly general design concept integrating problem formulation and problem solution is shown. The designer is supported in his decision by several different multi–criteria optimization strategies searching for optimal solutions. An important feature of this approach is the possibility of including several different models specific to special requirements. This offers a high flexibility of problem formulation, especially for choosing the model as simple as possible in order to save computational time and to get good insight in the dynamic behavior and the effects of design changes on the performance of the technical system.

References

[1] Schiehlen, W. (ed.): *Multibody Systems Handbook*. Berlin: Springer, 1990.

[2] Grübel, G. et al.: ANDECS – A Computation Environment for Control Applications of Optimization. In: *Computational Optimal Control*, by R. Bulirsch and D. Kraft (eds.). Basel: Birkhäuser, 1994, pp. 237–254.

[3] Bestle, D. and Eberhard, P.: Analyzing and Optimizing Multibody Systems. *Mech. Struct. and Mach.* **20** (1992) 67–92.

[4] Bestle, D.: *Analyse und Optimierung von Mehrkörpersystemen*. Berlin: Springer, 1994.

[5] Stadler, W. (ed.): Multicriteria Optimization in Engineering and in the Sciences. New York: Plenum Press, 1988.

[6] Bestle, D. and Eberhard, P.: Automated Approach for Optimizing Dynamic Systems. In: *Computational Optimal Control*, by R. Bulirsch and D. Kraft (eds.). Basel: Birkhäuser, 1994, pp. 225–235.

[7] Bestle, D. and Eberhard, P.: *NEWOPT / AIMS2.2 – Ein Programmpaket zur Analyse und Optimierung von mechanischen Systemen*. User's Manual. Stuttgart: University, Institute B of Mechanics, 1994.

[8] Eberhard, P.: *Zur Mehrkriterienoptimierung von Mehrkörpersystemen*. PhD–Thesis. Stuttgart: University, Institute B of Mechanics, 1995.

ON THE AUTOMATIC DIFFERENTIATION OF COMPUTER PROGRAMS AND AN APPLICATION TO MULTIBODY SYSTEMS

CHRISTIAN H. BISCHOF

Mathematics and Computer Science Division
Argonne National Laboratory
9700 S. Cass Avenue, Argonne, IL 60439
bischof@mcs.anl.gov

Abstract. Automatic differentiation (AD) is a methodology for developing sensitivity-enhanced versions of arbitrary computer programs. In this paper, we provide some background information on AD and address some frequently asked questions. We introduce the ADIFOR and ADIC tools for the automatic differentiation of Fortran 77 and ANSI-C programs, respectively, and give an example of applying ADIFOR in the context of the optimization of multibody systems.

1. Introduction

Assume that we have a code for the computation of a function f and $f : x \in \mathbf{R}^n \mapsto y \in \mathbf{R}^m$, and we wish to compute the derivatives of y with respect to x. We call x the *independent variable* and y the *dependent variable*.

In computing derivatives, we should keep the following issues in mind:

Reliability: The computed derivatives should ideally be accurate to machine precision.

Computational Cost: In many applications, the computation of derivatives is the dominant computational burden. Hence, the amount of memory and runtime required for the derivative code should be minimized.

Scalability: Whatever method we choose should be applicable to a 1-line formula as well as a 50,000-line code.

D. Bestle and W. Schielen (eds.), IUTAM Symposium on Optimization of Mechanical Systems, 41–48.
© *1996 Kluwer Academic Publishers.*

Human Effort: Derivatives are a means to an end. Hence a user should not spend much time in computing derivatives, in particular in situations where computer models are bound to change frequently.

Handcoding, divided-difference approximations, and symbolic methods traditionally have been used for the computation of derivatives. However, these methods fall short with respect to the previously mentioned criteria. The main drawbacks of divided-difference approximations are their numerical unpredictability and their computational cost. In contrast, both the handcoding and symbolic approaches suffer from a lack of scalability and require considerable human effort.

In this paper, we discuss another approach for computing derivatives, based on automatic differentiation (AD). AD techniques rely on the fact that every function, no matter how complicated, is executed on a computer as a (potentially very long) sequence of elementary operations such as additions, multiplications, and elementary functions such as **sin** and **cos** (see, for example, [10, 16]. By applying the chain rule of derivative calculus over and over again to the composition of those elementary operations, one can compute, in a completely mechanical fashion, derivatives of f that are correct up to machine precision [12].

In the next section, we give a brief overview of automatic differentiation. Section 3 introduces the ADIFOR and ADIC AD tools for Fortran 77 and ANSI-C, respectively, and Section 4 answers some commonly asked questions. In Section 5, we report on the application of ADIFOR in the context of the optimization of a multibody system. Lastly, we summarize our results.

2. Automatic Differentiation

Traditionally, two approaches to automatic differentiation have been developed: the so-called forward and reverse modes. These modes are distinguished by how the chain rule is used to propagate derivatives through the computation. We briefly summarize the main points about these two approaches; a more detailed description can be found in [4] and the references therein.

The forward mode propagates derivatives of intermediate variables with respect to the independent variables and follows the control flow of the original program. By exploiting the linearity of differentiation, the forward mode allows us to compute arbitrary linear combinations $J \cdot S$ of columns of the Jacobian

$$J = \begin{pmatrix} \frac{\partial y(1)}{\partial x(1)} & \cdots & \frac{\partial y(1)}{\partial x(n)} \\ \vdots & & \vdots \\ \frac{\partial y(m)}{\partial x(1)} & \cdots & \frac{\partial y(m)}{\partial x(n)} \end{pmatrix}. \tag{1}$$

For an $n \times p$ matrix S, the effort required is roughly $O(p)$ times the runtime and memory of the original program. In particular, when S is a vector s, we compute the directional derivative $J * s = \lim_{h \to 0} \frac{f(x+h*s)-f(x)}{h}$.

In contrast, the reverse mode of automatic differentiation propagates derivatives of the final result with respect to an intermediate quantity, so-called adjoint quantities. To propagate adjoints, one must be able to reverse the flow of the program, and remember or recompute any intermediate value that nonlinearly affects the final result. In particular, one must remember the intermediate values taken by variables that are overwritten, and keep a log of the branch directions taken. Also, changing a "+" to a "*" in the computer code can have profound ramifications for the complexity of the generated reverse mode code, while it does not have much effect for the forward mode.

For a $q \times m$ matrix W, the reverse mode allows us to compute the row linear combination $W \cdot J$ with $O(q)$ times as many floating-point operations as required for the evaluation of f. In a straightforward implementation, however, the storage requirements may be proportional to the number of floating-point operations required for the evaluation of f, as a result of the tracing required to make the program "reversible." When W is a row vector w, we compute the derivative $\frac{\partial (w^T * J)}{\partial x}$. The reverse mode is particularly attractive for the computation of long gradients, as its floating-point complexity does not depend on the number of independent variables.

In either case, automatic differentiation produces code that computes derivatives accurate to machine precision [12]. The techniques of automatic differentiation are directly applicable to computer programs of arbitrary length containing branches, loops, and subroutines.

3. Automatic Differentiation Tools

We are involved in the development of the ADIFOR (jointly with Rice University) and ADIC tools, which provide automatic differentiation functionality for Fortran 77 and ANSI-C, respectively, The ADIFOR 2.0 system is mature, and reference [4] lists 25 references reporting on the use of ADIFOR in various application domains, on codes of up to 60,000 lines. ADIC, in contrast, is in the prototype phase, but has been successfully applied to codes of up to 10,000 lines. ADIFOR and ADIC employ a source transformation approach directly rewriting the source code. This approach requires considerable compiler infrastructure, and ADIFOR and ADIC employ the ParaScope [8] and Sage++ [7] compiler environments developed at Rice and Indiana University, respectively. For references to other automatic differentiation tools, see [4].

ADIFOR and ADIC employ a hybrid forward/reverse-mode approach

$$
\left.\begin{array}{l}
\text{r\$1} = \text{x}(1) \ast \text{x}(2) \\
\text{r\$2} = \text{r\$1} \ast \text{x}(3) \\
\text{r\$3} = \text{r\$2} \ast \text{x}(4) \\
\text{r\$4} = \text{x}(5) \ast \text{x}(4) \\
\text{r\$5} = \text{r\$4} \ast \text{x}(3) \\
\text{r\$1bar} = \text{r\$5} \ast \text{x}(2) \\
\text{r\$2bar} = \text{r\$5} \ast \text{x}(1) \\
\text{r\$3bar} = \text{r\$4} \ast \text{r\$1} \\
\text{r\$4bar} = \text{x}(5) \ast \text{r\$2}
\end{array}\right\}
$$

Reverse Mode for computing $\dfrac{\partial y}{\partial x(i)}$:

$$
\text{r\$jbar} = \frac{\partial y}{\partial x(i)}, \; i = 1, \dots, 4
$$

$$
\text{r\$3} = \frac{\partial y}{\partial x(5)}
$$

```
do g$i$ = 1, g$p$
   g$y(g$i$) = r$1bar * g$x(g$i$,1)
             + r$2bar * g$x(g$i$,2)
             + r$3bar * g$x(g$i$,3)
             + r$4bar * g$x(g$i$,4)
             + r$3 * g$x(g$i$, 5)
enddo
y = r$3 * x(5)
```

} Forward Mode:
Assembling ∇y
from $\dfrac{\partial y}{\partial x(i)}$ and $\nabla x(i)$,
$i = 1, \dots, 5$.

} Computing function value

Figure 1. Sample Segment of an ADIFOR-generated Code

to generating derivatives. For each assignment statement, they use the reverse mode to generate code that computes the partial derivatives of the result with respect to the variables on the right-hand side and then employ the forward mode to propagate overall derivatives. For example, ADIFOR transforms the Fortran statement

$$
\text{y} = \text{x}(1) \ast \text{x}(2) \ast \text{x}(3) \ast \text{x}(4) \ast \text{x}(5)
$$

into the code segment shown in Figure 1.[1] Note that none of the common subexpressions $x(i) \ast x(j)$ are recomputed in the reverse-mode section for $\dfrac{\partial y}{\partial x(i)}$. The variable g\$p\$ denotes the number of (directional) derivatives being computed. For example, if g\$p\$ = 5, and g\$x(1:5,1:5) is the 5 × 5 identity matrix (i.e., g\$x(i,j) = $\frac{\partial x(i)}{\partial x(j)}$), then upon execution of these statements, g\$y(1:5) equals $\frac{dy}{dx}$. On the other hand, assume that we wished only to compute derivatives with respect to a scalar parameter s, so g\$p\$ = 1, and, on entry to this code segment, g\$x(1,i) = $\frac{\partial x(i)}{\partial s}$, $i = 1, \dots, 5$. Then the do-loop in Figure 1 implicitly computes $\frac{dy}{ds} = \frac{dy}{dx}\frac{dx}{ds}$ without ever forming $\frac{\partial y}{\partial x}$ explicitly.

ADIFOR and ADIC provide the directional derivative computation possibilities associated with the forward mode of automatic differentiation. We also mention that both ADIFOR and ADIC can transparently exploit sparsity in derivative computations by replacing the dense vector loop in Figure 1 with a call to a SparsLinC routine [4, 5], which, as a byproduct of

[1] The dollar sign indicates ADIFOR-generated variables. ADIFOR 2.0 could use any other character instead, taking care not to generate duplicate names.

the computation, will automatically compute the sparsity pattern of large sparse Jacobians.

None of these AD tools require any knowledge of the application domain. Hence, unlike handcoding or symbolically assisted approaches, automatic differentiation enables derivatives to be updated easily when the original code changes. Information on these tools as well as application highlights and reports can be found on the world-wide web at

http://www.mcs.anl.gov/autodiff/index.html.

4. Frequently Asked Questions

Given the mathematical underpinnngs of the concept of derivatives, the "ignorance" with which one can apply an AD tool usually provokes some of the questions that we try briefly to address here.

Question: How do you know that the code represents a globally differentiable function?

Answer: We don't. AD computes the derivative defined by the sequence of assignment statements executed in the course of a function evaluation. Hence, for a branch (if-statement), which potentially introduces a nondifferentiability, AD will compute a one-sided directional derivative. This problem is further discussed in [9].

Question: How do you deal with intrinsics?

Answer: Some intrinsics functions, such as abs() and sqrt(), are not differentiable in all points of their domain. At these points, ADIFOR invokes the ADIntrinsics system [4] to provide a (user customizable) default value, and prints a warning message. The ADIC prototype uses a similar, although less refined, mechanism.

Question: What happens when you differentiate through iterative processes?

Answer: It depends. AD generates a new iteration, and it is not clear a priori whether the new iteration will converge and what it will converge to, although empirically, AD leads to the desired result. However, derivative convergence may lag, or derivatives may diverge. For some commonly used approaches for solving nonlinear systems of equations, this issue is discussed in [11]. This problem clearly requires more research, but the emergence of robust AD tools has made it possible to tackle this problem for sophisticated numerical methods.

5. An Example: The Iltis All-Terrain Vehicle

The dynamic and kinetic behavior of vehicles can be modeled through multibody systems. Optimization techniques can then be employed to improve the design of such a vehicle with respect to comfort, ride, and handling. For an overview of this field as well as the methods employed, see [1].

In general, the motion of a multibody system can be described as follows:

$$\left.\begin{array}{rcl} \dot{y} & = & v(t,y,z,p) \\ M(t,y,p)\,\dot{z} + k(t,y,z,p) & = & q(t,y,z,p) \end{array}\right\}, \qquad (2)$$

where $\dot{y} = \frac{\partial y}{\partial t}$ is the derivative with respect to the time t, M is the mass matrix, k are the coriolis forces, q the external forces, y generalized position coordinates, z generalized velocity coordinates, and p the design parameters.

An efficient method for optimizing a multibody system is the adjoint variable method developed by Bestle and Eberhard [2], which requires the derivatives $\frac{\partial M_{mn}}{\partial t}$, $\frac{\partial M_{mn}}{\partial y_i}$, $\frac{\partial M_{mn}}{\partial p_k}$, $\frac{\partial (k_m - q_m)}{\partial y_i}$, $\frac{\partial (k_m - q_m)}{\partial z_j}$, and $\frac{\partial (k_m - q_m)}{\partial p_k}$. In [13], Häußermann applied the first version of ADIFOR [3] to several multibody systems and compared it with symbolic approaches and with approximations of derivatives via divided differences.

However, application of ADIFOR 1.0 to the so-called Iltis problem, a benchmark problem modeling an all-terrain vehicle [15], proved to be somewhat laborious. ADIFOR 1.0 was unable to process the subroutine of several thousand lines describing the equations of motion that had been generated with the NEWEUL [14] package. The problem had to be split by hand, a somewhat laborious and error-prone process.

With the new ADIFOR 2.0 system, however, one can now process the code as is. Differentiating with respect to 20 parameters, one obtains the results shown in Table 1. Computations were performed on a Silicon Graphics Indigo with 32 MB RAM and a 100 Mhz MIPS R4000 microprocessor. Here "Iltis" refers to the original code, and "Iltis.AD" refers to the

TABLE 1. Results of Applying ADIFOR 2.0 to the Iltis Problem

	Iltis.AD	Iltis	Ratio
Memory (MB)	3.52	0.52	6.7
Runtime (sec)	42.6	2.13	20.0
Lines of code	71,887	11,172	6.4

code generated by ADIFOR 2.0. We see that the memory required by the ADIFOR-generated code increases by a factor of 6.7, whereas runtime increases by a factor 20, the same cost increase one would also experience with divided-difference approximations. In most cases, however, ADIFOR-generated code outperforms one-sided divided-difference approximations, typically by a factor 1.5 to 3, and by a factor of 7.4 in the best case so far [6]. Code expansion is considerable because of the somewhat unusual structure of the NEWEUL-generated code. The number of lines of code increases by a factor of 6.4, and the resulting length of the .AD versions of the NEWEUL-generated files prevented compilation on an HP workstation. In our experience, code expansion by a factor 2 to 3 is typical. The generated code accurately computes the desired derivatives, whereas the study by Häußerman shows that this is not necessarily the case for divided difference approximations.

6. Conclusions

This paper gave a brief introduction into automatic differentiation. We reviewed the forward and reverse mode of automatic differentiation, answered some commonly asked questions, and introduced the ADIFOR and ADIC automatic differentiation tools. We also presented results on applying ADIFOR 2.0 to the Iltis multibody benchmark problem, which showed that reliable and efficient derivatives can be computed by using AD with minimal recourse to laborious and error-prone hand coding.

Acknowledgments

We thank Peter Eberhard for providing us with the Iltis code and for performing the benchmark runs. We also thank Peter Eberhard and Dieter Bestle for introducing us to multibody system optimization. Lastly, we thank Ralf Knösel for processing the Iltis code with ADIFOR 2.0.

This work was supported by the Office of Scientific Computing, U.S. Department of Energy, under Contract W-31-109-Eng-38; by the National Aerospace Agency under Purchase Order L25935D; and by the National Science Foundation, through the Center for Research on Parallel Computation, under Cooperative Agreement No. CCR-9120008.

References

1. Dieter Bestle. *Analyse und Optimierung von Mehrkörpersystemen*. Springer, Berlin, 1994.
2. Dieter Bestle and Peter Eberhard. Analyzing and optimizing multibody systems. *Mechanical Structures and Machinery*, 20(1):67–92, 1992.

48 C. H. BISCHOF

3. Christian Bischof, Alan Carle, George Corliss, Andreas Griewank, and Paul Hovland. ADIFOR: Generating derivative codes from Fortran programs. *Scientific Programming*, 1(1):11–29, 1992.
4. Christian Bischof, Alan Carle, Peyvand Khademi, and Andrew Mauer. The ADIFOR 2.0 system for the automatic differentiation of Fortran 77 programs, 1994. Preprint MCS-P481-1194, Mathematics and Computer Science Division, Argonne National Laboratory, and CRPC-TR94491, Center for Research on Parallel Computation, Rice University.
5. Christian Bischof and Andrew Mauer. ADIC – A tool for the automatic differentiation of C programs. Preprint MCS-P499-0295, Mathematics and Computer Science Division, Argonne National Laboratory, 1995.
6. Christian Bischof, Greg Whiffen, Christine Shoemaker, Alan Carle, and Aaron Ross. Application of automatic differentiation to groundwater transport models. In Alexander Peters et al., editors, *Computational Methods in Water Resources X*, pages 173–182. Kluwer Academic Publishers, Dordrehct, 1994.
7. Francois Bodin, Peter Beckman, Dennis Gannon, Jacob Goutwals, Srinivas Narayana, Suresh Srinivas, and Beata Winnicka. SAGE++: An object-oriented toolkit and class library for building Fortran and C++ restructuring tools. In *Proceedings of the Second Annual Object-Oriented Numerics Conference*. IEEE, 1994.
8. D. Callahan, K. Cooper, R. T. Hood, K. Kennedy, and L. M. Torczon. ParaScope: A parallel programming environment. *International Journal of Supercomputer Applications*, 2(4):84–99, December 1988.
9. Herbert Fischer. Special problems in automatic differentiation. In Andreas Griewank and George F. Corliss, editors, *Automatic Differentiation of Algorithms: Theory, Implementation, and Application*, pages 43 – 50. SIAM, Philadelphia, Penn., 1991.
10. Andreas Griewank. On automatic differentiation. In *Mathematical Programming: Recent Developments and Applications*, pages 83–108. Kluwer Academic Publishers, Amsterdam, 1989.
11. Andreas Griewank, Christian Bischof, George Corliss, Alan Carle, and Karen Williamson. Derivative convergence of iterative equation solvers. *Optimization Methods and Software*, 2:321–355, 1993.
12. Andreas Griewank and Shawn Reese. On the calculation of Jacobian matrices by the Markowitz rule. In Andreas Griewank and George F. Corliss, editors, *Automatic Differentiation of Algorithms: Theory, Implementation, and Application*, pages 126–135. SIAM, Philadelphia, 1991.
13. Uli Häußermann. Automatische Differentiation zur Rekursiven Bestimmung von Partiellen Ableitungen. STUD-102, Institut B für Mechanik, Universität Stuttgart, 1993.
14. E. Kreuzer and G. Leister. Programmsystem NEWEUL'90. Technical Report Anleitung AN-24, Institut B für Mechanik, Universität Stuttgart, 1991.
15. G. Leister and W. Schiehlen. Benchmark-beispiele des DFG-schwerpunktprogramms dynamic von mehrkörpersystemen. Technical Report Zwischenbericht ZB-64, Band 2, Institut B für Mechanik, Universität Stuttgart, 1991.
16. Louis B. Rall. *Automatic Differentiation: Techniques and Applications*, volume 120 of *Lecture Notes in Computer Science*. Springer Verlag, Berlin, 1981.

LAYOUT OF LINEAR AND NONLINEAR STRUCTURES BY SHAPE AND TOPOLOGY OPTIMIZATION

K.-U. BLETZINGER, K. MAUTE, R. REITINGER, E. RAMM
Institut für Baustatik, Universität Stuttgart,
D-70550 Stuttgart, Germany

1. Introduction

Structural optimization of mechanical systems has evolved to provide powerful tools during all stages of structural design [1]. They are characterized by decisions on the structural layout, the shape, and the dimensions of the structure. Equivalent structural optimization problems deal with optimization of topology, shape, and cross sections, respectively. The result depends on interactions between these three different types of optimization problems, as well as on the underlying physical assumptions of the simulation of structural response. Even if linear static or dynamic structural response is assumed structural optimization problems are usually non-linear and in the case of topology optimization usually non-convex. Additionally, powerful optimization procedures have to combine methods of different disciplines: geometric modeling, computational mechanics, and mathematical programming [2]. Consequently, due to the inherent complexity, optimization problems tend to be formulated in a way that neither the full potential of the concerned methods are exploited nor the effects of interactions between the different types of optimization problems are considered. This is e.g. the case for the interactions of topology and shape optimization, and the shape optimization of imperfection sensitive structures with large deflections and instability response. The contribution focuses on these two distinct areas of structural optimization which are of actual importance.

2. Shape optimization of buckling sensitive structures

It is well known for certain structures, in particular shells, that small deviations from the ideal shape may cause a dramatic reduction of the buckling load. This imperfection sensitivity is in particular pronounced if structural optimization is used to increase the load carrying capacity and therefore the efficiency of the structural design. It is essential to include the effects of shape imperfections into the optimization procedure. Stiffened panels, as they are used in many fields of engineer-

49

D. Bestle and W. Schielen (eds.), IUTAM Symposium on Optimization of Mechanical Systems, 49–56.
© 1996 *Kluwer Academic Publishers.*

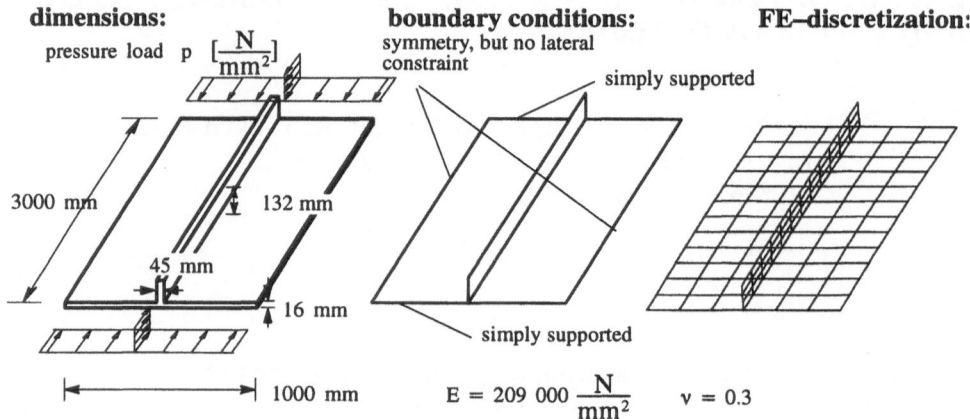

Figure 1. Stiffened panel: model for structural analysis.

ing, are typical examples of these kinds of structures. Consider e.g. the panel depicted in Figure 1, which was also investigated by Tvergard [3], among others. The panel is loaded by a constant axial pressure load scaled by the load parameter λ. The finite element method is applied for structural analysis. The structure is discretized by eight–noded isoparametric degenerated shell elements.

A geometrically nonlinear analysis shows that the ideal structure will fail by buckling into a global Euler type buckling mode, Figure 2. The corresponding critical load parameter is determined to be $\lambda_{cr} = 256.6$ N/mm^2. If a shape imperfection is added the critical point converts to a limit point and the critical load factor drops to $\lambda_{imp} = 194.7$ N/mm^2. The imperfection was assumed to be affine to the first eigenmode with an amplitude of 10mm.

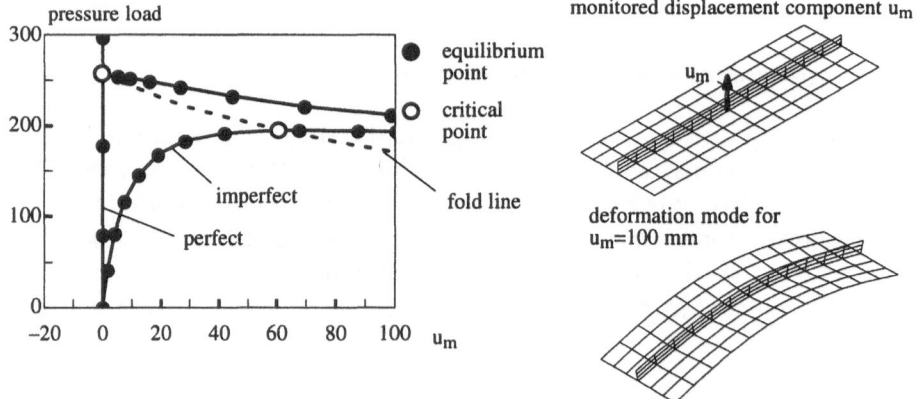

Figure 2. Load–deflection–diagrams of perfect and imperfect initial structure.

The objective of optimization is to maximize the critical load factor. Additionally, the total volume of the structure has to remain constant and is prescribed by an equality constraint. The proposed general optimization procedure [4,5] is controlled by an SQP method [6] and allows to define any combination of shape and thickness variables. In the special case of the mentioned panel the thicknesses of the panel skin and the stiffener as well as the height of the stiffener are variables. Two different optimization problems have been considered: a) maximize the critical load of the perfect structure, and b) maximize the minimum of the critical load factors of the perfect and imperfect structure, Figure 3. For case a) the critical load of the perfect panel could be increased to $\lambda_{cr} = 295.4$ N/mm^2, However, the structure remains to be sensitive to shape imperfections, as it is shown by a subsequent analysis of the optimized structure. For case b) the critical load of the imperfect structure is considerably improved to $\lambda_{cr} = 273.4$ N/mm^2, as desired.

The optimization procedure is based on the idea to directly trace the critical points of the load deflection relation as the structural shape converges to the optimal solution. After the critical load factor and the corresponding deflections of the initial structure have been evaluated these data are used as starting values to determine directly the critical load of the next iterate. This is repeated with any transition from one optimization step to the next. That means, that during the whole opti–

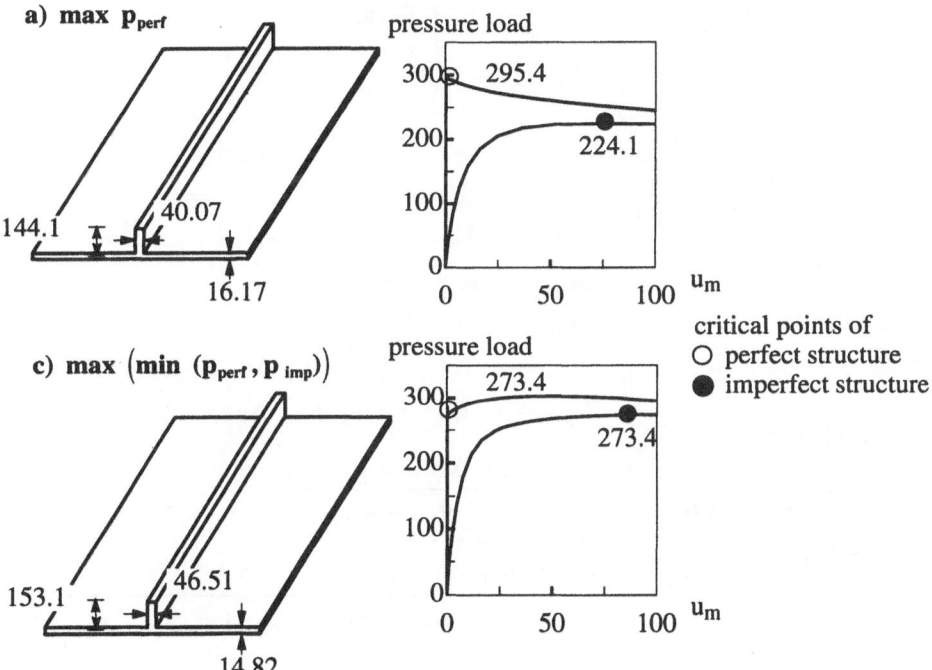

Figure 3. Optimization of stiffened panel with three variables.

mization procedure a complete nonlinear structural analysis starting from the un-
deflected shape has to be carried out only for the initial structure [4,5].

The underlying analytical concepts are the methods a) of numerical continuation or path–following to trace the nonlinear relations of load and deflections, and
b) for the direct calculation of the singular points of interest. For a detailed list of
literature see [4]. In either case the non–linear equations of equilibrium are augmented by additional equations which are a) an additional control equation in the
n+1 dimensional space of deflections $\mathbf{u}$ and load parameter λ, and b) the singularity
condition and a length constraint on the eigenmode which act in the 2n+1 dimensional space of deflections $\mathbf{u}$, eigenmode ϕ, and load factor λ. Refer to Figure 4
for further details on the formulation. The sensitivity of λ is consistently derived
from theses formulae. Since the stiffness matrix $\mathbf{K}_T$ is singular at the critical point,
$\mathbf{K}_T^{-1}$ is to be understood as a symbolic notation. The problem of inverting a singular
matrix is solved by a penalty formulation. The two types of methods are applied
to either the perfect or the imperfect structure, depending on which critical load has
to be optimized. In the latter case the shape of imperfection is evaluated by an atten-
dant eigenvalue analysis of the perfect structure. Further applications of the optimization procedure are reported in [5] e.g. form finding of free form shells.

Path–Following

$$\left\{ \begin{array}{c} G\,(\mathbf{u},\,\lambda) \\ c\,(\mathbf{u},\,\lambda) \end{array} \right\} = 0$$

Linearization:

$$\begin{bmatrix} \mathbf{K}_T & -\mathbf{P} \\ c_{,\mathbf{u}}{}^T & c_{,\lambda}{}^T \end{bmatrix} \begin{bmatrix} \Delta\mathbf{u} \\ \Delta\lambda \end{bmatrix} = - \begin{bmatrix} G \\ c \end{bmatrix}$$

Partitioning:

$$\Delta\mathbf{u}_P = \mathbf{K}_T^{-1}\,\mathbf{P}$$

$$\Delta\mathbf{u}_G = -\mathbf{K}_T^{-1}\,G$$

Iterative correction:

$$\Delta\lambda = \frac{c + c_{,\mathbf{u}}{}^T \Delta\mathbf{u}_G}{c_{,\lambda} + c_{,\mathbf{u}}{}^T \Delta\mathbf{u}_P}$$

$$\Delta\mathbf{u} = \Delta\mathbf{u}_P\,\Delta\lambda + \Delta\mathbf{u}_G$$

Extended System Method for critical point

$$\left\{ \begin{array}{c} G\,(\mathbf{u},\,\lambda) \\ \mathbf{K}_T\,(\mathbf{u},\,\lambda)\,\phi \\ 1\,(\phi) \end{array} \right\} = 0$$

$$\begin{bmatrix} \mathbf{K}_T & 0 & -\mathbf{P} \\ (\mathbf{K}_T\phi)_{,\mathbf{u}} & \mathbf{K}_T & (\mathbf{K}_T\phi)_{,\lambda} \\ 0^T & 1_{,\phi} & 0 \end{bmatrix} \begin{bmatrix} \Delta\mathbf{u} \\ \Delta\phi \\ \Delta\lambda \end{bmatrix} = - \begin{bmatrix} G \\ \mathbf{K}_T\,\phi \\ 1 \end{bmatrix}$$

$$\Delta\mathbf{u}_P = \mathbf{K}_T^{-1}\,\mathbf{P}$$

$$\Delta\mathbf{u}_G = -\mathbf{K}_T^{-1}\,G$$

$$\Delta\phi_P = -\mathbf{K}_T^{-1}\,(\,(\mathbf{K}_T\,\phi)_{,\mathbf{u}}\,\Delta\mathbf{u}_P + (\mathbf{K}_T\phi)_{,\lambda}\,)$$

$$\Delta\phi_G = -\mathbf{K}_T^{-1}\,(\,(\mathbf{K}_T\,\phi)_{,\mathbf{u}}\,\Delta\mathbf{u}_G)$$

$$\Delta\lambda = \frac{\phi^T\,\Delta\phi\,_G + \|\phi\|\,1_0}{\phi^T\,\Delta\,\phi\,_P}$$

$$\Delta\mathbf{u} = \Delta\mathbf{u}_P\,\Delta\lambda + \Delta\mathbf{u}_G$$

$$\phi = \Delta\phi_P\,\Delta\lambda + \Delta\phi_G$$

- Approximation of directional derivative of $\mathbf{K}_T$

$$(\mathbf{K}_T\,\phi\,)_{,\mathbf{u}}\,\Delta\mathbf{u} \approx \tfrac{1}{\varepsilon}((\mathbf{K}_T(\mathbf{u} + \varepsilon\phi,\lambda)\,\Delta\mathbf{u} - \mathbf{K}_T(\mathbf{u},\lambda)\,\Delta\mathbf{u}\,)$$

Figure 4. Path–following and direct calculation of critical points.

3. Adaptive topology optimization

During the preliminary design of structures the definition of layout or topology is one of the most important tasks. During the last years numerous methods have been suggested and applied to a broad range of problems. An overview can be found in Bendsøe and Mota Soares [7]. The most important and widely adapted approaches to the optimization of topology are those which are based on an optimal redistribution of material within a certain design space, Figure 5. In most cases, they are formulated as the minimization of structural compliance with respect to a volume equality constraint. Further extensions applied the methods to the optimum design of structures with respect to natural frequencies and other objectives [8–10]. Commonly, a special optimality criterion method is used for the solution. However, mathematical programming schemes have also been used. The field of material density is unknown and a non–linear relation between material density and stiffness is assumed. This relation is either artificially defined or derived from homogenization theory. The discretization of the design space is in general assumed to be constant during optimization and identical to the finite element discretization which is used for structural and sensitivity analyses.

The main advantages of material based topology optimization methods are their simplicity and stability. However, they have a number of shortcomings: (1) a large number of design variables since the discretization of design space and analysis model are identical, (2) void areas are analyzed and 'optimized', (3) in regions of the design space, where the parametrization is too coarse, the structure cannot be clearly identified, (4) because of the fixed discretization the optimum layout exhibits jagged boundaries which have to be smoothed in an extra post processor procedure [11], and (5) local quantities like stresses cannot be controlled during the optimization process.

The modifications of adaptive topology optimization are intended to overcome the above mentioned shortcomings [12]. The main idea is to separate design and analysis models. The design model is conventionally defined by a fine and regular discretization of the design space into design patches. It is used as a kind of back–

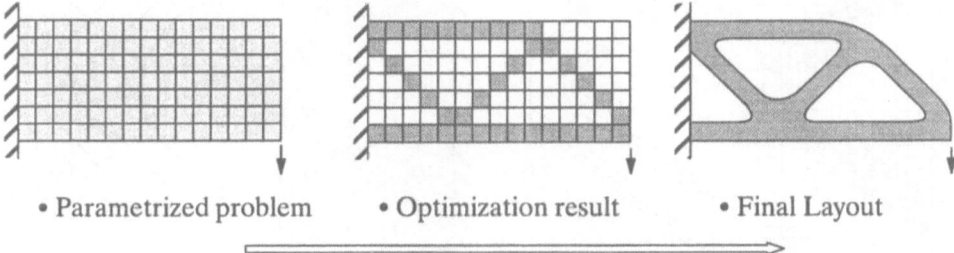

• Parametrized problem   • Optimization result   • Final Layout

Figure 5. Material based topology optimization.

Design Model   Analysis Model

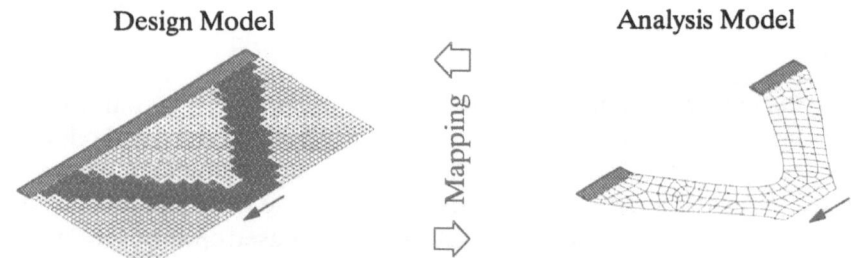

Figure 6. Alternative design modelling.

ground mesh to store the material distribution information and always spans the entire design space allowing the analysis model to shrink and swell. The analysis model, however, is adaptively adjusted to cover the non–void regions of the design space, Figure 6. It describes the active areas of the design process by smooth boundaries and with just as many elements as necessary. With the generation of the analysis model the material properties are transferred from the design model. Now, the active variables of the optimization procedure are the material parameters of the actual analysis model. Their distribution is modified by a standard material re-distribution technique and is mapped back into the design model to formulate the next analysis model and to proceed with the overall optimization procedure.

The smooth adaptation of boundaries has consequences on several other aspects of the technique. Because of the irregular shape of the active domain the mesh of the analysis model is generated by a free mesh generator [13]. The mesh varies drastically with every optimization step. That in turn means that the material law used in the redistribution technique has to be mesh independent. Therefore, a powered orthotropic law with three variables (two directional properties and one angle) is used [12].

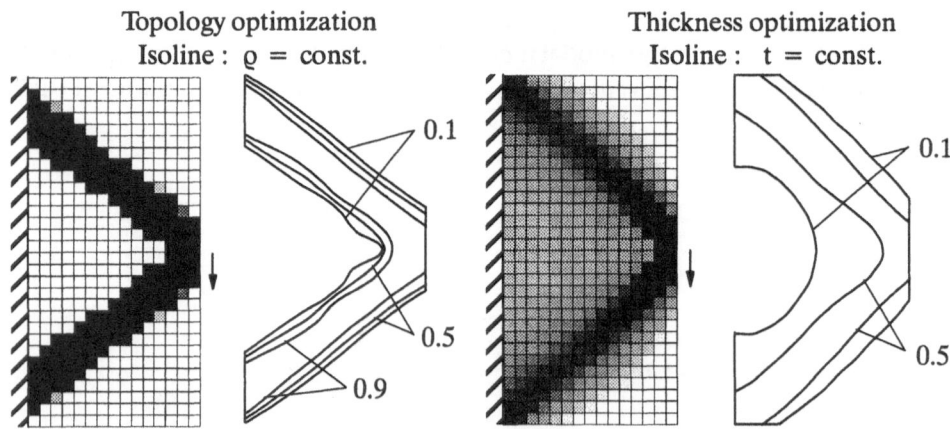

Figure 7. Material distribution for topology and thickness optimization.

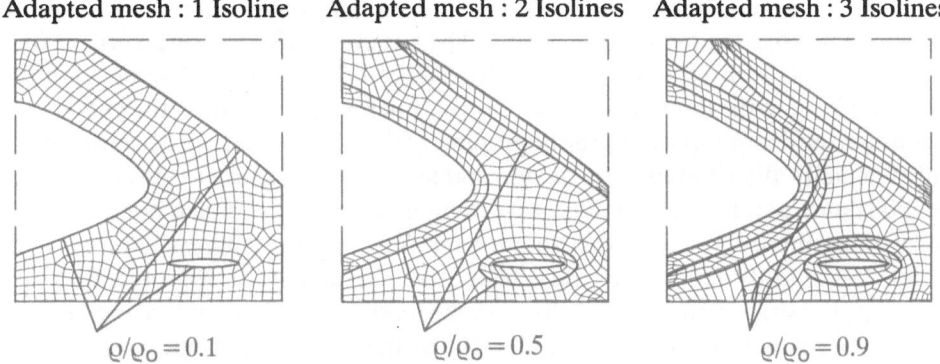

Figure 8. Mesh refining by multiple isolines.

Isolines of the material density distribution within the design model are used to define the boundaries of the analysis model. Regions with a density below a certain threshold value are neglected. Since the density gradient is very steep at the transition zones from void to active areas the choice of proper threshold values is not crucial, compared to the gradient of thickness distribution if thicknesses are variable, Figure 7. Several isolines can be used to define a layered analysis mesh at the boundary to further improved the response quality of the mesh, Figure 8.

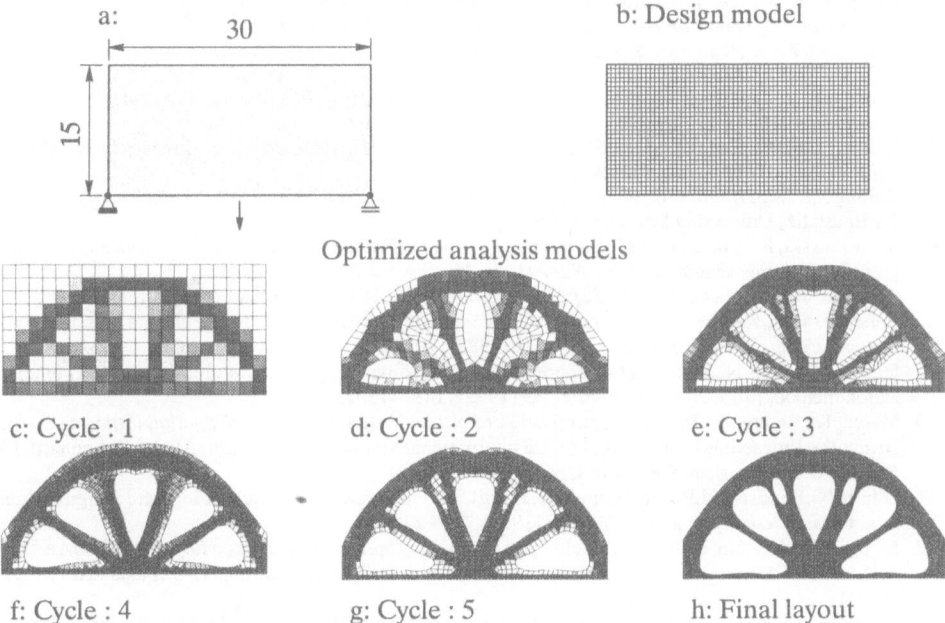

Figure 9. Topology optimization of a beam–like structure.

A rectangular wall structure, Figure 9, is fixed at the lower left corner and vertically supported at the lower right corner. The structure is loaded by a vertical load in the center of the lower edge. The objective is to find the structural layout of maximum stiffness with the mass restricted to 40 percent of the maximum possible amount in the design space. At the beginning of the optimization process the material is equally distributed. The design model of design space is discretized by 2x900 square patches. The linear finite element analysis is carried out by 2x2 reduced integrated, eight–noded, isoparametric plane stress elements. The topology which was roughly determined in cycle one does not change until cycle four. Since topology and shape depend on each other, the topology of the structure is changing in cycle 5, further improving the objective of the design problem.

4. Conclusions

Two challenging fields of structural optimization have been presented. They are intended to extend the application to problems like the consideration of buckling phenomena, shape imperfections, and the interactions of shape and topology. Procedures have been developed which combine most sophisticated methods like numerical continuation, extended system methods, and free mesh generation.

5. Literature

1. Ramm, E., Bletzinger, K.–U., Reitinger, R., and Maute, K.: The challenge of structural optimization, 2. *International Conference on computational structures technology*. Athens, Greece, 1994.
2. Bletzinger, K.–U., Kimmich, S. and Ramm, E.: Efficient modeling in shape optimal design. *Comp. Systems in Eng.* 2 (1992), 483–495.
3. Tveergard, V.: Imperfection–sensitivity of a wide integrally stiffened panel under compression. *Int. J. Solids Structures* 9 (1973), 177–192.
4. Reitinger, R., Bletzinger, K.–U. and Ramm, E.: Shape optimization of buckling sensitive structures, *Computing Systems in Engineering* 5 (1994), 65–75.
5. Reitinger, R.: *Stabilität und Optimierung imperfektionsempfindlicher Tragwerke*, Ph.D. dissertation, Institut für Baustatik, Universität Stuttgart, 1995.
6. Schittkowski, K.: The nonlinear programming method of Wilson, Han and Powell with an augmented Lagrangian type line search function. *Numerische Mathematik* 38 (1981), 83–114.
7. Bendsøe, M.P., Mota Soares, C. (Ed): *Topology Design of Structures*, Kluwer, Amsterdam, 1991.
8. Bendsøe, M.P., Kikuchi, N.: Generating optimal topologies in structural design using a homogenization method, *Comp. Meth. Appl. Mech. Eng.*, 71 (1988), 197–224.
9. Diaz, A., Kikuchi, N.: Solutions to shape and topology eigenvalue optimization problems using a homogenization method, *Int. J. Num. Meth. Eng.*, 35 (1992), 1487–1502.
10. Maute, K., Ramm, E.: Topology optimization of plate and shell structures, in *Spatial, lattice and tension structures*, Proceedings of the IASS–ASCE International Symposium 1994, Eds. Abel, J.F., Leonard, J.W., Penalba C.U., American Society of Civil Engineers, 1994.
11. Olhoff, N., Bendsøe, M.P., and Rasmussen, J.: On CAD–integrated structural topology and design optimization, *Comp. Meth. Appl. Mech. Eng.*, 89 (1991), 259–279.
12. Maute, K. and Ramm, E.: Adaptive techniques in topology optimization, *Proceedings of the 5. AIAA/USAF/NASA/ISSMO Symposium on Multidisciplinary Analysis and Optimization*, Sep. 7.–9., Panama City, Florida (1994).
13. Rehle, N. und Ramm, E.: Generieren von FE–Netzen für ebene und gekrümmte Flächentragwerke, *Bauingenieur*, 70 (1995), in press.

CONSTRAINED CONTROL IN A MECHANICAL SYSTEM WITH TWO DEGREES OF FREEDOM

F.L.CHERNOUSKO and I.S.DOBRYNINA
Institute for Problems in Mechanics,
Russian Academy of Sciences,
pr.Vernadskogo 101, Moscow 117526, Russia
E-mail: chern@ipm.msk.su

We consider a linear controlled system of the fourth order that has a pair of imaginary eigenvalues and a pair of zero eigenvalues. The system subjected to a scalar bounded control is a simplified model of various oscillatory mechanical and electromechanical systems such as elastic systems, cranes carrying pending loads, arms of flexible manipulators, systems containing fluid in a tank, etc. We propose a control satisfying the imposed constraint and steering the system from an arbitrary initial state to the prescribed terminal state of equilibrium in finite time. The obtained control law is based on the well-known Kalman [1] approach, but also takes into consideration the imposed constraint as proposed in [2]. The control law is presented in an explicit form via symbolic and numerical calculations. The results are illustrated by computer simulation.

1. Problem Statement

We will consider a two-mass system that contains an oscillatory link and is controlled by a scalar control

$$\ddot{\xi}_1 + \omega^2 \xi_1 = u, \qquad \ddot{\xi}_0 = u. \tag{1.1}$$

Here, ξ_i, $i = 0$, 1, are generalised coordinates, $\omega > 0$ is the natural frequency of the oscillator, and u is a bounded scalar control

$$|u(t)| \leq a \qquad a = const > 0 \tag{1.2}$$

Two-mass systems, consisting of a carrying body that moves horizontally with acceleration u and a mass connected with the body by a spring or a pendulum suspended to the body, can serve as mechanical models of system (1.1). In these examples ξ_0 is the displacement of the body and ξ_1 is the elongation of the spring or a small linear angular deviation of the pendulum.

D. Bestle and W. Schielen (eds.), IUTAM Symposium on Optimization of Mechanical Systems, 57–64.
© 1996 *Kluwer Academic Publishers.*

We pose the problem of determining a control $u(t)$ that satisfies (1.2) and steers system (1.1) from an arbitrary initial state at $t = 0$

$$\xi_i(0) = \xi_i^0, \qquad \dot{\xi}_i(0) = \eta_i^0, \qquad i = 0, 1 \qquad (1.3)$$

to the final rest state

$$\xi_i(T) = 0, \qquad \dot{\xi}_i(T) = 0, \qquad i = 0, 1. \qquad (1.4)$$

We simplify relations (1.1), (1.2) by substituting in them the variables

$$\xi_1 = \omega^{-2} a y, \qquad \xi_0 = \omega^{-2} a z, \qquad t = \omega^{-1} t', \qquad u = a u' \qquad (1.5)$$

Then, system (1.1), (1.2) assumes the form

$$\ddot{y} + y = u, \qquad \ddot{z} = u, \qquad (1.6)$$
$$|u| \leq 1. \qquad (1.7)$$

In what follows, we examine the system in the form of (1.6), (1.7), denoting the derivatives in the new time t' by dots and dropping the primes in t' and u'. The variables substitution (1.5) brings conditions (1.3) and (1.4) to the form

$$y(0) = x_1^0, \qquad \dot{y}(0) = x_2^0, \qquad z(0) = x_3^0, \qquad \dot{z}(0) = x_4^0, \qquad (1.8)$$
$$y(T) = 0, \qquad \dot{y}(T) = 0, \qquad z(T) = 0, \qquad \dot{z}(T) = 0. \qquad (1.9)$$

Here, x_i^0, $i = 1,...,4$, are some specified constants, and $T > 0$ in (1.9) is the as yet unknown time at which the process terminates.

This problem reduces to the construction of a control $u(t)$ that satisfies constraint (1.7) and steers system (1.6) from the specified initial state (1.8) to the specified final state (1.9).

2. Control Design.

We use Kalman's general approach [1] for constructing a control as a linear combination of the natural motions of the non-controlled system. This approach was extended in [2] to the case of control constraints. Let $x = (y, \dot{y}, z, \dot{z})^T$ be the state vector of system (1.6). We represent (1.6) in the general form

$$\dot{x} = Ax + Bu, \qquad (2.1)$$

where A and B are matrices of dimensions 4×4 and 4×1, respectively. We write the initial, (1.8), and final, (1.9), conditions in the general form

$$x(0) = x^0 = (x_1^0, x_2^0, x_3^0, x_4^0)^T, \qquad x(T) = 0. \qquad (2.2)$$

In the considered case, we have

$$A = \begin{pmatrix} 0 & 1 & 0 & 0 \\ -1 & 0 & 0 & 0 \\ 0 & 0 & 0 & 1 \\ 0 & 0 & 0 & 0 \end{pmatrix} \qquad B = \begin{pmatrix} 0 \\ 1 \\ 0 \\ 1 \end{pmatrix} \qquad (2.3)$$

Following the approach indicated in the preceding, we look for a control in the form

$$u(t) = Q^T(t)C, \qquad Q(t) = \Phi^{-1}(t)B, \qquad (2.4)$$

where C is a constant four-dimensional vector and $\Phi(t)$ is the fundamental matrix of the homogeneous system (2.1), namely,

$$\Phi(t) = \begin{pmatrix} \cos t & -\sin t & 0 & 0 \\ \sin t & \cos t & 0 & 0 \\ 0 & 0 & 1 & -t \\ 0 & 0 & 0 & 1 \end{pmatrix} \qquad (2.5)$$

Then, by (2.3)-(2-5), we obtain

$$u(t) = -C_1 \sin t + C_2 \cos t - C_3 t + C_4, \qquad (2.6)$$

where the C_i, $i = 1,...,4$, are the components of the vector C. In the general case, this vector is determined by the equation

$$C = -R^{-1}(T)x^0, \qquad (2.7)$$

where the notation

$$R(T) = \int_0^T Q(t)Q^T(t)dt \qquad (2.8)$$

is introduced. Having calculated the elements of the matrix $R(T)$ in (2.8) by using equalities (2.3)-(2.5), we find

$$R(T) = \begin{pmatrix} (T - sc)/2 & -s^2/2 & s - Tc & c - 1 \\ -s^2/2 & (T + sc)/2 & 1 - c - Ts & s \\ s - Tc & 1 - c - Ts & T^3/3 & -T^2/2 \\ c - 1 & s & -T^2/2 & T \end{pmatrix} \qquad (2.9)$$

The notations $s = \sin T$ and $c = \cos T$ are introduced in (2.9). Let φ_{ij}, $i, j = 1,...,4$, be the element of the inverse matrix $R^{-1}(T)$ of (2.9). Then, by virtue of (2.7), the expression for control (2.6) assumes the form

$$u(t) = \sum_{i=1}^{4} (\varphi_{1i} x_i^0 \sin t - \varphi_{2i} x_i^0 \cos t + \varphi_{3i} x_i^0 t - \varphi_{4i} x_i^0). \qquad (2.10)$$

Thus, control (2.10) steers, for any $T > 0$, system (2.1) (or (1.6)) from the initial state (2.2) (or (1.8)) to the final rest state (2.2) (or (1.9)) in time T. However, this control does not, generally speaking, satisfy constraint (1.7). To take this constraint into consideration, we apply the Cauchy inequality to relation (2.10)

$$|u| \leq \left(\sum_{i=1}^{4} x_i^{02} \right)^{1/2} \left[\sum_{i=1}^{4} (-\varphi_{1i} \sin t + \varphi_{2i} \cos t - \varphi_{3i} t + \varphi_{4i})^2 \right]^{1/2}. \qquad (2.11)$$

We introduce the auxiliary functions

$$p(t,T) = \sum_{i=1}^{4} (-\varphi_{1i} \sin t + \varphi_{2i} \cos t - \varphi_{3i} t + \varphi_{4i})^2, \qquad (2.12)$$

$$r(T) = \left[\max_{0 \leq t \leq T} p(t,T) \right]^{-1/2}. \qquad (2.13)$$

Then, inequality (2.11) is rewritten in the form

$$|u| \leq [p(t,T)]^{1/2} \leq |x^0| / r(T). \qquad (2.14)$$

We choose the termination time T from the condition

$$|x^0| = r(T). \qquad (2.15)$$

It follows from (2.4) that when T is selected according to (2.15), the constraint imposed on the control, (1.7), is satisfied for all $t \in [0,T]$.

Thus, we arrive at the following procedure for constructing the control $u(t)$. First, we determine the elements $\varphi_{ij}(T)$ of the inverse matrix $R^{-1}(T)$ and calculate the functions $p(t,T)$ and $r(T)$ by equalities (2.9), (2.12), and (2.13). These constructions, which must be executed once for the given system, are presented in Section 3. When they have been executed, one can construct, for any initial vector x^0, the desired bounded control that steers the system to the coordinate origin. To do this, we first determine the time T from condition (2.15) and then find the control u from (2.10).

3. Determining the Function r(T)

We will use the REDUCE symbolic calculations language to find, using a computer, the analytical representations of the elements φ_{ij}, $i,j = 1,...,4$, and of the matrix $R^{-1}(T)$ which is the inverse of (2.9). The expressions for φ_{ij} turned out to be rather cumbersome. To illustrate, we present one element of the matrix $R^{-1}(T)$:

$$\varphi_{11} = \left[-2T^5 - T^4 \sin 2T + 16T^3 \sin^2 T - 48T^2 \sin T (1 - \cos T)\right.$$

$$\left.+48T (1 - \cos T)^2\right] / \left[-T^6 + T^4 (8 \cos T + \sin^2 T + 16) + 8T^3 \sin T (2 \cos T - 5)\right. \quad (3.1)$$

$$\left.+48T^2 (1 - \cos T - 2 \sin^2 T)+240T \sin T (1 - \cos T) + 192(2 - 2\cos T + \sin^2 T)\right].$$

Using formula (2.12) and the obtained expressions of the form (3.1) for φ_{ij}, one can calculate the maximum values of $p(t,T)$ for $t \in [0,T]$. The function $r(T)$ in (2.13) was determined in this way. Its graph is shown in Fig.1

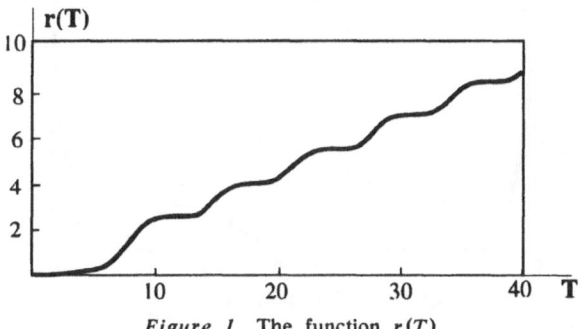

Figure 1. The function $r(T)$.

We will analyse the behaviour of the function $r(T)$ in the limiting cases.

Let the process duration T tends to 0. We expand the function $p(t,T)$ in (2.12) in a Maclaurin series (we use REDUCE) in the current time:

$$p(t,T) = (\varphi_{12} + \varphi_{14})^2 + (\varphi_{22} + \varphi_{24})^2 + (\varphi_{23} + \varphi_{34})^2 + (\varphi_{24} + \varphi_{44})^2 -$$

$$-2t [\varphi_{12} (\varphi_{11} + \varphi_{13} + \varphi_{22} + \varphi_{24}) + \varphi_{13} (\varphi_{14} + \varphi_{23} + \varphi_{34}) + \quad (3.2)$$

$$+\varphi_{14} (\varphi_{11} + \varphi_{24} + \varphi_{44}) + \varphi_{23} (\varphi_{22} + \varphi_{24} + \varphi_{33}) + \varphi_{34} (\varphi_{24} + \varphi_{33} + \varphi_{44})]+....$$

We then expand the numerators and denominators of the elements of the symmetric matrix $R(T)$ in series in T. We obtain

$$\varphi_{11} = (T^9 / 180 - T^{11} / 630+...) / (T^{16} / 18144000-...). \quad (3.3)$$

The other elements have similar representations. Estimates of the orders of the expansions in T in the numerators and denominators of the functions $\varphi_{ij}(t,T)$ in (3.3) show that, to obtain the principal term of the expansion of the function $p(t,T)$ in accordance with (3.2), it suffices to retain only the principal term (of the order T^{16}) in the denominators of formulas (3.3). In the numerators of these expressions, one has to take into account terms of various orders, and by collecting terms of like powers, one obtains the representation

$$p(t,T) = 1411200 T^{-8} f(\tau). \qquad (3.4)$$

The notation

$$f(\tau) = 1 - 24\tau + 204\tau^2 - 760\tau^3 + 1380\tau^4 - 1200\tau^5 + 400\tau^6, \quad \tau = t/T \in [0,1] \quad (3.5)$$

is used here. The graph of the polynomial $f(\tau)$ is shown in Fig.2.

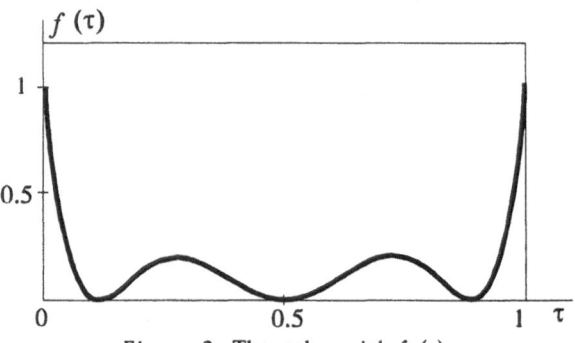

Figure 2. The polynomial $f(\tau)$.

One can easily see that $f(\tau)$ attains its largest value at $\tau = 0$ and $\tau = 1$: $f(0) = f(1) = 1$. Then, it follows from (3.4) and (3.5) that $\max p(t,T) = 1411200\, T^{-8}$. We substitute this result into (2.13) and obtain

$$r(T) = 8.4 \cdot 10^{-4} T^4, \qquad T \to 0. \qquad (3.6)$$

Let the control process duration $T \to \infty$. We substitute into (2.12) the expressions for φ_{ij}, $i,j = 1,...,4$, calculated similarly to (3.1), and expand the function $p(t,T)$ in a series of negative powers of T. Transformations made by using the REDUCE language yield the expansion

$$p(t,T) = 4T^{-2}\left\{9(t/T)^2 - 12(t/T) + 5 + T^{-1}\left[12(T - 2t)T^{-1}\sin(T - t) - \right.\right.$$

$$\left.\left. - \sin 2(T - t) - \sin 2t - 12(2T - 3t)T^{-1}\sin t\right]\right\} + O(T^{-4}), \quad T \to \infty. \qquad (3.7)$$

We rewrite this expansion in the form

$$p(t,T) = 4T^{-2}\left[p_0(\tau) + T^{-1}p_1(\tau,T)\right], \qquad p_0(\tau) = 9\tau^2 - 12\tau + 5, \qquad (3.8)$$
$$p_1 = 12(1 - 2\tau)\sin T(1 - \tau) - \sin 2T(1 - \tau) - \sin 2T\tau - 12(2 - 3\tau)\sin T\tau.$$

Let us find the maximum of (2.13) as $T \to \infty$ using representation (3.8). The quadratic trinomial $p_0(\tau)$ attains its maximum on the interval $[0,1]$ at $\tau = 0$. Since the contribution of the second addend in (3.8) as $T \to \infty$ is small, we have, up to higher-order infinitesimals

$$\max_{0 \le t \le T} p(t,T) = p(0,T) = 20T^{-2} + 4T^{-3}(12\sin T - \sin 2T), \quad T \to \infty \quad (3.9)$$

We used here expansion (3.7). Substituting (3.9) into (2.13) and expanding the result in a series of powers of T^{-1}, we obtain

$$r(T) = (10T - 12\sin T + \sin 2T) / (20\sqrt{5}) + 0(T^{-1}), \quad T \to \infty. \quad (3.10)$$

We differentiate (3.10) in T: $r'(T) = (2 - \cos T)(1 - \cos T) / (5\sqrt{5}) \ge 0$. Consequently, $r(T)$ is a monotonically increasing function as $T \to \infty$. It follows from the presented calculations and analytical expansions that the function $r(T)$ increases monotonically from 0 to ∞ as T varies from 0 to ∞. Consequently, Eq. (2.15) has, for any $|x^0|$ a unique solution.

4. Calculation of Control and Numerical Simulation

The procedure for designing the control is described at the end of Section 2. We dwell first on a practical numerical solution of Eq. (2.15). We partition the entire semi-infinite interval of variation of T into three parts: $[0, T_0]$, $[T_0, T_1]$, and $[T_1, \infty)$, to which three intervals of variation of r correspond: $[0, r_0]$, $[r_0, r_1]$, and $[r_1, \infty)$. Here, $r_i = r(T_i)$, $i = 0,1$. On the interval $[0, T_0]$, one uses the asymptotic representation (3.6) for small T, on the interval $[T_0, T_1]$, one uses the table of numerical values of $r(T)$, and on the interval $[T_1, \infty)$, one uses the asymptotic representation (3.10) for large T. First, we determine, for the specified x^0, by comparing $|x^0|$ with r_0 and r_1, which of the three intervals includes the desired T. We then determine T as follows. If $T \in [0, T_0]$, then, by (3.6), we have $T = [|x^0|/0.00084]^{1/4}$. If $T \in [T_0, T_1]$, we find T by linear interpolation from the table of values of $r(T)$ that is stored in the computer memory. If $T \in [T_1, \infty)$, we use representation (3.10). In this case, it is convenient to look for T in the form

$$T = 2\sqrt{5}|x^0| + \theta. \quad (4.1)$$

Substituting (4.1) into (3.10), we obtain the equation

$$F(\theta) = 10\theta - 12\sin(2\sqrt{5}|x^0| + \theta) + \sin[2(2\sqrt{5}|x^0| + \theta)] = 0$$

for θ. This equation is solved by some numerical method, for example, the successive interval halving method.

When the duration T has been determined for the specified initial vector x^0, the control $u(t)$ at any instant t can be calculated from (2.10). Here, one uses the analytical expressions for the functions φ_{ij}, $i,j = 1,...,4$, see (3.1), which were obtained by analytical transformations. The control calculated in this manner can be substituted into the right-hand parts of system (2.1), which is integrated numerically or analytically (using the fundamental matrix (2.5)) for the initial conditions (2.2). Some results of simulation with $x^0 = (-1, 2, 0.5, 1)$ are presented in Fig.3. For this example, $T = 13.12$. The curves in Fig.5 represent the projection of the phase trajectory of $x(t)$ on the $(y,\dot{y})$ and $(z,\dot{z})$ hyperplanes.

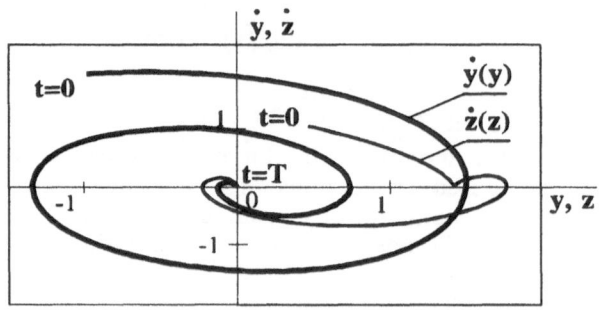

Figure 3. Results of simulation.

The paper is based on research supported by the International Science Foundation, Grant No M4F000, and by the Russian Foundation of Basic Research, Project No 93-01-16286.

References

1.   Kalman, R.E.: On the General Theory of Control Systems, in *Proc. 1st IFAC Congress*, v.1., London: Butterworths, 1960.

2.   Chernousko, F.L.: On the Construction of a Bounded Control in Oscillatory systems, *J. Appl.Math. and Mech. (PMM)* **52** No 4 (1988), 426-433.

3.   Dobrynina, I.S., and Chernousko, F.L.: Constrained Control of a Fourth-Order Linear System, *J. Computer and Systems Science International* **32**(4) (1994), 108-115.

A NEW APPROACH TO THE MIN-MAX DYNAMIC RESPONSE OPTIMIZATION

Dong-Hoon Choi and Min-Soo Kim

Department of Mechanical Design and Production Engineering,
Hanyang University, Haengdang-Dong 17, Sungdong-Ku, Seoul, Korea

ABSTRACT

For the treatment of a max-value cost function in a dynamic response optimization problem, we propose the approach of directly handling the original max-value cost function in order to avoid the computational burden of the previous transformation treatment. In this paper, it is theoretically shown that the previous treatment results in demanding an additional equality condition as a part of the Kuhn-Tucker necessary conditions. Also, it is demonstrated that the usability and feasibility conditions for the search direction of the previous treatment retard convergence rate. To investigate the numerical performance of both treatments, typical optimization algorithms in ADS are employed to solve a typical example problem. All the algorithms tested reveal that the suggested approach is more efficient and stable than the previous approach. Also, the better performing of the proposed approach over the previous approach is clearly shown by contrasting the convergence paths of the typical algorithms in the design space of the sample problem. Min-max dynamic response optimization programs are developed and applied to three typical examples to confirm that the performance of the suggested approach is better than that of the previous one.

1. Introduction

The optimal design problem of many mechanical systems and seismic-resistant structures is often mathematically modeled as a min-max dynamic response optimization problem. It is the time dependency and implicit nature of point-wise state variable constraint functions and a max-value cost function that make the min-max dynamic response optimization problem difficult and expensive to solve. While most works[1-4] have focused on efficiently dealing with point-wise state variable constraint functions, the topic of efficient treatment of the max-value cost function has not been sufficiently studied to the authors' knowledge.

For the treatment of the max-value cost function, all the previous works[1-4] have replaced it with an artificial design variable and imposed an additional point-wise state variable constraint function. Although the transformed problem appears to be mathematically equivalent to the original problem, the previous approach suffers from computational burden in the context of numerical optimization algorithm. In this paper, based on the Kuhn-Tucker necessary conditions, we theoretically show that an additional condition must be satisfied for the constrained optimality of the transformed problem. Also, we elucidate the condition on the components of the descent direction vector required for the search direction to be usable as well as feasible, which slows down the

65

D. Bestle and W. Schielen (eds.), IUTAM Symposium on Optimization of Mechanical Systems, 65–72.
© 1996 Kluwer Academic Publishers.

convergence rate of the previous approach.

In order to eliminate the computational burden of the previous approach, we propose the approach of directly handling the original max-value cost function. Typical optimization algorithms in ADS[5] are used to compare the numerical performance of the direct treatment with that of the transformation treatment. Also, dynamic response optimization programs are developed and applied to three typical examples to compare the performances of the two treatments for the max-value cost function in dynamic response optimization.

2. Min-Max Dynamic Response Optimization Problem

Let $\mathbf{b} \in R^n$ be a vector of design variables and $\mathbf{z} \in R^k$ be a vector of generalized velocities and displacements. A typical min-max dynamic response optimization problem can be presented as follows: find $\mathbf{b}$ to

$$\text{minimize} \quad \Psi_0 = \max_{t \in [0,T]} \; f_0(\mathbf{b}, \mathbf{z}, t), \tag{1}$$

satisfying the state equation

$$P(\mathbf{b})\dot{\mathbf{z}} = f(\mathbf{b}, \mathbf{z}, t), \; 0 \le t \le T, \tag{2}$$

$$\mathbf{z}(0) = \mathbf{z}^0,$$

and the constraints

$$\Psi_i(\mathbf{b}, \mathbf{z}, t) \le 0, \quad 0 \le t \le T, \quad i = 1, \dots, m, \tag{3}$$

$$\Phi_j \equiv \int_0^T h_j(\mathbf{b}, \mathbf{z}, t) dt \le 0, \quad j = 1, \dots, p, \tag{4}$$

$$g_l(\mathbf{b}) \le 0, \quad l = 1, \dots, q. \tag{5}$$

The differential equations of motion are presented in the general first-order form in Eq. (2), where $\dot{\mathbf{z}}$ denotes a vector of the time derivatives of state variables, $P(\mathbf{b})$ is a matrix whose elements depend on the design variables, $f(\mathbf{b}, \mathbf{z}, t)$ is a vector of forcing function and support reactions, $\mathbf{z}^0$ is a vector of given initial conditions, and T is the length of time interval of interest. The constraints of Eq. (3) are the point-wise state variable constraints. The constraints of Eq. (4) are the integral-type state variable constraints which can be readily incorporated into the state space formulation. The constraints of Eq. (5) are explicit functions of design variables which can be routinely handled in any nonlinear programming algorithms. In this paper, we concentrate on the treatment of the max-value cost function of (1).

3. Treatment of the Max-Value Cost Function

3-1. Previous Treatment. For the treatment of the max-value cost function of (1), the previous approach transforms it as follows:

$$\text{minimize} \quad b_{n+1}, \tag{6}$$

$$\text{subject to} \quad f_0(\mathbf{b}, \mathbf{z}, t) - b_{n+1} \le 0, \quad 0 \le t \le T, \tag{7}$$

where b_{n+1} is an artificial design variable.

Fig. 1 graphically shows the concept of the transformation treatment. As the optimization process proceeds, a line meaning the artificial design variable b_{n+1} is

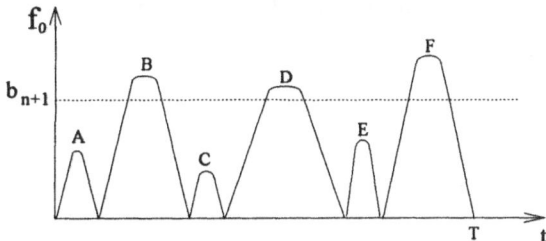

lowered while the chosen dynamic response f_0 in the entire time interval is kept below this line at the same time.

In investigating the previous treatment from the optimization theory point of view in the following sections, we will consider the transformed problem of (6) and (7) for a

Fig. 1 Graphical description of the previous treatment

clear presentation.

3-1-1. Kuhn-Tucker Necessary Conditions. The Lagrangian function of the transformed problem is

$$L(\tilde{\mathbf{b}},\mathbf{z}) = b_{n+1} + \int_0^T \mu(t)[f_0(\mathbf{b},\mathbf{z},t) - b_{n+1}]dt \qquad (8)$$

where the extended design variable vector $\tilde{\mathbf{b}}$ is newly defined as $\{b_1\ b_2\ \dots\ b_n\ b_{n+1}\}^T$, $\mu(t)$ is the Lagrange multiplier function for the constraint of (7), and $[\bullet]^+$ represents max $[\bullet,0]$. Now, we assume the optimal point $\tilde{\mathbf{b}}^*$ to be a regular point of the feasible region and $f_0(\tilde{\mathbf{b}}^*,\mathbf{z},t)$ to be a differentiable function. For $\tilde{\mathbf{b}}^*$ to be a relative minimum, the following conditions must be satisfied:

$$dL(\tilde{\mathbf{b}}^*,\mathbf{z})/d\tilde{\mathbf{b}} = 0, \qquad (9)$$
$$\mu^*(t)[f_0(\tilde{\mathbf{b}}^*,\mathbf{z},t) - b_{n+1}] = 0, \qquad 0 \le t \le T, \qquad (10)$$

The Kuhn-Tucker conditions of (9) and (10) will become equivalent to the necessary condition for optimality of the original problem of Eq. (1) if the following additional condition of (11) is satisfied, which can be derived from the $(n+1)$th component of Eq. (9).

$$1 - \int_0^T \mu^*(t)dt = 0 \qquad (11)$$

Thus, the transformation treatment requires an additional condition that the integration of the Lagrange multiplier function for the constraint of (7) over the time interval must be equal to 1 at the optimum. In the dual space, Eq. (11) can be interpreted as an equality constraint on dual variables. Therefore, the transformed problem is more difficult to solve than the original problem from the optimization theory point of view.

3-1-2. Condition on a Search Direction Vector. We now pay attention to the search direction vector to examine the numerical performance of the previous treatment during optimization process. Notice that the first derivative of the cost function of (6) with respect to b_{n+1} has the opposite sense to that of the constraint of (7). It algorithmically means that the original cost function is adversely split into two conflicting functions.

The conflict can also be interpreted in the context of feasible-direction methods. In the feasible-direction methods, we want to find a search direction $\mathbf{S}$ which will reduce the cost function without violating the active constraints for some finite move. This condition on the search direction for the transformed problem of (6) and (7) can be mathematically expressed as

$$\frac{db_{n+1}}{d\tilde{\mathbf{b}}} \cdot \mathbf{S} \leq 0, \tag{12}$$

$$\frac{d\{f_0(\mathbf{b}, \mathbf{z}, t_k) - b_{n+1}\}}{d\tilde{\mathbf{b}}} \cdot \mathbf{S} \leq 0, \qquad k \in \mathbf{J}, \tag{13}$$

where $\mathbf{J}$ is a set of time intervals where the constraint of (7) is active. Since $db_{n+1}/d\tilde{\mathbf{b}} = \{0, 0, \ldots, 0, 1\}^T$ and thus $(db_{n+1}/d\tilde{\mathbf{b}}) \cdot \mathbf{S} = S_{n+1}$, Eqs. (12) and (13) can be integrated as follows:

$$\sum_{i=1}^{n} \left(\frac{df_0(\mathbf{b}, \mathbf{z}, t_k)}{db_i} \cdot S_i \right) \leq S_{n+1} \leq 0, \qquad k \in \mathbf{J}. \tag{14}$$

The adverse effect of splitting the original cost function into two conflicting functions is reflected by the inequality relation of (14) among the components of a search direction vector. This restriction on the components $\{ s_1\ s_2\ \ldots\ s_n \}$ corresponding to the design variables, which does not exist if the original cost function is directly handled, may distort the search direction and lead to the small distance of travel in line search. We will illustrate that enforcing the condition of (14) may even stall the optimization process of the feasible direction method in Section 4-1.

3-2. The Direct Treatment of the Max-Value Cost Function. In order to alleviate the computational burden by eliminating the complexity of the previous approach, the original max-value cost function of (1) is directly handled in this study. Fig. 2 graphically shows the concept of the direct treatment, in which only the maximum dynamic response is minimized.

Once line search is used during optimization process, the cost value(pseudo-cost value with SUMT) of the current iteration is guaranteed to be always smaller than those of the previous iterations. Thus, we propose to use only optimization algorithms which employ line search in order to prevent the convergence difficulty due to the possible oscillations in the extreme dynamic response from iteration to iteration. We believe that it is not a severe restriction since line search is embedded in most of the efficient optimization algorithms in use these days.

The gradient of the max-value cost function of (1) can be

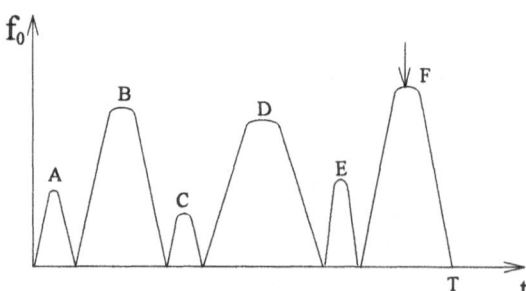

Fig. 2 Graphical description of the direct treatment

analytically obtained with exploiting the fact of $df_0(\mathbf{b}, \mathbf{z}, t_p)/dt = 0$ as follows:

$$\frac{d\Psi_0}{d\mathbf{b}} = \frac{\partial f_0(\mathbf{b},\mathbf{z},t_p)}{\partial \mathbf{b}} + \frac{\partial f_0(\mathbf{b},\mathbf{z},t_p)}{\partial \mathbf{z}} \cdot \left[\frac{d\mathbf{z}}{d\mathbf{b}}\right]_{t=t_p}, \tag{15}$$

where t_p is the time point of the maximum response. It can be evaluated by using either the direct differentiation method or the adjoint variable methods for calculating the gradient of the dynamic response at a particular time developed by Hsieh and Arora[2].

4. Numerical Results and Discussion

In order to compare the numerical performance of the proposed direct treatment with that of the previous one, both are applied to solve one example problem by using the various algorithms in ADS[5] and three example problems by using the dynamic response optimizers in IDOL[6].

4-1. Performance Comparison with Various Optimization Algorithms. A two degree of freedom dynamic absorber is excited by a forcing function over the range of frequency ratio from 0.5 to 1.5. The design problem is to find the damping and stiffness constants that minimize the ratio of the maximum amplitude of the main mass subject to rattle space and side constraints. One may refer to Problem 2 in pp. 193-196 of Reference 1 for detailed information.

The problem is solved by using four algorithms in ADS to test whether the effectiveness of the proposed approach depends on the optimization algorithm employed. The four algorithms are Sequential Linear Programming (SLP), Sequential Quadratic Programming (SQP), Feasible Direction method (FDR) and Augmented Lagrange Multiplier method (ALM). Since ADS cannot handle parametric optimal design problem by using state space methods, one rattle space constraint is expanded to 30 discrete constraints over the exciting frequency ratio of [0.5, 1.5], and the finite difference method is used for design sensitivity analysis.

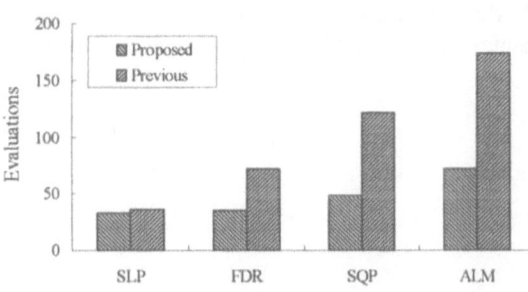

In Fig. 3, the numbers of function evaluations are compared for the four algorithms employed. Here, the number of function evaluations include those necessary for design sensitivity analysis. The proposed treatment can be seen more efficient

Fig. 3 Performance comparison with 4 algorithms

than the previous one for all four algorithms. Even though improvement in case of using SLP is not appreciable with the good initial guess for an artificial design variable (b_3^0) which we chose for the previous treatment, the improvement has been found to become significant when using a bad b_3^0.

In order to trace the source of computational enhancement obtained by using the proposed treatment, the convergence paths of both treatments in design space are depicted in Fig. 4 for the case of using the feasible direction method. Starting from the same initial design point, the proposed and previous treatments require 6 and 11 gradient evaluations, respectively, to reach the similar optimum design points. Fig. 4 clearly shows that the lengths of travel in line search of the previous treatment are much shorter than those of the proposed one. Specifically, optimization process stalls at the design points of 7 and 9, and several descent directions are almost univariate near the optimum. We have scrutinized the reason of stall at the points of 7 and 9 to find out that the search directions at those points cannot satisfy the condition of (14), which does not exist with the direct treatment.

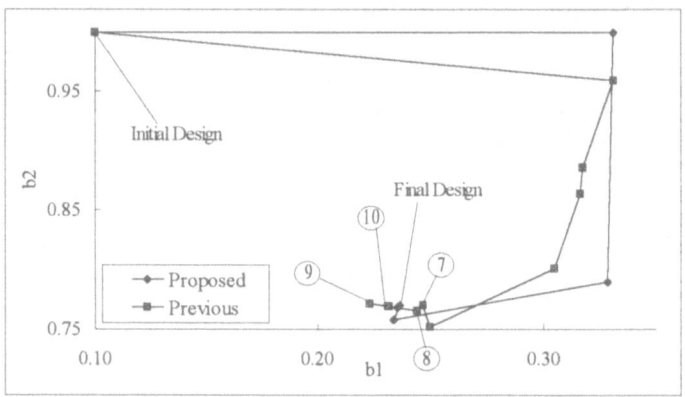

Fig. 4 Convergence paths of both treatments with feasible direction method

The convergence paths of both treatments in design space are also depicted in Fig. 4 for the case of using SQP. Starting from the same initial design point, the proposed and previous treatments require 6 and 17 gradient evaluations, respectively, to reach the optimum points. It is impressive that the proposed treatment results in nearly quadratic convergence to the optimum, while the previous one leads to small distance of travel in most of line searches and several almost univariate search directions at the points of 15-18.

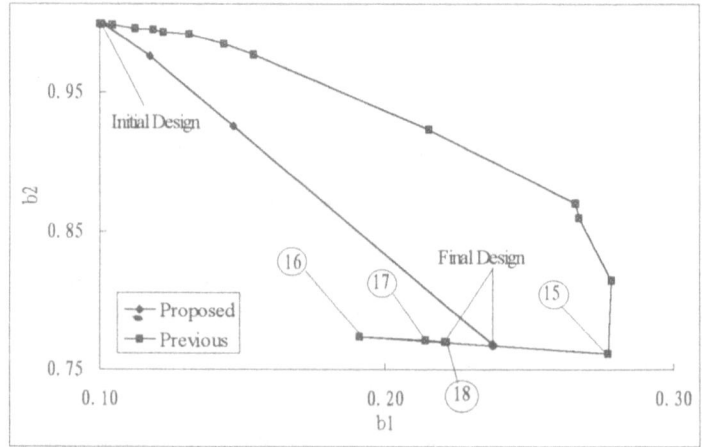

Fig.5 Convergence paths of both treatments with SQP

Tracing the convergence paths in Figs. 3 and 4 confirms that the previous treatment remarkably worsens the original min-max optimization problem from the algorithmic point of view and thus requires much more computational effort than the proposed treatment which eliminated unnecessary adverse effects.

4-2. Min-Max Transient Response Optimization Problems. Two dynamic response optimizers, IDOL 1.6D[6], are developed to evaluate the effectiveness of the proposed treatment for min-max transient response optimization problems. They employ the same numerical procedures except for the treatment of a max-value cost function and are run on the same computer to directly compare the efficiency of the proposed treatment with that of the previous one. Specifically, we use the ALM method, which can be effective for dynamic response optimization of large scale systems, with employing the BFGS method for unconstrained suboptimization and a sequential polynomial approximation method for line search. The Adams-Bashforth Adams-Moulton predictor-corrector method is used for dynamic analysis and the adjoint variable method for design sensitivity analysis.

These optimizer are used to solve three typical transient response optimization problems such as a single degree of freedom nonlinear impact absorber, a two degree of freedom dynamic absorber, and a five degree of freedom vehicle suspension system. One may refer to Reference 4 for detailed information. The optimization results of the three example problems obtained by using the proposed and previous treatments are listed in Table 1, where NG is the number of gradient evaluations and NF is the number of function evaluations.

Table 1 Performance comparison between both treatments

	Impact Absorber		Dynamic Absorber		Vehicle Suspension System	
	Proposed	Previous	Proposed	Previous	Proposed	Previous
b1	0.5978	0.5972	1.3491	1.3319	50.0	50.0
b2	0.5976	0.5972	0.0198	0.0231	200.11	221.54
b3		0.5969		2.3677	277.93	276.01
b4					50.0	49.94
b5					76.87	76.26
b6					80.0	80.0
b7						255.51
f*	0.5976	0.5969	2.3678	2.3677	255.13	255.51
NG	32	47	12	50	22	55
NF	136	168	36	165	83	211

As shown in Table 1, the proposed treatment has reduced the number of gradient evaluations by 30% - 75% and the number of function evaluations by 20% - 75%, while giving the similar optimization results for the three example problems. This reduction demonstrates the superiority of the proposed treatment to the previous one.

5. Conclusion

The treatment of a max-value cost function in a dynamic response optimization problem is explored in this paper. In the previous treatment, the original max-value cost function has been split into an artificial design variable and an additional point-wise state variable constraint. The adverse effects of this split are elucidated in the contexts of optimization theory and algorithm.

To eliminate the complexity of the previous approach, we propose the approach of directly handling a max-value cost function with using any optimization algorithms which employ line search. The analytical methods for design sensitivity analysis of a max-value cost function are also presented.

To compare the numerical performance of the suggested approach with that of the previous one, both are applied to solve four example problems. The comparison clearly reveals that the proposed direct treatment is more efficient and stable than the previous one.

Acknowledgement
This research was supported by the Agency for Defense and Development Grant No. ADD-90-5-09, and the Turbo and Power Machinery Research Center, KOSEF.

Reference

1. Haug, E. J. and Arora, J.S., 1979, Applied Optimal Design, Wiley-Interscience, New York
2. Hsieh, C.C. and Arora, J.S., 1984, "Design Sensitivity Analysis and Optimization of Dynamic Response", Computer Methods in Applied mechanics and Engineering, Vol. 43, pp. 195-219
3. Hsieh, C. C. and Arora, J. S., 1985, "Hybrid Formulation for treatment of Point-wise State Variable Constraints in Dynamic Response Optimization", Computer Methods in Applied Mechanics and Engineering, Vol. 48, pp. 171-189
4. Paeng, J. K. and Arora, J. S., 1989, "Dynamic Response Optimization of Mechanical Systems with Multiplier Methods", ASME Journal of mechanism, Transmission, and Automation in Design, Vol. 111, pp. 73-80
5. Vanderplaats, G. N., 1985, ADS-A FORTRAN Program for Automated Design Synthesis Version 1.10 , Engineering Design Optimization, Inc.
6. Kim. M.-S. and Choi, D.-H. 1993, IDOL 1.6D User's Guide, Technical Report AMOD 93-01, AMOD Lab., Mechanincal Design & Production Engineering, Hanyang University.

CONSIDERING LOCAL BUCKLING EFFECTS FOR THE OPTIMIZATION OF STIFFENED COMPOSITE PANELS

Hans A. Eschenauer, Christof M. Weber
*Research Center for Multidisciplinary Analyses
and Applied Structural Optimization FOMAAS
University of Siegen
D-57068 Siegen / GERMANY*

1. Introduction

The present paper addresses the optimal layout of stiffened fibre composite plates (Fig. 1) considering buckling constraints; these plates are increasingly applied in many fields of engineering (air- and spacecraft technology, automotive industries, boatbuilding etc.).

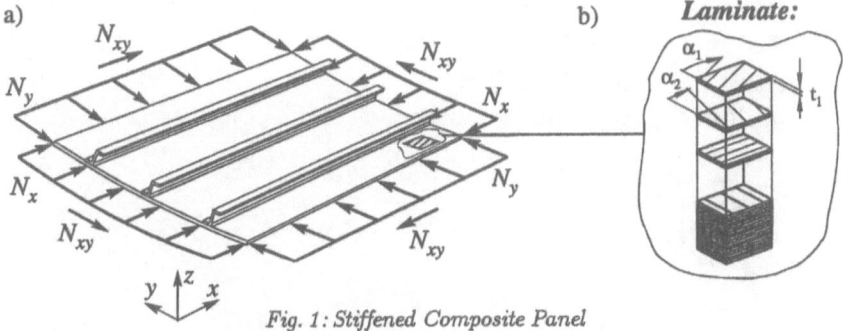

Fig. 1: Stiffened Composite Panel

This particular area of structural optimization still requires substantial investigations into its fundamentals. The structural analysis alone for the treatment of this type of problems may increase to such a degree that the complete optimization process requires extremely long computation times due to the processing of a high amount of data, a fact that calls for the development of "intelligent" procedures in order to reduce the computation effort to a tolerable measure and to maintain reduplicability of the whole process. For this purpose, a so-called "constructive design model" is introduced.

D. Bestle and W. Schielen (eds.), IUTAM Symposium on Optimization of Mechanical Systems, 73–80.
© 1996 *Kluwer Academic Publishers.*

2. Problem Definition

In the following, we will consider a plane or shallow panel which consists of a so-called base panel onto which stiffeners are applied according to Fig. 1a. This plate is assumed to be subject both to shear and compression loads. Plate and stiffeners can be made of either isotropic or CFRP-material (Fig. 1b). Plane structures of the above-described type are generally endangered by buckling; the buckling value can be maximized by choosing certain design influence parameters like the thickness distribution, the stacking sequence, the ply angles and ply thicknesses of the base panel, or the arrangement, shape or number of stiffeners. Thus, the optimal layout of such a panel shall be determined, where imperfections and the post-buckling behaviour shall not be considered in the layout.

The given optimization problem can be formulated in the following way:

- *Maximization of the buckling load*

$$\max_{x \in \mathbb{R}^n} \left\{ f(\boldsymbol{x}) \mid \boldsymbol{g}(\boldsymbol{x}) \geq 0 \right\} \tag{1}$$

with $f(\boldsymbol{x})$ buckling load to be maximized,

        $\boldsymbol{x}$ design variable vector,

        $g_1(\boldsymbol{x})$ given weight W,

        $g_{1+i}(\boldsymbol{x})$ upper and lower bounds of the design variables.

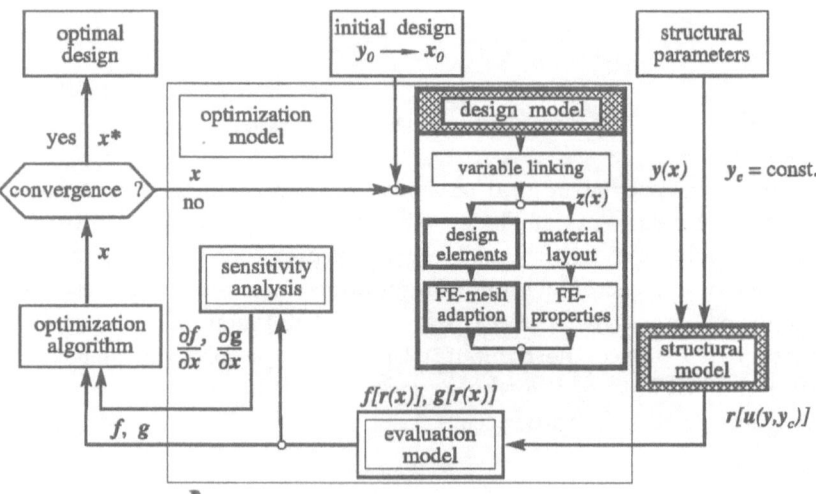

Fig. 2: Optimization loop

A fundamental solution procedure for general optimization problems - and thus also for Composite Structures - is presented by the "Three-Columns-Concept" [1]. Fig. 2 schematically illustrates the division of the optimization task into the three main parts "structural model", "optimization model" and "optimization algorithms" in the form of an optimization loop.

As already mentioned in the introduction, an increase of the optimization efficiency of structures endangered by buckling, like the stiffened panel treated here, requires a correspondingly structured design model (Fig. 2) to be formulated prior to the actual execution of the optimization. The single steps of a "constructive design modeling" are presented in the box. In the following, they shall be described in detail.

3. Augmented Design Model

3.1. DEFINITIONS – CONSTRUCTIVE DESIGN MODEL

The design model describes the relation between the design variables x and the variable parameters y of the analysis model required for the calculation of the component behaviour. By means of a linear transformation, where several analysis variables are assigned to one design variable, a so-called "variable linking", is achieved:

$$z_j = a_{ij}\, x_i + z_j^0 \qquad (2)$$

with: z_j, z_j^0 *j-th constructive variable, corresponding initial value* ,
       a_{ij} *allocation matrix* ,
       x_i *i-th design variable* .

The suitable definition of the design variables presents an important aspect within the optimization task. The simplest method to be realized in the scope of a shape optimization of components is to use the parameters of the structural analysis model as design variables, for instance by defining the FE-nodal coordinates of a FE-model as design variables. This procedure, however, has some decisive disadvantages illustrated in [2]. In the present case, the re-positioning of a stiffener would require a coupling of the components of the displacement vector (Δx_i^k) in order to secure the linkage between stiffener and base panel (Fig. 3a). In addition, the nodal displacements would have to be coupled to make sure that all nodes of the stiffener remain in the stiffener plane after the displacement. This procedure would thus not be sensible for practical applications.

A more suitable approach is achieved by applying "constructive design models"[3]. Their fundamental constituents are the geometrical modeling of the constructive layout, and the linking of the design variables x_i with the constructive

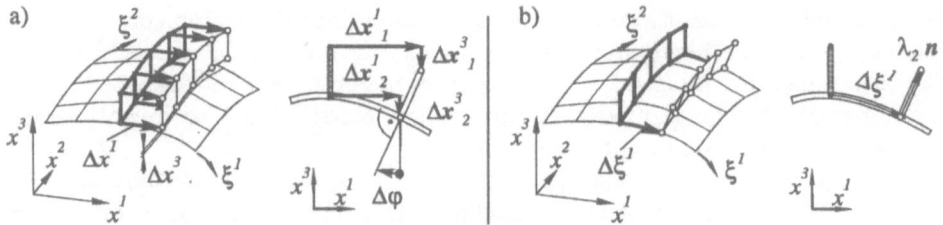

Fig. 3: Possibilities of positioning a stiffener on a curved panel

variables z_i (like dimensions and position of the stiffener) instead of directly with the analysis variables of the analysis model. When all constructive variables have been determined from the design variables, the new constructive layout of the component can be obtained. This procedure allows to use both the coefficients (e.g. for a variation of the component shape) and the independent parameters of the approach functions as design variables. The latter facilitate a re-location of the design variables on a prescribed contour and is used in our case for the positioning ($\Delta \xi^1$) of a stiffener on a given panel surface (Fig. 3b). Based upon the constructive layout, the analysis variables $\mathbf{y}$ are then calculated for the different analysis models (Fig. 2). Since in the present case only smooth and relatively small variations of the geometry occur during optimization, the necessary FE-mesh adaptation can be caried out by means of isoparametrical distortion rules [4].

Proceeding from the fundamentals of the constructive design models, the following sections shall introduce design models using Design Elements for stiffened panels.

3.2. DESIGN ELEMENTS CONCEPT

A suitable procedure for the shape optimization of structures is the Design Element Method introduced by IMAM [5], where a structure is divided into simple subelements like lines, surfaces, ruled bodies, denoted Design Elements. These areas are defined by keypoints; each design element is described by corresponding shape functions and is controlled by so-called "master nodes". A geometrical modeling of this type is used in many CAGD-procedures [6]. The master nodes to be varied during optimization are defined by a set of design variables.

For a plane structure the rule of interpolation within a Design Element can generally be given as follows:

$$\mathbf{r} = \mathbf{r}(x^k(\xi^\alpha)) = \mathbf{r}(\xi^\alpha) = \sum_{i=0}^{m} \sum_{j=0}^{n} b_i^k(\xi^1) \, b_j^k(\xi^2) \, a_{ij}^k \; ; \quad (\xi^1, \xi^2) \in [0,1] \qquad (3)$$

with x^k cartesian coordinates, $k = 1,2,3$,

         ξ^α independent surface parameters, $\alpha = 1,2$,

a_{ij}^{k} coefficients of the approach functions,
b_{i}^{k}, b_{j}^{k} parametrical approach functions.

In order to geometrically model the panel structures treated in this paper, section-wise defined bicubical BÉZIER-splines (patches) proved to be particularly suitable which are linked with each other C^2-steady [3].

In order to formulate a Stiffener Design Element, we require in addition to the position vector $r(\xi^\alpha)$ the tangential surfaces and the normal vectors. The combination and logical linking of surface elements allows for the definition of macro-elements used for the geometrical modeling of stiffener elements. The position of a stiffener can then be determined via a position vector to the initial and end point of the stiffener base (r_{A1}, r_{A2}); the shape variation is carried out through the stiffener parameters λ_1, λ_2, λ_3 (see Fig. 4).

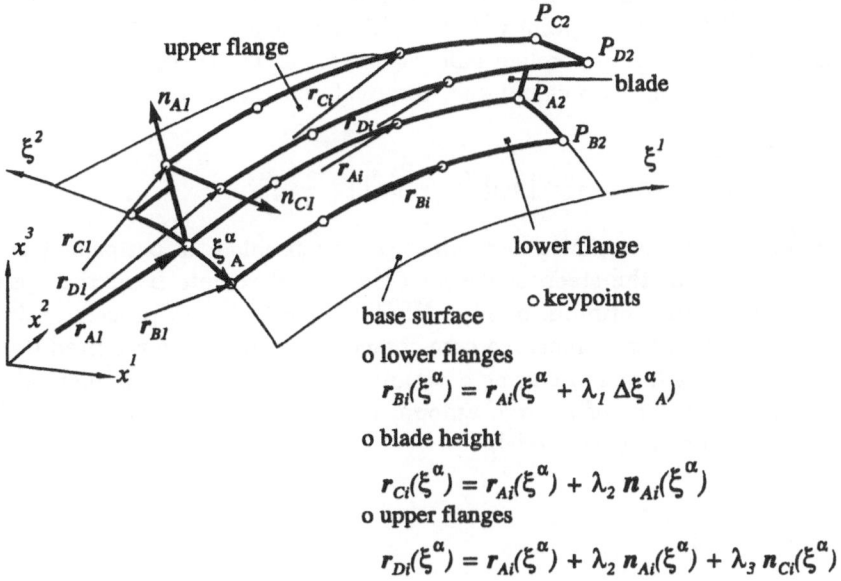

o lower flanges
$$r_{Bi}(\xi^\alpha) = r_{Ai}(\xi^\alpha + \lambda_1 \Delta\xi^\alpha_A)$$

o blade height
$$r_{Ci}(\xi^\alpha) = r_{Ai}(\xi^\alpha) + \lambda_2 \, n_{Ai}(\xi^\alpha)$$

o upper flanges
$$r_{Di}(\xi^\alpha) = r_{Ai}(\xi^\alpha) + \lambda_2 \, n_{Ai}(\xi^\alpha) + \lambda_3 \, n_{Ci}(\xi^\alpha)$$

Fig. 4: Stiffener design element

3.3. PARAMETRICAL DESCRIPTION OF THE CONSTRUCTIVE LAYOUT

By means of a parametrical description of the panel base in the form (3) the 3D surface can be transformed into a 2D unit area, where the stiffeners form the boundaries of the single sub-surfaces. Thus, the position of a stiffener in the 2D plane is also determined and can be moved on the plane by means of the surface parameters ξ^α (Fig. 5). Explicitly defined coupling relations for the displacement vectors as would be required in the 3-dimensional range are not necessary here.

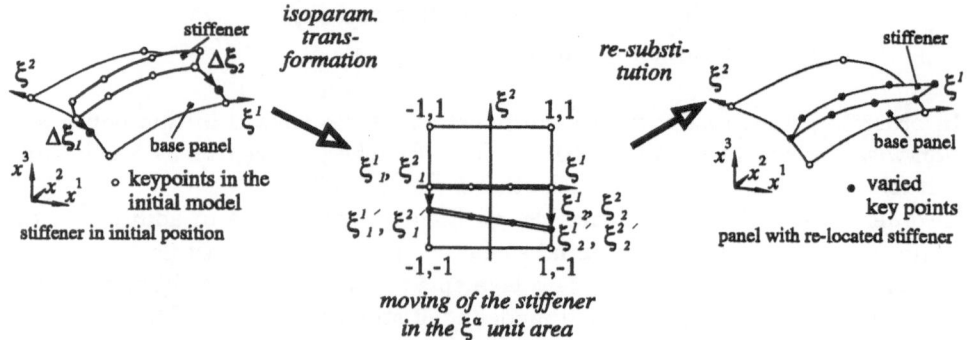

Fig. 5: Moving of a stiffener using Design Elements

For a transformation from the 3-dimensional into the 2-dimensional plane, the parameters $\xi^{\alpha *}$ for a given point $r^*(x^k)$ on the surface are to be determined. Since for this transformation no explicit rule can be given, the surface parameters $\xi^{\alpha *}$ for a given point $r^*(x^k)$ are calculated iteratively by means of a minimization of the distance vector:

$$min \; \frac{1}{2} \; | \, r_s(\xi^{\alpha}) - r^*(x^k) \, |^2 \; \Rightarrow \xi^{\alpha *} \qquad (4)$$

The solution of the unconstrained minimization problem is performed using a modified method of the steepest descend (1st order) remote from the optimum whereas, close to the optimum point, a NEWTON-RAPHSON procedure (2nd order) is applied in order to increase convergence rate. Since the required derivatives of the objective function can be determined analytically, the numerical effort is low even in the case of a large amount of discrete points $r^*(x^k)$ (e.g. all FE-nodal points of the analysis model).

4. Test Example

The following test example shall illustrate the efficiency of the developed design models. For this purpose, a simply supported composite panel is considered possessing six longitudinal stiffeners and which is subject to an uni-axial load in x-direction (see Fig. 6).

The material in the base panel is arranged in 12 and in the stiffeners in 10 symmetrical single layers. As design variables 11 parameters for the stiffener arangement, the stiffener dimensions and the material layout in the stiffeners and the base panel as well are considered. The objective function and the constraints for the test example are defined according to (1).

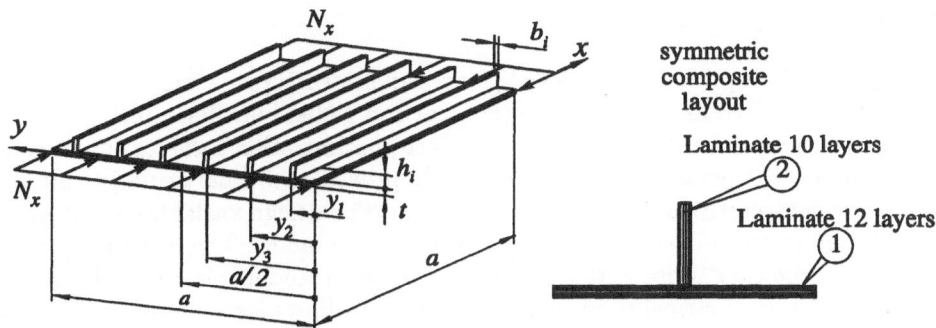

Fig. 6: Definition of the test panel to be optimized

The buckling analysis using FE is based on the eigenvalue equation obtained from the second variation of the total potential:

$$(K + \lambda K_G)\, u = 0 \qquad (5)$$

with $\quad K \qquad\qquad$ ordinary stiffness matrix,

$\qquad\quad K_G \qquad\qquad$ geometrical stiffness matrix,

$\qquad\quad \lambda = F_{crit}/F_{appl} \quad$ eigenvalue,

$\qquad\quad u \qquad\qquad$ eigenvector.

This equation is solved numerically and yields the buckling eigenvalue λ which is equivalent to the ratio of the critical buckling load F_{crit} and the applied load F_{appl}. For the modeling of the composite panel structures treated here, 8-nodes shell elements (Stif99) are used, each possessing three rotational and three translational degrees of freedom [7].

Table 1 shows the results of the optimization calculations. We have used a SQP-algorithm according to POWELL/ SCHITTKOWSKI [1]. It becomes obvious that the increase of the buckling load is caused by the enlargement of the stiffener elements by simultaneously reducing the wall thickness of the panels on the one hand. On the other hand, the increase is achieved by a substantial variation of the material layout of base panel and stiffeners. In the optimum point, the buckling values are very close to each other.

	objective function: 1st eigenvalue	2nd eigenvalue/ 3rd eigenvalue	figures
initial Design	0.986	1.703/ 2.355	
optimal Design	4.823	4.853/ 4.872	

Table 1: Results of the optimization

5. Conclusion

For finding the optimal layout of stiffend panel structures, a very accurate FE analysis is required in order to consider local and global buckling effects.

Due to the time-consuming structural analysis, optimization calls for "intelligent" design models; "constructive" design models proved to be particularly suitable, where the component is modeled geometrically. The design variables are then defined based upon the parameters of the constructive model. The mesh adaptation required in the case of discrete structural analysis procedures is here achieved on the basis of a geometrical mesh description which allows for a uniform distortion or adaption to the geometry variations during the optimization.

In a test example, an optimization increases the critical buckling load, achieved through a variation of the component geometry as well as of the material structure. It should be mentioned due to table 1 that in the optimal design there is a quasi multiple eigenvalue which means a clustering occurs. Further that means that such a optimal structure might be sensible to imperfections and it calls for further investigations.

ACKNOWLEDGMENT

The authors thank Mr. Michael Wengenroth for his committed assistance in the translation of this paper.

REFERENCES

1. Eschenauer, H.A.; Geilen, J.; Wahl, H.J.: *SAPOP - An optimization procedure for multicriteria structural design*. In: Hörnlein, H.R.E.M.; Schittkowski, K. (Eds.): Software systems for structural optimization. International Series of Numerical Mathematics, Vol. 110, Birkhäuser Verlag, Basel (1993) 207-227.

2. Braibant, V., Fleury C. and Beckers, P.: *Shape optimal design: An Approach matching C.A.D. and optimization concepts*, Report SA-109, Aerospace Laboratory of the University of Liège, Belgium, 1983.

3. Schuhmacher, G.: *Multidisziplinäre, fertigungsgerechte Optimierung von Faserverbund-Flächentragwerken*, Dissertation, Universität-GH Siegen, TIM Forschungsberichte T07-03.95, 1995.

4. Zienkiewicz, O. C., Phillips, D. V.: *An automatic mesh generation sheme for plane and curved surfaces by "isoparametric" coordinates*, Internat. J. for Numer. Meths. in Engrg. 3, (1971), 519-528.

5. Imam, M. H.: *Three-dimensional shape optimization*, Internat. J. for Numer. Meths. in Engrg. 18, (1982), 661-673.

6. Bletzinger, K. U.: *Formoptimierung von Flächentragwerken*, Dissertation, Universität Stuttgart, Bericht Nr.:11, 1990.

7. N.N.: *ANSYS User´s Manual for Revision 5.0*, Swanson Analysis Systems, Inc., Johnson Road, Houston, 1992.

OPTIMIZATION OF MULTIBODY SYSTEMS USING APPROXIMATION CONCEPTS

L.F.P. ETMAN, D.H. VAN CAMPEN AND A.J.G. SCHOOFS
Department of Mechanical Engineering
Eindhoven University of Technology
P.O. Box 513 Eindhoven, The Netherlands

Abstract.
Sequential approximate optimization is used to solve multibody optimum design problems. The transient optimization problem is formulated such that approximation concepts can be incorporated. Two multibody design examples illustrate the effectiveness of the approach.

1. Introduction

Optimum design of multibody systems is characterized by a specific kind of optimization problem. Generally, an optimization problem is formulated to determine the design variable values that will minimize an objective function subject to constraints. For multibody systems, the objective function and constraints are usually time dependent, which complicates the standard problem formulation. Additionally, for many engineering applications, multibody analysis routines are used to calculate the kinematic and dynamic behavior of the mechanical design. As a result, most objective function and constraint values follow from the numerical analysis. Therefore, to solve the optimization problem, the multibody code has to be coupled with a mathematical programming algorithm. Such a coupling may be difficult to implement and can lead to high computational costs.

Numerical design optimization tools have been mainly developed in the field of finite element structural analysis. Many tools use a suitable approximation concept as interface between analysis software and optimizer. Such an interface avoids programming difficulties, and is computationally more convenient than a direct coupling (Haftka and Gürdal, 1992). The basic

81

D. Bestle and W. Schielen (eds.), IUTAM Symposium on Optimization of Mechanical Systems, 81–88.
© 1996 *Kluwer Academic Publishers.*

idea is to separate the numerical analysis and optimization. This is established by building an explicitly known approximate optimization problem that can be easily handled by the optimizer.

Local function approximations are the most popular approximations used in optimization (Barthelemy and Haftka, 1993). Large numbers of design variables and constraints can be handled without great difficulty. Moreover, in many cases, design sensitivities can be obtained at reduced computational costs compared with finite difference sensitivity analysis. Local approximations of objective function and constraints are based on function values and derivative values with respect to the design variables in a single point of the design space. Since the approximations have a limited region of validity, a sequence of approximate optimization cycles has to be performed to reach the optimum solution.

Approximate optimization strategies can also be effectively applied to the design optimization of multibody systems. To illustrate this, an optimization tool has been developed based upon a local approximation concept. It has been coupled with the multibody analysis code MECANO (Samtech, 1994). Sensitivities have been calculated by finite differences. The effectiveness of the approach is demonstrated. Suitable intermediate design variables and responses are introduced with the aim to improve the approximations and the efficiency of the optimization process.

2. Optimization problem formulation

Within the time interval $t \in [0, T]$, the transient optimization problem is formulated as follows: find the set of design variable values $\mathbf{x} \in \Re^n$ that will minimize the objective function:

$$F(\mathbf{x}) = F(r_j(\mathbf{x}, t), \mathbf{x}) \qquad \begin{aligned} t &\in [0, T] \\ j &= 1, \ldots, n_r \end{aligned} \qquad (1)$$

subject to the constraints:

$$g_h(\mathbf{x}, t) = g_h(r_j(\mathbf{x}, t), \mathbf{x}) \le c_h \qquad \begin{aligned} \forall \ t &\in [0, T] \\ j &= 1, \ldots, n_r \\ h &= 1, \ldots, m \end{aligned} \qquad (2)$$

within the design space:

$$x_k^l \le x_k \le x_k^u \qquad k = 1, \ldots, n \qquad (3)$$

The scalar x_k is the k-th element of the design vector $\mathbf{x}$. The set of n_r functions $r_j(\mathbf{x}, t)$ represents the responses calculated by the multibody code.

The aforementioned formulation covers several different types of optimization problems that may occur for multibody systems. Most constraints will depend on the response values, although there is no restriction at all for constraints depending on design variables only. Response based constraints can be directly related to the response functions, such as displacements or bending stresses bounded to a maximum value. Other constraints may depend on a response at one specific time point, or be an integral function of a response on a certain time interval. Time point and integral type functions are applicable to the objective function as well.

If a response function $r(\mathbf{x}, t)$ directly determines the corresponding constraint value, a time dependent constraint follows:

$$g(\mathbf{x}, t) \leq c \qquad \forall\, t \in [0, T] \tag{4}$$

which must be satisfied during the entire simulation time. This time dependency has to be removed to reach an optimization problem that can be solved by a nonlinear programming algorithm. The most straightforward way is to discretize the time interval into n_t time points. Then, the original constraint is replaced by n_t constraints:

$$g_i(\mathbf{x}, t_i) \leq c \qquad i = i, \ldots, n_t \tag{5}$$

The time point distribution has to be dense enough to avoid too large constraint violations. This increases the number of constraints significantly.

Several equivalent constraint formulations have been proposed to remove time dependency without increasing the number of constraints. Generally, they can be written as an integral function of $g(\mathbf{x}, t)$ (Haftka and Gürdal, 1992). However, these constraints may suffer from a strong nonlinear behavior, non-differentiability, or even loss of information of the original constraint. Therefore, Hsieh and Arora (1984) took into account only the 'worst' time point constraints, i.e. all local maxima of the constraint functions.

3. Sequential approximate optimization

The optimization problem defined in the previous section is replaced by a sequence of explicitly known optimization problems using a local approximation concept. Vanderplaats (1993) described the basic program structure of the sequential approximate optimization process for finite element structural analysis. This structure is the same for multibody analysis.

Sequential approximate optimization starts with the analysis of the initially proposed design, followed by an evaluation of all constraint functions. Approximation models are generated for the critical and potentially critical constraints based upon analysis and design sensitivity data. The approximate optimization problem is built and the region of validity is bounded by so-called move limits. Within this search subregion, the approximate problem is solved by the optimizer. At the calculated optimum design, a new design cycle is started. This process is repeated until convergence.

The most simple approximation is a first-order Taylor series approximation of objective function and constraints. More fundamental implicit functions however are the responses $r_j(\mathbf{x}, t)$. Given a set of response values, objective function and constraint values can usually be easily obtained. So the response functions can be approximated instead of the objective function and constraints. This leads to the well-known intermediate response variables. From the design variable point of view, intermediate design variables $\mathbf{x}^I(\mathbf{x})$ can be formulated as well. The basic idea is the same: explicit nonlinear functional behavior can be used to improve the approximations, and thus the efficiency and reliability of the optimization process.

We define the approximate optimization problem of a particular cycle as: minimize the approximate objective function:

$$\tilde{F}(\mathbf{x}) = F(\tilde{r}_{ji}(\mathbf{x}_j^I(\mathbf{x}), t_i), \mathbf{x}) \qquad \begin{aligned} i &= 1, \ldots, n_t \\ j &= 1, \ldots, n_r \end{aligned} \qquad (6)$$

subject to the approximate active and potentially active constraints:

$$\tilde{g}_h(\mathbf{x}) = g_h(\tilde{r}_{ji}(\mathbf{x}_j^I(\mathbf{x}), t_i), \mathbf{x}) \le c_h \qquad \begin{aligned} i &= 1, \ldots, n_t \\ j &= 1, \ldots, n_r \\ h &= 1, \ldots, m \end{aligned} \qquad (7)$$

within the search subregion around the cycle start design $\mathbf{x}_0$:

$$x_{0k}^l - \alpha_k \le x_k \le x_{0k}^u + \alpha_k \qquad k = 1, \ldots, n \qquad (8)$$

It is assumed that the response functions $r_j(\mathbf{x}, t)$ are calculated at a set of n_t discrete time points: $r_{ji}(\mathbf{x}, t_i)$. Objective function and constraints follow from these discrete response values according to the selected type of functions. Responses $r_{ji}(\mathbf{x}, t_i)$ are linearly approximated at $\mathbf{x}_0$ with respect to the intermediate design variables:

$$\tilde{r}_{ji}(\mathbf{x}_j^I, t_i) = r_{ji}(\mathbf{x}_{0j}^I, t_i) + \sum_{k=1}^n (x_{kj}^I - x_{0kj}^I) \left(\frac{\partial r_{ji}}{\partial x_{kj}^I} \right)_{\mathbf{x}_{0j}^I} \qquad (9)$$

Every response may have its own intermediate design variables $\mathbf{x}_j^I(\mathbf{x})$.

The number of responses $r_{ji}(\mathbf{x}, t_i)$ to be approximated is highly influenced by the optimization problem formulation. Suppose a constraint is active that is an integral function of a response: $g(\mathbf{x}) = \int r(\mathbf{x}, t) dt$. Then, all discrete responses $r_i(\mathbf{x}, t_i)$ have to be approximated. However, for time point constraints $g_i(\mathbf{x}, t_i)$, a large amount of the constraints can be deleted. All constraints can be removed whose constraint value at $\mathbf{x}_0$ is smaller than e.g. 70% of the constraint bound. Additionally, for every local maximum of the constraint $g(\mathbf{x}, t)$, only a few time point constraints $g_i(\mathbf{x}, t_i)$ near the maximum have to be retained. This combines the worst time points of Hsieh and Arora (1984), and the constraint deletion of Vanderplaats (1993).

Constraint deletion gives the opportunity to reduce the costs of the sensitivity analysis. Gradients have to be calculated only for a limited set of constraints. For time point constraints this means that the multibody analysis code should be able to calculate gradients just at those time points for which the constraints remain in the set of active and potentially active constraints.

Usually, a local approximation concept requires an appropriate move limit strategy to ensure a good convergence of the optimization process. In this paper we adopt the move limit strategy described in Etman *et al.* (1994). So, move limits α_k are calculated from the absolute value of the design variable values x_{0k}, the maximum approximation error of the previous cycle, and the convergence behavior.

4. Test problems

4.1. DESIGN OF A VEHICLE SUSPENSION SYSTEM

A five degree of freedom vehicle suspension system, shown in Figure 1, is designed to have an optimal response to road undulation 'Profile No. 1' for a travelling speed of $11.4 \ ms^{-1}$ (Haug and Arora, 1979). Spring coefficients k_1, k_2 and k_3, and damper coefficients c_1, c_2, and c_3 have to be determined such that the maximum absolute acceleration of the driver's seat $|\ddot{z}_1|$ is minimized. Constraints are the maximum relative displacements between: chassis and drivers seat, chassis and front and rear wheels, and road surface and front and rear wheels. Every constraint is replaced by 600 time point constraints equally spaced on the time interval of 0 to 3 s. The constraints depend linearly on the responses $\ddot{z}_1$, z_1, z_2, z_3, z_4, and z_5. Therefore, a sequence of linear programming problems follows if no intermediate design variables are introduced.

Table 1 shows the convergence behavior of the approximate optimization process. For every design cycle, the maximum approximation error ($\max e_h$), the maximum constraint violation ($\max v_h$) and the objective function value (F) are tabulated. Additionally, the number of time points is given at which active and potentially active constraints occur (n_t^a), as well as the maximum design variable change ($\max \Delta x_k$).

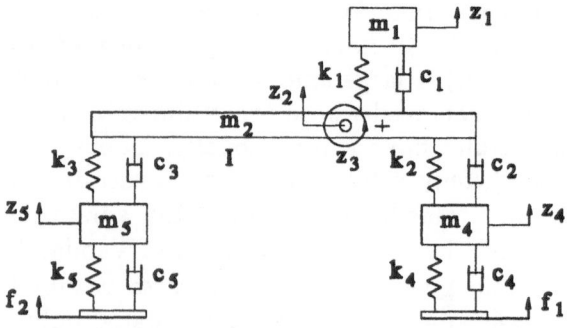

Figure 1.   Five degree of freedom vehicle model.

TABLE 1. Optimization history of the vehicle model.

Cycle	$\max \Delta x_k$ [%]	n_t^a	$\max e_h$ [%]	$\max v_h$ [%]	F [ms^{-2}]
0				51.0	8.43
1	15.0	26	23.3	35.5	7.73
2	15.0	37	19.0	20.1	7.39
3	15.0	37	20.4	5.62	7.37
4	15.0	39	3.40	0.612	7.22
5	20.0	27	4.76	1.24	6.99
6	26.7	30	7.39	0.968	6.81
7	35.6	20	3.84	0.808	6.69
8	36.1	16	3.06	0.511	6.54
9	5.03	19	0.515	-0.107	6.48
10	0.391	19	0.0219	0.00138	6.47

The optimum vehicle design was obtained within ten design cycles, starting from the infeasible initial design of Haug and Arora (1979). The final maximum acceleration and the active constraints correspond with Haug and Arora (1979), though the optimum design we found differs for k_3 and c_1: $k_3 = 35.0 \, kNm^{-1}$ and $c_1 = 4.05 \, kNsm^{-1}$, instead of $k_3 = 42.4 \, kNm^{-1}$ and $c_1 = 2.26 \, kNsm^{-1}$. This is probably caused by the different optimization problem formulation. Deviations with Etman *et al.* (1994) are explained by the more dense time discretization. Approximation models were built at 19 to 39 of the 600 time points. During the first three cycles the constraints were relaxed to obtain a feasible solution of the approximate optimization problem. Move limits were active for the first eight cycles.

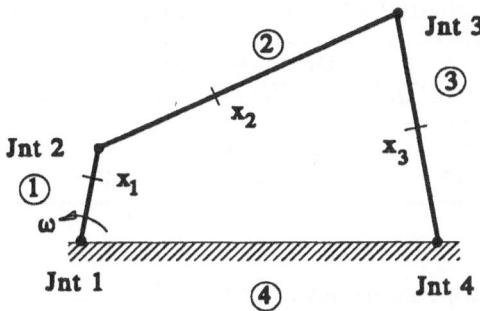

Figure 2.   Stress constrained four-bar mechanism.

4.2. STRESS CONSTRAINED DESIGN OF A FOUR-BAR MECHANISM

Figure 2 shows a four-bar mechanism that consists of three flexible links, connected to each other and the ground by revolute joints. The three mobile links have a circular cross section, an elasticity modulus of $6.895 \ 10^{10} \ Pa$, and a mass density of $2757 \ kgm^{-3}$. The lengths of the bars are $l_1 = 0.3048 \ m$, $l_2 = 0.9144 \ m$, $l_3 = 0.762 \ m$, and $l_4 = 0.9144 \ m$, respectively. The input crank of the mechanism rotates at a constant angular velocity of $10\pi \ rads^{-1}$. Due to the motion, bending stresses occur in the mobile links. The optimum design problem is now to minimize the mass of the mechanism by varying the cross sectional areas, with the bending stresses constrained to a maximum of $\sigma_a = 2.758 \ 10^7 \ Pa$.

Each link is modeled by six beam elements. MECANO computed the bending moments M_k^p in every p-th node of link k as a function of time. Stresses can then be calculated from:

$$\sigma_k^p = \frac{4\sqrt{\pi}}{A_k^{3/2}} M_k^p \qquad (10)$$

with A_k the cross sectional area of body k. The bending moments were treated as intermediate response variables. Sohoni and Haug (1982) defined the cross sectional areas A_k as design variables. We selected the diameters d_k instead, and used the cross sectional areas as intermediate design variables. A time interval of 0.3 to 0.5 s is considered, discretized into 201 time points. This exactly covers one period of steady state motion.

A good convergence behavior was found starting from the initial design $d_{1,2,3} = 357 \ mm$, with a maximum of 21 time points remaining in the active set. The optimum design is given in Table 2. The constraint violation, maximum approximation error and objective function accuracy were smaller than 0.1 %. The final objective function value reported in Sohoni

TABLE 2. Optimization of the stress constrained four-bar mechanism.

Reference	Optimum design			No.	Time	F
	d_1 [mm]	d_2 [mm]	d_3 [mm]	cycles	points	[kg]
This paper	37.3	23.6	19.8	13	0-21	2.67
Sohoni and Haug (1982)	36.2	28.1	12.2			2.69

and Haug (1982) is approximately the same. Deviations with respect to the design variable values are probably caused by the different method of stress analysis. The use of bending moments as intermediate response variables directly follows from equation (10). However, it is not clear yet which intermediate design variables correspond best with these bending moments. Maybe other intermediate design variables can be found that give a faster convergence than the cross sectional areas we used here.

5. Conclusion

Approximation concepts can be effectively used for the design optimization of multibody systems. The two examples demonstrate the effectiveness of separating the analysis and optimization routines. Time dependency can be removed by means of time discretization. The increase of the number of constraints is limited by only approximating the worst critical and potentially critical time point constraints. As a result, gradients are required only at a fraction of all time points. This gives the opportunity to reduce the costs of the design sensitivity analysis.

References

Barthelemy, J.-F.M. and Haftka, R.T. (1993) Approximation concepts for optimum structural design - a review, *Structural Optimization*, 5, pp. 129–144.

Etman, L.F.P., Thijssen, E.J.R.W., Schoofs, A.J.G. and Van Campen, D.H. (1994) Optimization of multibody systems using sequential linear programming, in B.J. Gilmore, D.A. Hoeltzel, D. Dutta and H.A. Eschenauer (eds.), *ASME Advances in Design Automation*, DE 69-2, pp. 525–530.

Haftka, R.T. and Gürdal, Z. (1992) *Elements of structural optimization*. Kluwer Academic Publishers, Dordrecht.

Haug, E.J. and Arora, J.S. (1979) *Applied optimal design*. John Wiley & Sons, New York.

Hsieh, C.C and Arora, J.S. (1984) Design sensitivity analysis and optimization of dynamic response, *Computer Methods in Applied Mechanics and Engineering*, 43, pp. 195–219.

Samtech (1994) *Samcef Mecano version 5.1*. Liège.

Sohoni, V.N. and Haug, E.J. (1982) A state space technique for optimal design of mechanisms, *ASME Journal of Mechanical Design*, 104, pp. 792–798.

Vanderplaats, G.N. (1993) Thirty years of modern structural optimization, *Advances in Engineering Software*, 16, pp. 81–88.

OPTIMAL ROCKING AND DAMPING OF A SWING

A.M.FORMAL'SKY and E.K.LAVROVSKY

Institute of Mechanics, Moscow Lomonossov State University,
Michurinsky Prospect 1, 119899, Moscow, Russia
e-mail: Formal@inmech.msu.su

The problem of maximising or minimising the inclination of a swing to the vertical at its highest point is considered. The swing control problem is of interest from the viewpoint of theoretical mechanics, control theory, oscillations theory. It is closely related to the problem of using extensible rods to damp the oscillations of a satellite around its centre of mass in gravitational field, and with control problems for some sporting motions. The system of a swing and a person on it is modelled by a pendulum and a material point. This point can be shifted within some limits along the straight line, passing through the pivot and the mass centre of the pendulum. The viscous damping because of air resistance opposing the motion of the swing, and also in the pivot are taken into account together with dry friction in the pivot.

1. The Motion Equations

We model the system of a swing and a person on it by planar compound pendulum of mass m and a material point of mass M that can be shifted along the line OC (figure 1).

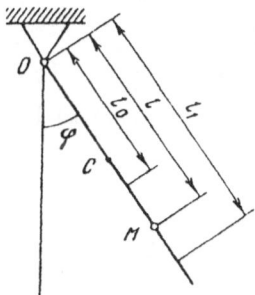

Figure 1. Model of a swing and a person on it.

89

D. Bestle and W. Schielen (eds.), IUTAM Symposium on Optimization of Mechanical Systems, 89–96.
© 1996 *Kluwer Academic Publishers.*

Here O is the pendulum suspension point, C is its mass centre. Let I be the moment of inertia of the pendulum around point O and $OC = \rho$. The distance OM is denoted by l. Let $l_0 \leq l \leq l_1$, where $l_0, l_1 = \text{const}$, $l_1 > l_0$. The motion equation of this system is

$$\frac{d}{dt}\left[(I + Ml^2)\frac{d\varphi}{dt}\right] + (cl^2 + c_1)\frac{d\varphi}{dt} + v + \zeta = 0, \quad \zeta = (Ml + m\rho)g\sin\varphi \qquad (1)$$

Here φ is the counter clockwise deviation angle of the pendulum from the vertical, c is positive coefficient of viscous damping because of air resistance opposing the motion of the mass M [1], c_1 is positive coefficient of viscous damping because of air resistance opposing the motion of the pendulum without the mass M, and also in the pivot, g is the gravity acceleration, and v is the torque of the dry friction forces in the pivot whose threshold is v_0

$$v = v_0\text{sgn}\,\dot{\varphi} \text{ for } \dot{\varphi} \neq 0 \qquad\qquad v = -\zeta \text{ for } \dot{\varphi} = 0 \text{ and } |\zeta| \leq v_0$$

$$\qquad\qquad\qquad\qquad\qquad\qquad\qquad\qquad\qquad\qquad\qquad\qquad\qquad\qquad\qquad (2)$$

$$v = v_0 \text{ for } \dot{\varphi} = 0 \text{ and } \zeta \leq -v_0 \qquad v = -v_0 \text{ for } \dot{\varphi} = 0 \text{ and } \zeta \geq v_0$$

We introduce the dimensionless time τ, moment of inertia j, coefficients of viscosity χ and χ_1, the threshold of dry friction forces torque δ, the gravitational torque of the pendulum μ, and distance OM denoted by u

$$\tau = \frac{t\sqrt{g}}{\sqrt{l_0}}, \quad j = \frac{I}{Ml_0^2}, \quad \chi = \frac{c\sqrt{l_0}}{M\sqrt{g}}, \quad \chi_1 = \frac{c_1}{Mgl_0\sqrt{l_0 g}},$$

$$\delta = \frac{v_0}{Mgl_0}, \quad \mu = \frac{m\rho}{Ml_0}, \quad u = \frac{l}{l_0}$$

The dimensionless distance $1 \leq u \leq U$ ($U = l_1 / l_0$, $U > 1$) from the pivot to the moving point is the controllable parameter. The equation (1) can be rewritten in the form

$$\dot{\varphi} = K / (j + u^2)$$

$$\qquad\qquad\qquad\qquad\qquad\qquad\qquad\qquad\qquad\qquad\qquad\qquad\qquad\qquad\qquad (3)$$

$$\dot{K} = -(\chi u^2 + \chi_1)K / (j + u^2) - \delta\text{sgn}K - (u + \mu)\sin\varphi$$

Here K is the total angular momentum of the system. The introduction of the variable K instead of angular velocity $\dot{\varphi}$ enables to avoid the differentiation of the control $u(t)$, which can change discontinuously in time together with the velocity $\dot{\varphi}$. The control u and the angle φ enter non-linearly in system (3). Only the threshold value of the torque due to dry friction forces is represented in (3).

2. Statement of the Problem

We shall assume that $\delta < 1 + \mu$. Then the dead spaces of system (1), (2) for all $l_0 \le l \le l_1$ ($1 \le u \le U$) do not occupy the entire range $-\pi \le \varphi < \pi$. Let the initial state of system (3)

$$\varphi(0) < 0, \quad K(0) = 0 \tag{4}$$

be situated outside these dead spaces.

The problem of the optimal swing rocking is to find a control u for which $\max \varphi(\theta)$ is obtained, where $\theta > 0$ is the first time instant when $K(\theta) = 0$. We will write this down as follows:

$$\max_{1 \le u \le U} \left[\varphi(\theta) \right], \quad K(\theta) = 0, \quad \theta > 0 \tag{5}$$

We write the problem of the optimal swing damping in the form

$$\min_{1 \le u \le U} \left[\varphi(\theta) \right], \quad K(\theta) = 0, \quad \theta > 0 \tag{6}$$

The initial state (4) is assumed to be such that a time $\theta > 0$ exists at which $K(\theta) = 0$. Problems (5) and (6) are actually Bulgakov's problem for accumulated perturbations [2]. But here the time θ at the right end is determined by the condition $K(\theta) = 0$. The modified statement of the Bulgakov's problem when the terminal time is not specified, was examined by V.V.Alexandrov and V.N.Zhermolenko [3].

Under condition (4) the momentum $K > 0$ in the interval $0 < \tau < \theta$. Therefore, in equations (3) only the first of relations (2) is used.

3. Method of the Problem Solving

For any admissible control $u(\tau)$ the momentum $K > 0$, and the angle $\varphi(\tau)$ increases strictly monotonically as τ increases from 0 up to θ. Therefore, when considering the phase trajectories of system (3) in the semi-plane $K > 0$, one can prove that problem (5) is solved by a control u which *maximises the derivative dK/dφ at each point* (φ, K) *of the phase trajectory*. A control u which *minimises this derivative*, solves the problem (6).

The extremum of this derivative is reached at some $u \in [1, U]$ which *yields the extremum* of the function

$$F(u) = -\chi K u^2 - \left[\delta + (u + \mu) \sin \varphi \right] (j + u^2) \tag{7}$$

Using this proposition we can describe the optimal control $u(\varphi, K)$ analytically and synthesise the optimal control.

4. System without Air Resistance and Dry Friction

Analysis of function (7) in the case when $\chi = \delta = 0$ shows that the optimal rocking control is of relay form.

$$u = U \text{ for } \varphi K < 0 \ (\varphi\dot{\varphi} < 0), \quad u = 1 \text{ for } \varphi K \geq 0 \ (\varphi\dot{\varphi} \geq 0) \tag{8}$$

This control means that the mass centre of the system is raised in the lowest point of the swing trajectory where $\varphi = 0$, and extended in its highest point where $\dot{\varphi} = 0$. It is well known that the swing oscillations increase under this control. The control (8) was examined by K.Magnus [1], but for a massless swing $(j = \mu = 0)$ and without investigations of its optimality.

The optimal damping control is the "opposite" of (8)

$$u = 1 \text{ for } \varphi K > 0 \ (\varphi\dot{\varphi} > 0), \quad u = U \text{ for } \varphi K \leq 0 \ (\varphi\dot{\varphi} \leq 0) \tag{9}$$

Control (8) maximises the deviation of the swing from the vertical over any specified number of half-oscillations, whereas control (9) minimises this deviation.

5. Rocking of the Swing with all Dissipative Forces

In the presence of all dissipative forces optimal control differs from (8) and (9). Optimal control structure turns out to depend on the relation between dimensionless moment of pendulum inertia j and the upper control limit U. Analysis of function (7) shows that for the problem (5) there are three possible cases

$$(1) \ j \leq 2U + 1, \quad (2) \ 2U + 1 < j < 3U^2, \quad (3) \ 3U^2 \leq j$$

In the case 1 optimal control is of relay form, as is (8)

$$u = U, \text{ if } \delta + \chi K + \Lambda \sin \varphi \leq 0$$
$$\left(\Lambda = U + \mu + \frac{j+1}{U+1} \right) \tag{10}$$
$$u = 1, \text{ if } \delta + \chi K + \Lambda \sin \varphi > 0$$

The switching line for control (10) is given by the equation

$$\delta + \chi K + \Lambda \sin \varphi = 0 \tag{11}$$

The switching of control (10) from $u=U$ to $u=1$ occurs not at $\varphi = 0$, as for control (8), but earlier, at $\varphi < 0$. As δ and χ increase while μ and j decrease, the switching point on each optimal trajectory moves away from the $\varphi = 0$ axis.

Figure 2 shows the synthesis picture for optimal rocking control (10) in the semi-plane $K>0$ (the dead spaces are shown hatched). The curve (11) intersects the semi-axis $K=0$, $\varphi<0$ at two points

$$\varphi = \varphi_0 = -\arcsin(\delta/\Lambda), \quad \varphi = -\pi - \varphi_0 \tag{12}$$

At $\chi=0$ the curve (11) becomes the two straight lines (12).

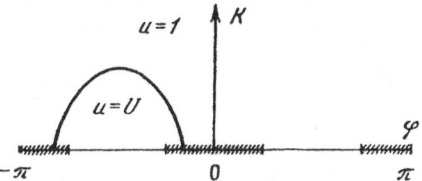

Figure 2. The synthesis of optimal rocking control (10) in the semi-plane $K>0$.

The synthesis of optimal control enables to solve the optimisation problem not only for initial states (4).

In the case 2 the optimal control is

$$
\begin{aligned}
u &= U, & &\text{if } \delta + \chi K + \lambda(U)\sin\varphi \le 0 \\
u &= u_1(\varphi, K), & &\text{if } -\lambda(U)\sin\varphi \le \delta + \chi K \le -\eta(1)\sin\varphi \\
u &= 1, & &\text{if } \delta + \chi K + \eta(1)\sin\varphi \ge 0
\end{aligned}
\tag{13}
$$

Here we denote

$$u_1(\varphi, K) = -\frac{1}{3}(A + \sqrt{A^2 - 3j}), \quad A = A(\varphi, K) = \frac{\delta + \chi K}{\sin\varphi} + \mu,$$

$$\lambda(v) = 2\sqrt{j + v^2} - v + \mu, \quad \eta(v) = \frac{1}{2v}(j + 3v^2) + \mu$$

The value $u_1(\varphi, K)$ is the smaller of the two roots of quadratic equation

$$dF / du = 3u^2 + 2Au + j = 0 \tag{14}$$

During optimal motion the control (13) at $\varphi < 0$ gets off the saturation level U, experiencing a jump, then strictly monotonically and continuously decreases down to 1. It reaches 1 at $\varphi < 0$ and remains equal to 1 up to $\varphi(\theta)$. At $j = 2U + 1$ the "relay-continuous" control (13) becomes the purely relay control (10).

In the case 3 the optimal control is

$$u = U, \qquad \text{if } \delta + \chi K + \eta(U)\sin \varphi \leq 0$$

$$u = u_1(\varphi, K), \quad \text{if } -\eta(U)\sin \varphi \leq \delta + \chi K \leq -\eta(1)\sin \varphi \qquad (15)$$

$$u = 1, \qquad \text{if } \delta + \chi K + \eta(1)\sin \varphi \geq 0$$

Figure 3 shows the synthesis picture for optimal control (15) in the semi-plane $K>0$. Unlike the case 2, in the case 3 the optimal control is continuous. Having got off the level U at $\varphi < 0$, it strictly monotonically decreases down to 1, reaches 1 at $\varphi < 0$ and remains equal to 1 up to $\varphi(\theta)$. At $j = 3U^2$ control (15) is identical with (13).

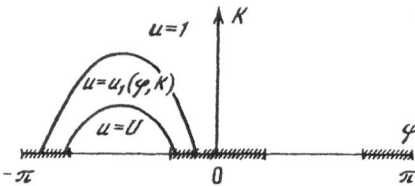

Figure 3. The synthesis picture for optimal rocking control (15) in the semi-plane $K>0$.

The results obtained here show that when there is dry friction at the pivot and (or) viscous friction due to air resistance to the motion of the mass M, in optimal motion the centre of mass of the system is displaced upwards not when the swing passes through its lowest point, as in the cases of control (8), but earlier.
The optimal control synthesis picture for swing rocking is symmetric relative to the origin of coordinates. Hence it is easy to extend it to the semi-plane $K < 0$.

6. Damping of the Swing

Analysis of function (7) shows that for the problem (6) there are three possible cases:
$$(1)\ j \leq 3, \quad (2)\ 3 < j < U(U+2), \quad (3)\ U(U+2) \leq j$$
In the case 1 optimal control has the form

$$u = 1, \qquad \text{if } \delta + \chi K + \eta(1)\sin \varphi \leq 0$$

$$u = u_2(\varphi, K), \quad \text{if } -\eta(1)\sin \varphi \leq \delta + \chi K \leq -\eta(U)\sin \varphi \qquad (16)$$

$$u = U, \qquad \text{if } \delta + \chi K + \eta(U)\sin \varphi \geq 0$$

Here

$$u_2(\varphi, K) = \frac{1}{3}\left(-A + \sqrt{A^2 - 3j}\right)$$

is the largest of two roots of quadratic equation (14).

Control (16) is a continuous function of its arguments. Having got off the level 1, it strictly monotonically increases up to U.

In the case 2 the optimal control is

$$u = 1, \qquad \text{if } \delta + \chi K + \lambda(1)\sin\varphi \leq 0$$

$$u = u_2(\varphi, K), \quad \text{if } -\lambda(1)\sin\varphi \leq \delta + \chi K \leq -\eta(U)\sin\varphi \qquad (17)$$

$$u = U, \qquad \text{if } \delta + \chi K + \eta(U)\sin\varphi \geq 0$$

During optimal motion the "relay-continuous" control (17) gets off the saturation level 1 experiencing a jump, then strictly monotonically and continuously increases up to U.

In the case 3 optimal control is of relay form, as is (9)

$$u = 1, \quad \text{if } \delta + \chi K + \Lambda\sin\varphi \leq 0$$
$$\qquad\qquad\qquad\qquad\qquad\qquad\qquad (18)$$
$$u = U, \quad \text{if } \delta + \chi K + \Lambda\sin\varphi > 0$$

In all the damping controls (16) - (18) the centre of mass of the system is lowered not when the swing passes through the lowest point, as under control (9), but earlier.

7. Rocking and Damping of the Swing within Prescribed Time

We consider problems (5) and (6) for which the time θ is given in advance. These problems can not be solved by extremising the function (7). We design their solutions numerically using the Pontryagin's maximum principle [4]. Equations (3) must be supplemented by the following relations

$$\dot\psi_1 = \psi_2(u + \mu)\cos\varphi, \qquad \dot\psi_2 = [\psi_2(\chi u^2 + \chi_1) - \psi_1 \mp 1]/(j + u^2)$$

$$H(u) = (\psi_1 \pm 1)K/(j + u^2) - \psi_2[(u + \mu)\sin\varphi + \delta + (\chi u^2 + \chi_1)K/(j + u^2)] \qquad (19)$$

$$\max_{1 \leq u \leq U} H(u) = C, \quad u = \arg\left[\max_{1 \leq u \leq U} H(u)\right], \quad \psi_1(\theta) = 0, \quad K(0) = K(\theta) = 0$$

where $\psi_1(t)$ and $\psi_2(t)$ are conjugate variables and C is unknown constant. The upper sign corresponds to the swing rocking and lower sign to the swing damping problem.

If the time θ is free, then $C=0$ and $\psi_2(\theta)=0$. However, it is simpler to solve the problem for free time θ by finding extremum of function (7) as above.

If the time θ is given in advance, then $C \neq 0$. Choosing different values of $\varphi(\theta)$ and $\psi_2(\theta)$ at the final time θ and integrating the equations (3), (19) from right to left, one can construct a two-parameter family of optimal trajectories. Thus we can select parameters $\varphi(\theta)$ and $\psi_2(\theta)$ so that the initial value of the angle φ be equal to the specified $\varphi(0)$ and the terminal time to the specified θ.

In figure 4 we show the results of numerical investigation of the swing rocking problem for $\delta = \chi = \chi_1 = 0$ (there is no dissipation), $\mu = 0.5$, $U = 2$ and $\varphi(0) = -0.4248$. Here is presented the dependence of the maximum of the rocking amplitude $\varphi(\theta)$ on the specified time θ for $j = 1/3; 6; 15$. The maximum value of $\varphi(\theta)$ is obtained if time θ is free.

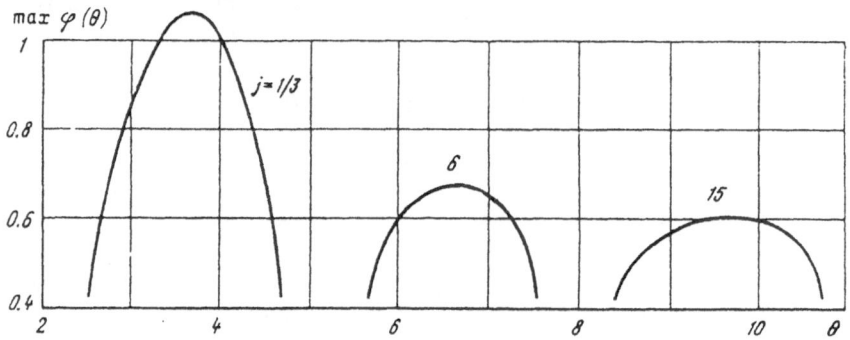

Figure 4. The dependencies of the maximum of the rocking amplitude $\varphi(0)$ on the specified time θ for $j=1/3;6;15$.

This maximum is larger for smaller moments of inertia j. It is clear that for a given value of $\varphi(0)$ the first positive time θ, at which $K(\theta)=0$, can not be chosen arbitrarily. It lies within certain limits. Figure 4 illustrates this.

References

1. Magnus. K. (1976) *Schwingungen*, B.G.Teubner, Stuttgart.
2. Bulgakov, B.V. (1954) *Oscillations*, Gostechizdat, Moscow.
3. Alexandrov, V.V. and Zhermolenko, V.N. (1972) On the absolute stability of second-order systems, *Bull. Moscow State Univ., Ser. 1 Math. and Mech.* 5, 102-109.
4. Pontryagin. L.S.. Boltyanskii, V.G.. Gamkrelidze, R.V. and Mischenko, Ye.F. (1961) *Mathematical Theory of Optimal Processes*, Fizmatgiz, Moscow.

AN INTEGRATED STRATEGY FOR THE CONTROL OF COMPLEX MECHANICAL SYSTEMS BASED ON SUB-SYSTEM OPTIMALITY CRITERIA

T.J. GORDON

University of Technology
Department of Aeronautical and Automotive Engineering
Loughborough
Leics. LE11 3TU
England

1. Introduction

Complex mechanical (or mechatronic) control systems often comprise a number of sub-systems, which may interact dynamically and yet be designed and implemented in isolation from one another - chassis control systems on motor vehicles are a case in point. At the simplest level, this can be wasteful in terms of duplication of transducers or electronic circuitry, and cost savings can be found from allowing systems to 'talk' to one another. More importantly, the sharing of dynamic information between different sub-systems can be expected to benefit the overall mechanical performance of the complex system - see for example Mastinu *et al.* (1994). In this paper, a high level of system integration will be considered, on-line dynamic control being optimised with respect to predictions and cost criteria derived from the separate sub-systems.

It is worth emphasising at the start, that the approach involves a genuine synthesis of sub-system designs. In particular

- No global dynamic system model is used, except for simulation. This makes the controller design tractable for large or complex systems.

- The approach is capable of dealing with non-linear subsystems.

- The integrated approach can be made tolerant to unexpected events, such as an actuator failure.

This paper covers a general form of the integrated control strategy in Section 2, and its more specific formulation in the case of multi-system LQR optimization in Section 3. An illustrative application to the control of a full-vehicle active suspension system is then considered in Sections 4 and 5.

97

D. Bestle and W. Schielen (eds.), IUTAM Symposium on Optimization of Mechanical Systems, 97-104.
© 1996 *Kluwer Academic Publishers.*

2. General Formulation

A global dynamic system Σ is modelled as a set of sub-systems Σ^a $(a = 1, ..., \sigma)$ which are defined so as to describe all of the significant system dynamics. There is no requirement here that the Σ^a be mutually exclusive - the choice of resolution into sub-systems, their degrees of freedom etc. can be regarded as a standard problem for dynamics modelling, which of course requires engineering skill and intuition on the part of the modeller. As a guideline however, it is desirable that each sub-system has identifiable variables and optimization criteria that lie predominantly within its domain. For example, in the active suspension example described below, we shall define "wheel control" sub-systems together with a "body control" sub-system; there each wheel control sub-system incorporates a simplified description of the vehicle body motion, and hence there is some 'overlap' in the dynamic modelling.

Suppose Σ is described by the (generally non-linear) state equations

$$\dot{\xi} = \Phi(\xi, \eta, \zeta) \tag{1}$$

where ξ is an N-component state-vector, η is an M-component vector of control inputs, and ζ is an R-component vector of disturbance inputs. The corresponding equations for Σ^a are

$$\dot{x} = f^a(x, u, z) \tag{2}$$

with vector dimensions (n^a, m^a, r^a). It is assumed that each set of local variables (x, u, z) has a known relationship to the corresponding global variables:

$$x = X^a(\xi) \quad u = U^a(\eta) \quad z = Z^a(\zeta) \tag{3}$$

For linear systems, these become matrix relationships, e.g. $X^a(\xi) \equiv \overline{X}^a \cdot \xi$, where $\overline{X}^a$ is a corresponding $n^a \times N$ matrix of coefficients. The functions in equations (2) and (3) are assumed known; however the global system equations (1) are not directly applied, and may be unknown.

The systems integration strategy requires that each subsystem design supplies *cost criteria* for the subsequent application of control. This study will employ deterministic optimal control (see for example Kwakernaak and Sivan 1972), and the criteria then take the form of Hamiltonian functions

$$H^a(x, u, p) = p^T f^a(x, u, 0) + L^a(x, u) \tag{4}$$

Here $L^a(x, u)$ is an underlying cost function and p is the costate vector obtained from local design optimization. In the normal implementation, an optimal control is synthesised by the algebraic minimisation of H^a with respect to u:

$$u = \arg\min\{H^a\} \tag{5}$$

Clearly this must be replaced by an alternative condition in the integrated control - otherwise the individual sub-systems will provide different forms of the 'optimal' control.

Integration of the local controls will be based on the following two principles, where in both cases we distinguish between 'internal' and 'external' variables, denoted ($'$) and ($''$) respectively.

(a) Predictive Model Integration. From equation (3)

$$\dot{\mathbf{x}} = \left(\partial \mathbf{X}^a / \partial \xi\right)\dot{\xi} \tag{6}$$

Using any local model (2), dynamic relations can be deduced for ξ. To avoid conflict between different local models, a subspace of internal states is required in each case:

$$\mathbf{x}' = \mathbf{P}^a(\mathbf{x}) \tag{7}$$

The resulting relations for ξ are then of the form

$$P'^a\dot{\xi} = Q'^a \tag{8}$$

where

$$P'^a(\xi) \overset{def}{=} \left(\partial \mathbf{P}^a / \partial \mathbf{x}\right)\cdot\left(\partial \mathbf{X}^a / \partial \xi\right), \quad Q'^a \overset{def}{=} \left(\partial \mathbf{P}^a / \partial \mathbf{x}\right)\cdot \mathbf{f}^a(\mathbf{X}^a(\xi), \mathbf{U}^a(\eta), 0) \tag{9}$$

An integrated 'model' is then obtained by assembling the sub-system equations to give

$$P\dot{\xi} = Q \tag{10}$$

where

$$P \overset{def}{=} \left[\left(P'^1\right)^T\left(P'^2\right)^T \cdots \left(P'^\sigma\right)^T\right]^T, \quad Q \overset{def}{=} \left[\left(Q'^1\right)^T\left(Q'^2\right)^T \cdots \left(Q'^\sigma\right)^T\right]^T \tag{11}$$

As a final condition, the integrated predictive model be *complete*; in other words equation (9) is required to provide a unique prediction for $\dot{\xi}$:

$$\sum n'^a = N , \quad \det P \neq 0 \tag{12}$$

(b) Hamiltonian Criterion
As well as providing localised predictions, each sub-system is assumed to define an internal component of dynamic cost:

$$L^a(\mathbf{x}, \mathbf{u}) = L'^a(\mathbf{x}, \mathbf{u}) + L''^a(\mathbf{x}, \mathbf{u}) \tag{13}$$

The external component $L''^a(\mathbf{x}, \mathbf{u})$ is some approximation to the dynamic costing associated with the other sub-systems, and may be required to ensure the local optimization is properly defined. Using local dynamic optimization, the costates and Hamiltonian functions can be split similarly:

$$H'^a(\mathbf{x}, \mathbf{u}, \mathbf{p}) = \mathbf{p}'^T\mathbf{f}^a(\mathbf{x}, \mathbf{u}, 0) + L'^a(\mathbf{x}, \mathbf{u})$$
$$H''^a(\mathbf{x}, \mathbf{u}, \mathbf{p}) = \mathbf{p}''^T\mathbf{f}^a(\mathbf{x}, \mathbf{u}, 0) + L''^a(\mathbf{x}, \mathbf{u}) \tag{14}$$

This leads to the defining relation for integrated system control:

$$\eta = \arg\min\left\{\sum\nolimits_a H'^a\right\} \tag{15}$$

which is a natural generalisation of the earlier condition (5).

3. LQR Realisation

In the Linear Quadratic Regulator (LQR) formalism, the cost function takes the following form

$$L = \begin{bmatrix} \mathbf{x}^T & \mathbf{u}^T \end{bmatrix} \begin{bmatrix} Q & S \\ S^T & R \end{bmatrix} \begin{bmatrix} \mathbf{x} \\ \mathbf{u} \end{bmatrix} \tag{16}$$

and each of the sub-system coefficient matrices (Q, S, R) may be split into internal and external components, e.g. $Q = Q' + Q''$. The 'optimal' control for the sub-system is

$$\mathbf{u}_{opt}(\mathbf{x}) = -R^{-1}(B^T P + S^T)\mathbf{x} \overset{def}{=} K\mathbf{x} \tag{17}$$

This is not to be applied in closed-loop, but is useful for intermediate analysis. P is the usual Riccati matrix for the sub-system optimization, which may be split into internal and external components, $P = P' + P''$, via the following pair of equations:

$$P'(A + BK) + A^T P' + Q' + S'K = 0 \tag{18a}$$

$$P''(A + BK) + A^T P'' + Q'' + S''K = 0 \tag{18b}$$

These equations are linear in P' and P'' and may be solved easily . Note that although P is a symmetric matrix, in general P' and P'' are non-symmetric.

4. Suspension Control Systems

We consider the example of an active vehicle suspension system. These systems are well-known in the literature, and many authors have applied various optimal control methods in design and analysis (see for example Abdel Hady, 1989, Gordon et al., 1994). The full-vehicle system can be resolved into five dynamic control sub-systems; one sub-system is associated with each wheel, and one with the body - the cost terms and degrees of freedom are outlined in Table 1, and model parameters are given in Table 2. The body control sub-system contains no representation of the wheel inertias, and the wheel control systems only use simplified body inertia representations. There is also a sixth sub-system employed, which is purely kinematic; it has no associated cost function, and simply defines the relationships between displacements and velocities. From these six sub-systems, the completeness condition (12) is met, and the above integrated control scheme may be implemented.

Four basic control systems have been considered, and the following labelling is used consistently in Figures 1 - 6.

   (a) Reference passive control

   (b) Isolated control - where each vehicle corner is controlled locally, based on the LQR optimization of its wheel control subsystem model.

   (c) Integrated control - as described above.

   (d) Global control - where the design is based on a single multivariable system model.

Quadratic cost functions can be constructed quite easily for the variables given in Table 1. The performance objectives are to maintain minimum absolute pitch and roll deflections of the body, as well as reducing the accelerations in vehicle body-bounce, these criteria being chosen to emphasise the integrated nature of the control task. The quadratic cost functions also contain terms in tyre and suspension workspace variables, these acting as constraint functions in the optimisation.

TABLE 1. Active Suspension Sub-Systems

Sub-system	Degrees of Freedom	Costed Internally	Costed Externally
Wheel control	body height	dynamic tyre load variations	body acceleration
	wheel height		actuator effort
	'road' surface		suspension deflection
Body control	bounce	suspension deflections	actuator effort
	pitch	pitch angle	
	roll	roll angle	
	'road' surface	bounce acceleration	

TABLE 2. Vehicle system parameters

body mass	1400 kg	wheelbase	2.7 m
body roll moment of inertia	500 kg m^2	track	1.5 m
body pitch moment of inertia	2500 kg m^2	tyre vertical stiffness	2×10^5 N/m
wheel masses	25 kg - front	passive spring stiffness	2×10^4 N/m
	35 kg - rear		
mass centre location behind front axle	1.2 m	passive damping rate	10^3 N/ms^{-1}

Cost function weighting parameters were roughly set to match peak and rms tyre and suspension deflections to the passive system, under free motion resulting from impulsive initial disturbances. These were as follows:

Condition 1:   Initial vertical velocity $v_0 = 2.51$ m/s of a single front wheel
Condition 2:   Initial vertical velocity $v_0 = 2.03$ m/s of a single rear wheel
Condition 3:   Initial vertical velocity $v_0 = 0.936$ m/s of the body (pure bounce)
Condition 4:   Initial angular velocity $\omega_0 = 1.75$ rad/s of the body in roll
Condition 5:   Initial angular velocity $\omega_0 = 0.717$ rad/s of the body in pitch

The initial velocities for these conditions were chosen to give roughly 'equal' effects on the passive system, with peak tyre deflections of 20 mm resulting from Conditions 1 and 2, and peak suspension deflections of 100 mm resulting from Conditions 3 - 5. These conditions were then used as the basis for cost function 'tuning', the active systems (b) - (d) being required to remain within the passive peak limits, and to maintain rms deflections of roughly similar magnitudes. Figures 1 and 2 show normalised tyre and suspension responses for two cases - clearly the 'tuning' conditions are being respected here.

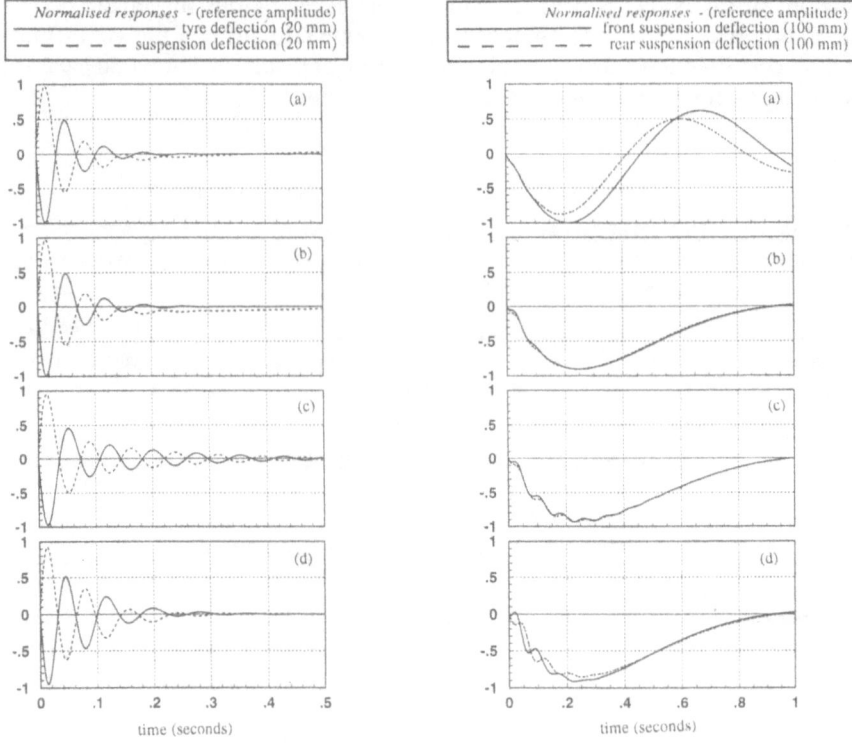

Figure 1. Responses to Condition 1   *Figure 2.* Responses to Condition 3

In Figure 3, the vehicle settles onto an uneven road surface, the left front wheel falling into a 50 mm depression. The responses in Figure 4 are for a low speed run over a road surface which is sinusoidal on the left track (vehicle speed = 7 m/s, wavelength = 5.4 m chosen to emphasise pitch inputs, amplitude = ±50 mm) and flat on the right track. Systems (c) and (d) both give excellent control over the body roll and pitch attitudes. All three active systems also give much improved control over bounce acceleration (compared to the passive system) though the global design (d) is clearly best in this respect. The isolated active control is only capable of controlling the oscillations in the attitude angles; steady-state errors are poorly controlled.

The conclusion thus far is that the integrated design (c) gives an excellent compromise solution to the global optimisation problem - the use of local optimality criteria hardly degrades performance compared to design (d). A stronger conclusion is apparent from Figures 5 and 6 however, again using the test conditions of Figures 3 and 4, but *with a mechanical failure introduced in one of the actuators*: the left front active control unit has degenerated to act as an undamped linear spring (rated in accordance with the passive system data). With the Hamiltonian minimisation (15) carried out *on-line*, (c) provides the best of the four controls; the global control (d) becomes nearly unstable in body bounce, as evidenced by the very high accelerations.

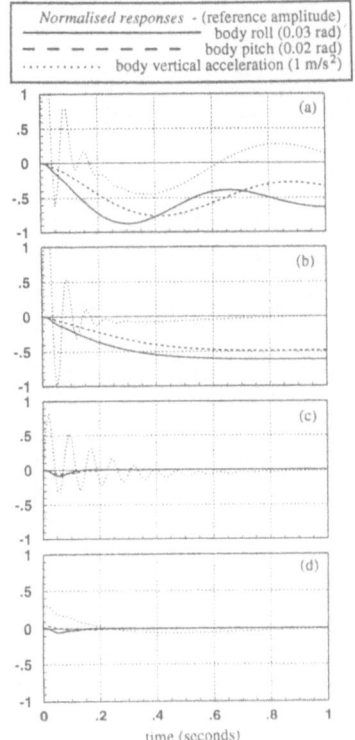

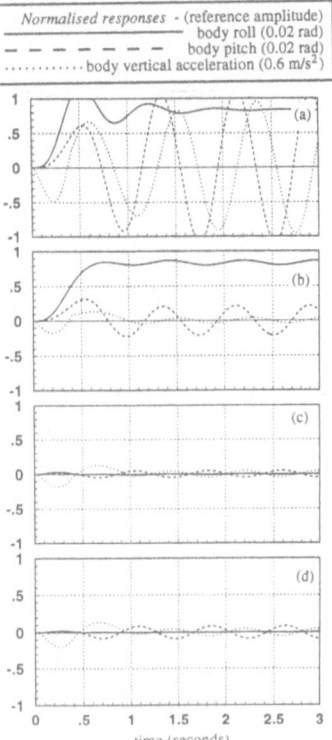

Figure 3. Settling on an Uneven Surface   *Figure 4.* Sinusoidal Road Input

5. Discussion and Conclusions

In this brief paper, it has not been possible to consider the many detailed steps involved in the synthesis of the integrated optimal control strategy for the example active suspension system. However, the broad reasoning is intuitively straightforward, and can be summarised quite simply:

- each dynamic sub-system is modelled and optimised in isolation, but with simplistic representations of 'external' interactions included
- each sub-system supplies prediction and cost evaluation data to an integrating controller
- the integrating controller uses a Minimum Hamiltonian principle to emulate classical optimal control system design. Since the minimisation is purely algebraic, it should be feasible to carry this out on-line in many applications

The resulting integrated control is potentially far superior to that provided by isolated sub-system designs, and potentially more robust than controllers designed on the basis of global system models - this in addition to the greatly increased flexibility of the overall synthesis process.

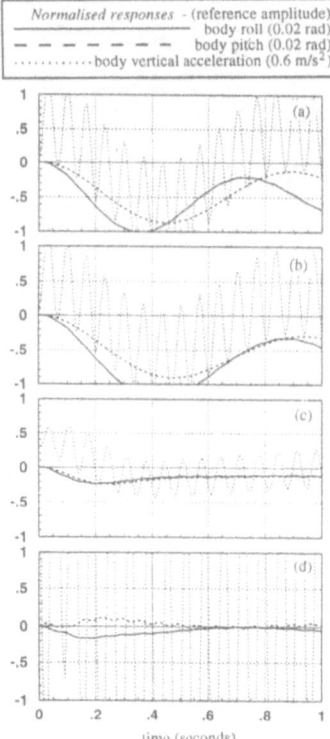

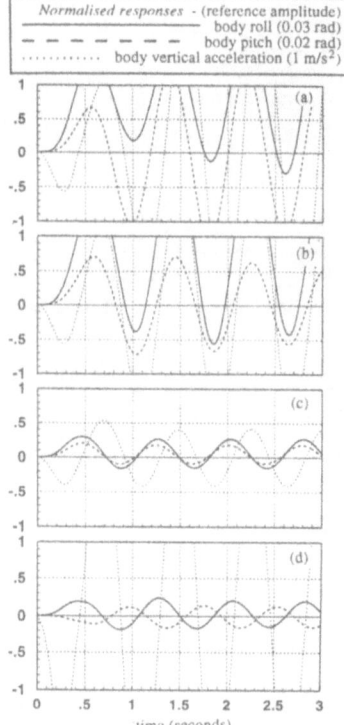

Figure 5. Settling Responses
- with faulty actuator

Figure 6. Sinusoidal Response
- with faulty actuator

Of course, this study has been limited to one simple example, and the approach is as yet restricted to the optimal deterministic regulator (linear or non-linear). Formal stability properties have yet to be proved, and links with other forms of optimal control have yet to be made. However, the results obtained are extremely encouraging, both from the point of view of performance and of flexibility in design procedure.

References

Abdel Hady, M.B.A. and Crolla, D.A. (1989) Theoretical analysis of active suspension performance using a four-wheel vehicle model, *Proc. Instn. Mech. Engrs. Part D*, **203**, 125 - 135.

Gordon, T.J., Palkovics, L., Pilbeam, C. and Sharp, R.S. (1994) Second generation approaches to active and semi-active suspension control system design, *The Dynamics of Vehicles on Roads and Tracks*, Proceedings of the 13th IAVSD Symposium, Chengdu, China, 1993, *Supplement to Vehicle System Dynamics*, **23**, 158 - 171.

Mastinu, G., Babbel, E., Lugner, P., Margolis, D., Mittermayr, P. and Richter, B. (1994) Integrated control of lateral vehicle dynamics, *The Dynamics of Vehicles on Roads and Tracks*, Proceedings of the 13th IAVSD Symposium, Chengdu, China, 1993, *Supplement to Vehicle System Dynamics*, **23**, 358 - 377.

Kwakernaak, H. and Sivan, S., *Linear Optimal Control Systems*, Wiley, New York, 1972.

DESIGN OPTIMIZATION OF A SOLID SURFACE DEPLOYABLE ANTENNA

S.D. GUEST AND S. PELLEGRINO
Cambridge University Engineering Department
Trumpington Street, Cambridge, CB2 1PZ, U.K.

Abstract. This paper describes the design optimization of a new rigid-panel deployable antenna, where the reflective surface is split into wings, and each wing is subdivided into panels which are connected together by revolute joints. The surface folds by wrapping the wings around a central hub. In order to deploy, small deformations of the antenna are required. The design has been optimized to minimize this deformation. Based on the optimization results, a model of the antenna has been built and successfully tested.

1. Introduction and Antenna Concept

Deployable antennas are an essential element of space technology. The maximum diameter of launch vehicles limits the diameter of satellites to about 4 m, but antennas larger than this are required for earth observation, astronomy and communications. Many previous deployable antennas have been furlable, but deployable antennas with a solid surface are required for frequencies above about 5 GHz (Rogers *et al.*, 1993).

The Solid Surface Deployable Antenna (SSDA) is a new concept for foldable antennas with a solid reflector surface. The concept took its inspiration from a wrapping fold pattern for thin flat membranes described by Guest and Pellegrino (1992). The antenna surface is split into *wings*, and each wing is further subdivided into *panels* which are connected together by revolute joints. The SSDA is folded by wrapping the wings around a central *hub*. *Connecting bars* are used to couple the motion of different wings.

Many different versions of the SSDA have been described by Guest (1994), but this paper concentrates on one particular variant, which is shown in Fig. 1. This SSDA has 6 wings, each wing is made up of 5 panels, and the connecting bar is attached by revolute joints to panel 5 on one wing, and panel 4 on the next wing. This arrangement was chosen because a simple consideration of the

105

D. Bestle and W. Schielen (eds.), IUTAM Symposium on Optimization of Mechanical Systems, 105–112.
© 1996 *Kluwer Academic Publishers.*

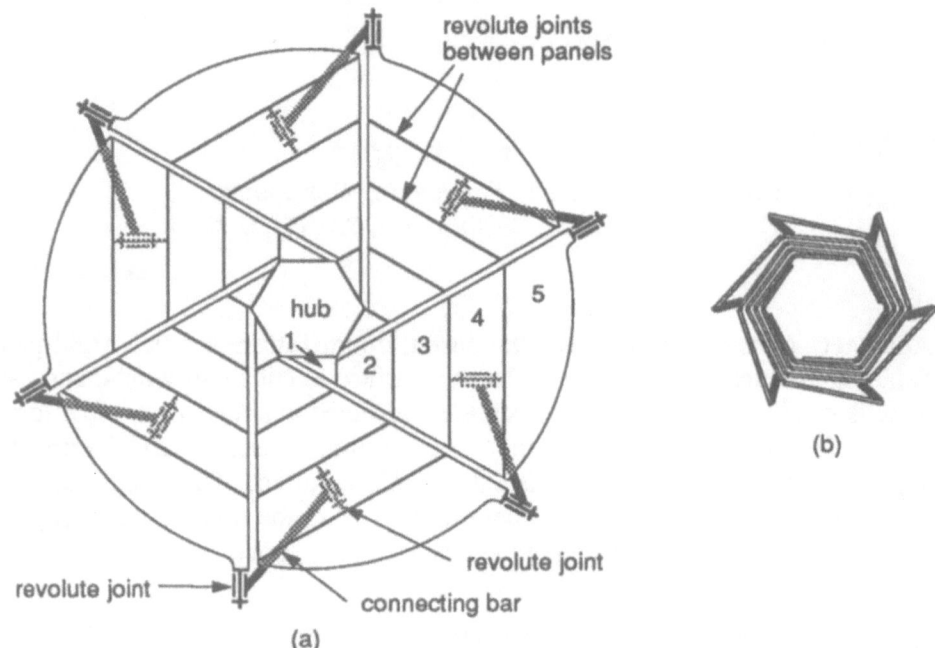

Figure 1. The SSDA: (a) Fully deployed, showing panel numbers;
(b) Fully folded, the panels are shown flat for clarity.

mobility, M, of the mechanism shows that it has no less than 1 degree of freedom (dof). This can be shown by noting that the SSDA consists of 36 bodies, each having 6 dof (the hub is taken as a reference), and 42 revolute joints, each imposing 5 constraints. Deployment will be activated by driving at the same rate 6 symmetrically placed hinges, thereby imposing 5 additional constraints.

$$M \geq 6 \times 36 - 5 \times 42 - 5 = 1 \tag{1}$$

If all constraints are independent, then Eq. 1 is an equality. However, the existence of one or more mechanisms does not guarantee that there will be a continuous motion, without any parts being disconnected, between the fully folded and fully deployed configurations. Indeed no such motion has been found for the SSDA. A practical way around this difficulty is to optimize the design of the SSDA so that, even though some deformation is required during deployment, its magnitude is sufficiently small not to damage the antenna.

This paper describes the optimization processes which went into the design of a working model of the SSDA. The objective of the design optimization was to minimize the deformation of the antenna during deployment.

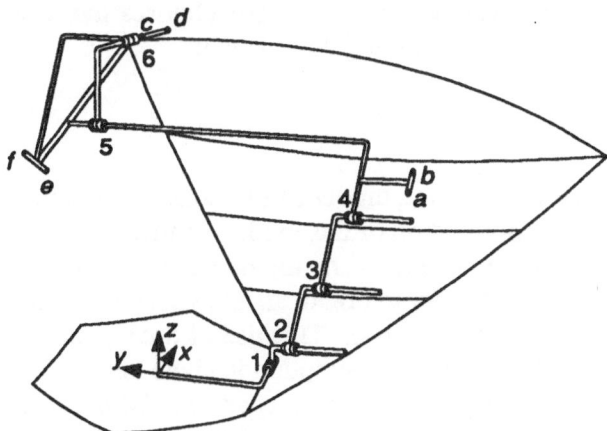

Figure 2. Simplified SSDA model, showing hinge numbers. The position of
the connecting bar is defined by points a–f.

The paper is laid out as follows. Section 2 briefly describes the model of
the antenna that was used in the optimization process, and Section 3 describes
the analytical tools that were used in the mathematical formulation of this
model. Section 4 describes the optimization process itself, and shows the results
of this process. Section 5 describes a physical model based on this optimum
design which was designed and tested.

2. Optimization Model

The objective of the optimization is to minimize the deformation of the
SSDA during deployment. Any optimization technique requires that, for any
given trial design, the maximum deformation level be estimated. An accurate
model of the SSDA could be obtained using finite elements, but then estimating
the deformation associated with each trial design would require a complex nonlinear simulation. Therefore a simpler model of the SSDA is set up. First, it is
assumed that the configuration of the SSDA is perfectly symmetric at all stages
of deployment. Second, all elements of the antenna are modelled as rigid, apart
from the joint at one end of each connecting bar. The design is optimized to
minimize the deformation of these flexible joints.

The model used is shown in Fig. 2. Because of symmetry, only one wing is
considered. The panels are modelled as rigid bars, connected together by
revolute hinges 1–6. Hinge 7 is replaced by points e and f on this wing, and by
points a and b on the next wing. Points e and a are connected together by a
deformable link, as are points f and b. The deformation of the system, Δ, is
defined as the sum of the squares of the lengths of these links. The

optimization process described in this paper chooses the design that has the *smallest maximum value* of Δ during deployment.

3. Analytical Tools

Having introduced a simplified model of the antenna, it is necessary to find a simple and effective formulation of this model, so that Δ can be easily found for any configuration of the antenna. The method used was to model the antenna using dual quaternions in order to obtain an analytical expression for Δ in terms of the rotation angles at hinges 1–6. This use of dual quaternions is described fully in Guest (1994), but a brief description is given here.

A general displacement of a body in space can be defined by a distance d and an angle θ along a line in space called the screw axis. This axis can be defined by its 6 *Plücker coordinates* $(\mathbf{w}, \mathbf{p} \times \mathbf{w})$, where $\mathbf{w}$ is a unit vector in the direction of the screw axis, and $\mathbf{p}$ is the position vector of a point on that axis. Using the notation of dual quantities, this is written as a displacement by a dual angle $\hat{\theta} = \theta + \varepsilon d$ along the dual vector $\hat{\mathbf{w}} = \mathbf{w} + \varepsilon(\mathbf{p} \times \mathbf{w})$, where ε is an operator having the property that $\varepsilon^2 = 0$.

A simple way to calculate the net effect of two general displacements on a body is to define those displacements using *dual quaternions* (McCarthy, 1990). Using the quantities defined above, a dual quaternion is defined as

$$\hat{A} = \cos\frac{\hat{\theta}}{2} + \sin\frac{\hat{\theta}}{2}\left(\hat{w}_x i + \hat{w}_y j + \hat{w}_z k\right)$$

where the quantities i, j and k multiply together according to the rule

$$i^2 = j^2 = k^2 = -1$$

$$ij = k, ji = -k, \text{ and cyclic permutations}$$

If a dual quaternion $\hat{A}$ represents the first spatial displacement of a body, and a dual quaternion $\hat{B}$ represents a second spatial displacement, the dual quaternion $\hat{C}$, representing the net effect of the two displacements, can be calculated by the dual quaternion multiplication $\hat{C} = \hat{B}\hat{A}$.

To calculate the position of a point at the end of a chain of bodies, the first operation is to form the dual quaternion $\hat{P}$ that represents the initial position and orientation of that point. This can be thought of as a displacement from the origin of coordinates to the point of interest. $\hat{P}$ is then multiplied by the dual quaternions $\hat{T}_i$ which represent the displacement of each joint in the chain, working from the point of interest back towards the reference. For instance, the position and orientation of point a, shown in Fig. 2, after hinges 1–4 have rotated by specified amounts, can be calculated from

$$\hat{P}_a' = \hat{T}_1\,\hat{T}_2\,\hat{T}_3\,\hat{T}_4\,\hat{P}_a$$

where $\hat{P}_a$ represents the position and orientation of a in the initial configuration, $\hat{P}_a'$ represents the position and orientation of a in the final configuration, and $\hat{T}_i$ represents the displacement due to hinge i, a pure rotation about its hinge axis. Thus it is possible to obtain the position of any point on the antenna as an analytical expression in terms of the hinge rotations. From this it is trivial to find Δ.

4. Optimization Process

The final objective was to build a model antenna using an existing axisymmetric parabolic reflector surface of 1.48 m diameter, with a focal length/diameter ratio of 0.42. The design of this antenna was optimized in three stages. First, a trial design was generated. Second, a simulation was carried out to find the maximum Δ during deployment. Third, the results of this simulation were used to find the optimum design.

4.1. TRIAL DESIGN

The shape of the panels and the rotation of hinges 1–5 between the deployed and folded configurations can be completely determined by considering how the antenna folds, i.e. by ensuring that it wraps compactly around the hub, as shown in Fig. 1(b). This, however, leaves undecided both the position of the connecting bar and the total rotation of hinge 6. In order to generate a trial design of the SSDA, the position of the connecting bar is fixed by varying the 4 points a–d defined in Fig. 2. Points e and f are simply points a and b rotated about the z-axis by $\pi/3$. This appears to give 13 free parameters, but actually there are only 3 because of the following 10 practical constraints. The distance between points a and b, and points c and d, is fixed. Points a and c lie on the antenna surface. Also, for the mechanism to operate correctly, point c must lie on the outer edge of the antenna. Finally, in the folded configuration, points e and f are again obtained by rotating points a and b about the z-axis by $\pi/3$.

The 3 parameters chosen to specify each trial design are the y-coordinate of point c, c_y, the x-coordinate of point a, a_x, and the angle that line c-d makes with the x-y plane, ϕ. All 3 parameters are defined in the fully-deployed configuration. The remaining 10 parameters are obtained by solving the system of 10 constraint equations outlined above. This system of non-linear equations is solved using the routine C05NJF (NAG, 1990). A similar approach has been followed by Hayman *et al.* (1994) in the design optimization of a coilable mast.

4.2. SIMULATION

For each trial design, a deployment simulation is performed in order to calculate the maximum deformation, Δ_{max}, required for the antenna to deploy. The simulation starts with the antenna fully folded. In this configuration it is strain-free, and the orientation of each of the hinges is known. Hinge 6 is then opened in 20 equal steps. At each step, the angles of hinges 1–5 are calculated using the routine E04JAF (NAG, 1990), which finds the optimal value of the 5 angles by minimising Δ. Five inequality constraints are imposed on the angles to simulate the hinge stops that in reality would prevent the hinges from turning past their fully deployed position. The simulation thus gives Δ, and also the hinge angles, at every stage in the deployment of a particular trial design.

4.3. OPTIMUM DESIGN

It was found that the plot of Δ produced by the simulation always has two peaks, one soon after the start of the simulation, and one close to the end. It was also found that, if the values of a_x and c_y are kept constant, Δ_{max} could be minimized by varying the parameter ϕ until the two deformation peaks have the same size.

Because of this simplification, the overall optimization was carried out using a hand search. For each point on a grid of points (a_x, c_y), the minimum value of Δ_{max} was found by varying ϕ. Using this method, it soon became clear that the optimal design of the SSDA has c_y as large as possible and a_x as small as possible, subject to points a and c remaining on the 4th and 5th panels, as shown in Fig. 2.

Typical results for the optimization are shown in Fig. 3. Figure 3(a) shows the plots of Δ for three simulations. Plot (i) is for $a_x = 0.500$ m, $c_y = 0.000$ m and $\phi = 0.02$ rad; ϕ is optimal for this value of a_x and c_y, and hence the two peaks of deformation are the same size. Plot (ii) is for $a_x = 0.382$ m, $c_y = 0.251$ m and $\phi = 0.10$ rad; a_x and c_y are optimal but ϕ is non-optimal, and hence the two deformation peaks are of very different size. Plot (iii) is for optimal values of all three parameters, $a_x = 0.382$ m, $c_y = 0.251$ m and $\phi = 0.34$ rad. For this case, $\Delta_{max} = 5 \times 10^{-4}$ m^2, and this corresponds to a misfit at the hinge of only 16 mm, compared with the 1.48 m diameter of the antenna.

The variation of the hinge angles during deployment is shown for the optimal design in Fig. 3(b). Figure 4 shows four views from the deployment simulation of this design. Similar views revealed some interference between panels during the early stages of deployment, and so the spacing between panels in the folded configuration was increased slightly and a new optimization process was carried out. The revised design was not substantially different

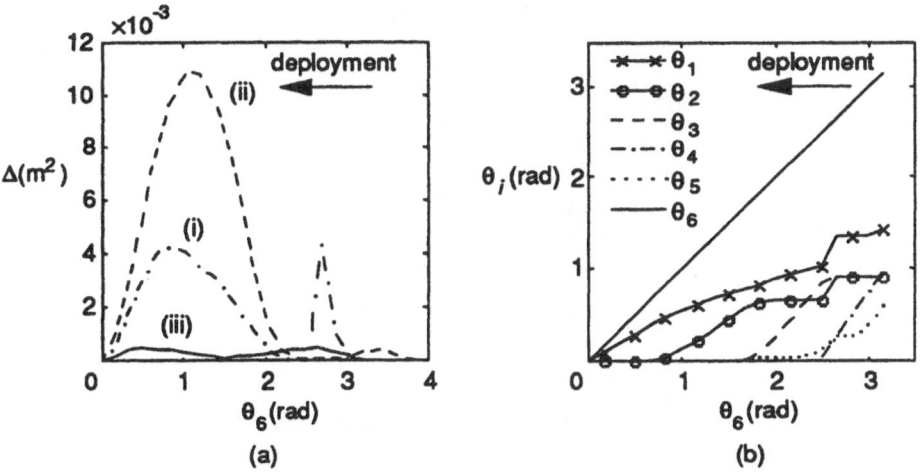

Figure 3. Simulation results: (a) Deformation during 3 simulations; (b) Hinge angles during deployment of the optimum configuration.

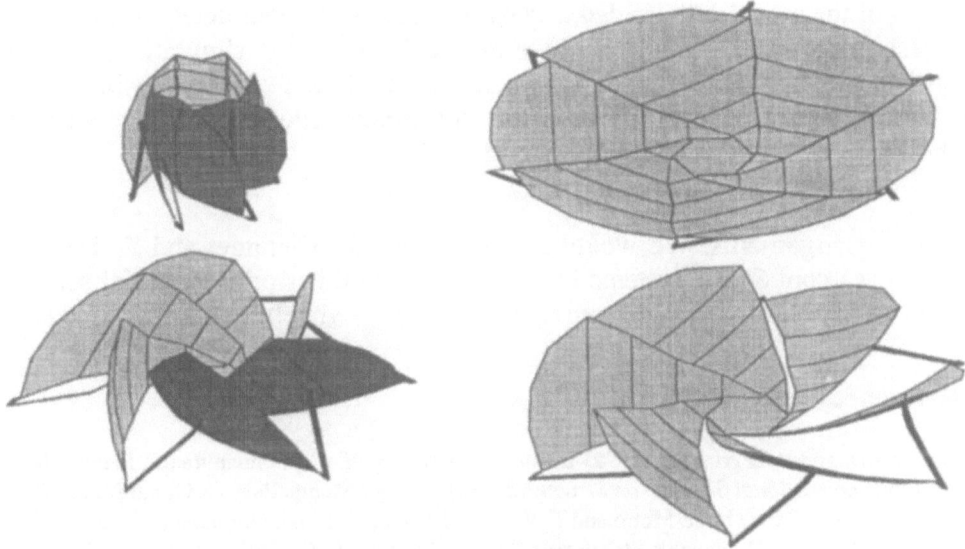

Figure 4. Four views of the antenna unfolding.

from the original one, and, in fact, all results given in this section refer to the revised design.

5. Conclusion

Following the optimization process described, a model of the antenna was made. It consisted of a glass fibre surface, cut into panels. The hinges between

112

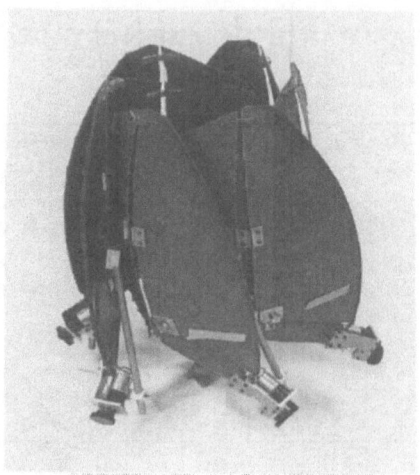

Figure 5. The model antenna fully folded.

panels were simple brass hinges. The connecting bar were Al-alloy tubes, and hinge 6 of each wing was driven by a heavily geared down electrical motor. A photo of the model fully folded is shown in Fig. 5. Further details are given in Guest (1994). The model deploys as predicted by the simulation, thereby justifying the simplified modelling technique used for the design optimization. There is some evidence of both the deformation peaks, with the antenna showing snap-through behaviour at each peak.

Acknowledgements. We would like to thank J.F. Clemmet and R. Dace of Matra Marconi Space Systems for their help with this work. Financial support from EPSRC and Matra Marconi Space Systems is gratefully acknowledged.

References

Guest, S.D. and Pellegrino, S. (1992) Inextensional wrapping of flat membranes. *Proceedings, First International Seminar on Structural Morphology,* Montpellier, La Grand Motte, 7–11 September (Edited by R. Motro and T. Wester), LMGC, Universite Montpellier II, 203–215.

Guest, S.D. (1994) *Deployable Structures: Concepts and Analysis.* PhD dissertation, University of Cambridge.

Hayman, G.J., Hedgepeth, J.H. and Park, K.C. (1994) Design freedoms of articulating Astromast and their optimization for improved performance. *Proceedings, 35th SDM Conference,* Hilton Head, 18–20 April, 811–817.

McCarthy, J.M. (1990) *An Introduction to Theoretical Kinematics.* The MIT Press.

NAG (1990) *The Nag Fortran Library Manual, Mark 14.* The Numerical Algorithms Group Ltd.

Rogers, C.A., Stutzman, W.I., Campbell, T.G. and Hedgepeth, J.M. (1993) Technology assessment and development of large deployable antennas. *ASCE Journal of Aerospace Engineering* **6,** 34–54 .

PERIMETER CONSTRAINED TOPOLOGY OPTIMIZATION OF CONTINUUM STRUCTURES

R. B. HABER
Department of Theoretical & Applied Mechanics
University of Illinois at Urbana-Champaign, Urbana, IL 61801, USA

M. P. BENDSØE and C. S. JOG
Mathematical Institute
Technical University of Denmark, DK-2800 Lyngby, Denmark

Abstract

The *perimeter method* for variable-topology shape optimization enforces an upper-bound constraint on the perimeter of the solid part of the structure. The perimeter constraint ensures a well-posed design problem and allows the designer to control the number of holes in the optimal design and to establish their characteristic length scale. Thus single-step procedures for topology design and detailed shape design are possible.

1. Introduction

In recent years, new techniques for *variable-topology* shape optimization problems have been introduced. In these methods, holes may be added or deleted to modify the connectivity of the structure during the course of the optimization process, yielding significant weight reductions or improved performance relative to fixed-topology designs. Figure 1 illustrates a typical variable-topology shape design problem. A fixed amount of elastic material is to be distributed within a candidate design domain. No restrictions are placed on the connectivity of the solid region. It is now well-known that this design problem is not well posed in general. Typically, for a given design, a design with the same volume and lower compliance can be obtained by increasing the number of holes. Eventually, this leads to "chattering" designs with microscopic perforations.

There are two alternatives for generating a well-posed, variable-topology optimization problem; one must either directly address the possibility of chattering solutions in the problem formulation (*relaxation*) or take steps to ensure that they do not occur. In the former option accommodates the chattering designs by expanding the design space, as in the *homogenization method* for topology optimization[*]. The second option involves restricting the design space to exclude chattering designs [6]. This is the basis of the new variable-topology design method presented in this paper.

---

[*] We will not here try to review the literature on topology design methods, but refer the reader to the recent review paper [9] as well as to other recent papers cited in the references.

D. Bestle and W. Schielen (eds.), IUTAM Symposium on Optimization of Mechanical Systems, 113–120.
© 1996 *Kluwer Academic Publishers.*

When there are no a priori restrictions on the configuration of the microstructure and in the absence of an explicit penalty on intermediate volume fractions, homogenization methods commonly generate optimal designs with perforated microstructures. This is consistent with the expected form of continuum solutions to the relaxed problem. Unfortunately, it may be impractical to manufacture designs with perforated micro-structures. When the design is required to include only macroscopic holes and if use of for example fiber-reinforced composite materials is excluded, the relaxed optimization problem does not lead directly to useful designs. Nonetheless, the relaxed solution still provides a useful bound on the compliance that can be achieved through variable-topology design and it might also suggest an effective macroscopic configuration to an experienced designer.

Although relaxed solutions containing perforated microstructures might convey useful information, they do not directly address the macroscopic design problem illustrated in Figure 1. Therefore, it is worthwhile to consider modifications to the relaxed formulation that suppress perforated material in the optimal design. The suppression can be achieved either by introducing an explicit penalty on intermediate values of the bulk density [1] or by restricting the microstructure configuration to for example square holes in square, periodic cells [3, 8, 11]. We refer to such methods as *penalized homogenization methods*.

This paper presents another approach to topology optimization called the *perimeter method*. In contrast to homogenization methods which expand the design space to include chattering designs, the perimeter method restricts the solution space to exclude chattering designs. An upper-bound constraint on the perimeter of the solid region enforces this restriction. The perimeter constraint leads to a well-posed continuum optimization problem for which solutions, comprised exclusively of solid material and void, are guaranteed to exist. There is no need to consider perforated microstructures or to apply homogenization theory, since geometric features with microscopic length scales are specifically excluded in the statement of the optimization problem.

The finite element solution procedure based on the new formulation offers a number of advantages: 1) the new method is convergent with respect to grid refinement, so it can be used for detailed shape design as well as preliminary conceptual design; 2) the designer is able to establish a characteristic length-scale for holes, thereby controlling the number of holes in the design; 3) the method is compatible with any design objective (it is not limited to compliance minimization) and general forms of design and response constraints; 4) the new procedure can be implemented with standard finite element analysis and optimization software.

Another approach to the variable-topology optimization problem that operate on the macroscopic scale is the *bubble method* [4], which iteratively introduces new holes and applies fixed-topology shape optimization algorithms in a hierarchical algorithm. Also, in recent work [10], ideas from image processing are introduced on the level of the optimization algorithm for a variable density model to enforce designs which only have variations above a specified scale.

2. A Topology Optimization Problem with a Constraint on Perimeter

Consider the problem of finding the stiffest (minimum compliance) structure under a single loading condition that can be obtained by distributing a fixed volume V^* of ho-mogeneous, isotropic, linear elastic material within a two-dimensional candidate domain

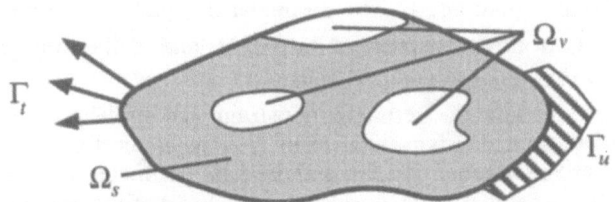

Figure 1. Domain diagram for the variable-topology optimization problem

Ω with boundary $\Gamma = \overline{\Gamma_u \cup \Gamma_t}$ (see Figure 1). No restriction is placed on the connectivity of $\Omega_s \subset \Omega$, the solid part of Ω where material is present. Let $\overline{C}$ designate the elasticity tensor of the solid isotropic material. We introduce the indicator function,

$$\chi : \Omega \to \{0,1\} \text{ where } \chi(\mathbf{x}) = 1 \text{ for } \mathbf{x} \in \Omega_s \text{ and } \chi(\mathbf{x}) = 0 \text{ for } \mathbf{x} \in \Omega_v \equiv \Omega \setminus \Omega_s$$

to represent arbitrary configurations of Ω_s. The region Ω_v is the portion of Ω occupied by void. The volume of the solid part of the structure is given by $\int_\Omega \chi\, d\Omega$, and the elasticity tensor in Ω is given by $C = \chi\overline{C}$.

We will in analogy with recent work use the potential energy formulation of the stiffness version of the variable-topology, compliance optimization problem on Ω, in a small-deformation elasticity setting. The optimization problem is thus stated as

$$\sup_{\chi \in V_\chi} \inf_{\mathbf{u} \in V_u} \Pi \tag{1}$$

where

$$V_u \equiv \left\{ \mathbf{u} : u_i \in H^1(\Omega); \mathbf{u} = \overline{\mathbf{u}} \text{ on } \Gamma_u \right\} \text{ and } V_\chi \equiv \left\{ \chi \,\middle|\, \forall \mathbf{x} \in \Omega : \chi(\mathbf{x}) = 0 \text{ or } 1; \int_\Omega \chi\, d\Omega \le V^* \right\}$$

are the sets of admissible displacements and admissible indicator functions, respectively. In (1), Π denotes the potential energy given as

$$\Pi = \int_\Omega \left(\frac{1}{2} \varepsilon(\mathbf{u})' \left[\chi\overline{C} \right] \varepsilon(\mathbf{u}) - \mathbf{b} \cdot \mathbf{u} \right) d\Omega - \int_{\Gamma_t} \left(\overline{\mathbf{t}} \cdot \mathbf{u} \right) d\Gamma$$

in terms of displacement vector $\mathbf{u}$, linearized strain $\varepsilon(\mathbf{u})$, body force $\mathbf{b}$ and prescribed surface traction $\overline{\mathbf{t}}$ on the traction boundary.

In problem (1) the inner problem enforces equilibrium; the outer problem seeks an optimal design for minimum compliance. It is a *macroscopic* topology design problem, since it does not include microstructure. It is an ill-posed problem, as explained above, due to a tendency to develop chattering designs.

2.1. THE PERIMETER METHOD

The new approach seeks to exclude chattering functions from the design space by limiting the design's perimeter. We define the perimeter of a design as the measure of the boundary of the solid region: $|\partial\Omega_s|$. The perimeter is equal to the total surface area of the boundary of Ω_s in three-dimensional problems, and to the total arc length of the boundary of Ω_s in two-dimensional problems.

Figure 2 illustrates how the perimeter constraint accomplishes the objective of excluding chattering designs. Designs with identical areas are perforated with holes of two different sizes. Both the number of holes and the total perimeter of the holes increase as the radius of the holes decreases. Thus, designs with low perimeter measures have fewer and larger holes than designs with high perimeter measures. Chattering designs have an infinite number of infinitely fine perforations, and are characterized by an infinite perimeter measure. Therefore, an upper bound constraint on the perimeter effectively excludes microscopic perforations and chattering solutions from the feasible design space. Further, the number and sizes of a finite set of holes can be controlled by limiting the perimeter without otherwise restricting the shape or layout of the holes.

In order to obtain a well-posed macroscopic topology design problem that excludes perforated material, we append to problem (1) an upper bound constraint on the perimeter; this assures existence of solutions as well [2]:

$$\sup_{\chi \in V_\chi;\ |\partial\Omega_s| \le P^*} \inf_{u \in V_u} \Pi \tag{2}$$

Here, P^* is a designer-specified value for the upper bound on the perimeter.

The use of the indicator function to represent the design configuration in problem (2) suggests that a raster geometry model, similar to a television image, could be used in numerical methods for solving the topology design problem. This is in contrast to the boundary-representation geometry models that are commonly used in finite element procedures for shape optimization. In the raster approach, a refined (often uniform) finite element grid covers the candidate design domain. A χ-value of either 0 or 1 is assigned to each element to define the design geometry. The boundaries of the solid region are represented implicitly by the element edges across which the value of χ changes. This raster geometry approach is also an integral feature of the homogenization methods.

The raster approximation of the continuum problem (2) leads to an integer programming problem that could be solved, in principle, by an appropriate global optimization strategy such as simulated annealing. However, raster geometry models generate optimization problems that are very large and in the case of (2) function calls are extremely 'expensive'. Therefore, we here approximate problem (2) with a continuous programming problem to obtain a practical numerical method. The remainder of this paper deals with continuous approximations to (2). Since the form of the approximation

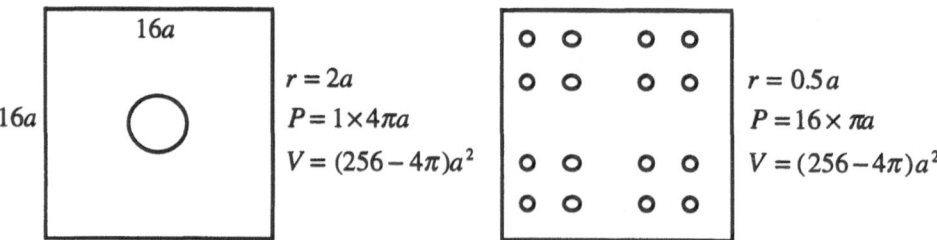

Figure 2. Two designs of equal volume. The design with small holes has the largest perimeter.

is not unique, it is important to keep in mind that problem (2) represents the basic statement of the new perimeter method for topology design. In particular, the existence proof of Ambrosio and Buttazzo [2] provides the mathematical support for the perimeter method and any numerical method that derives from it.

2.2. CONTINUOUS APPROXIMATIONS TO THE INTEGER PROBLEM

We replace the indicator function χ with a distributed *interpolation parameter*, $\rho : \Omega \rightarrow [0,1]$ to achieve a continuous problem that is amenable to discretized solution methods. We have solid material where $\rho = 1$, and void wherever $\rho = 0$. Intermediate values of ρ indicate "transitional" material between solid and void. There is no particular physical significance to the transitional material, its only purpose is to provide a continuous optimization problem. Ultimately, we will introduce a penalty term to force ρ to its extreme values to approximate the integer problem (2). Thus, transitional material is suppressed in the optimal design. New expressions in terms of ρ for the volume, perimeter and elasticity tensor are required. The volume is simply $V(\rho) = \int_{\Omega} \rho \, d\Omega$.

We also need a substitute for the perimeter measure $|\partial \Omega_s|$ that is compatible with the continuous interpolation model. The *total variation of* ρ is a suitable measure which approaches the perimeter, in the limit, as the amount of transitional material is forced to zero [5]. In anticipation of a piecewise-continuous finite element discretization, we partition Ω into a set of open, disjoint regions Ω_α so that $\Omega = \overline{\bigcup \Omega_\alpha}$ and consider densities in the set $V_\rho = \left\{ \rho \in L_\infty(\Omega) : \rho|_{\Omega_\alpha} \in H^1(\Omega_\alpha); 0 < \rho_{\min} \leq \rho \leq 1 \, \forall x \in \Omega \right\}$. The lower bound on the interpolation parameter ensures that elastic energy is positive definite (a requirement for the numerical scheme). We then use the following smooth perturbation of the total variation ([12]) to express the "perimeter" of $\rho \in V_\rho$, accounting for the possibility that ρ may be discontinuous across sets of measure zero (e.g., the element boundaries).

$$P \equiv \int_{\Omega \backslash \Gamma_J} g_h(\nabla \rho, \xi) d\Omega + \int_{\Gamma_J} j(\langle \rho \rangle, \xi) d\Gamma$$

where $\Gamma_J = \Omega \backslash \bigcup \Omega_\alpha$ is the jump set of ρ, $\langle \rho \rangle$ is the jump in ρ across Γ_J. The functions $g_h(\mathbf{w}, \xi)$ and $j(r, \xi)$ are smooth approximations to $|\mathbf{w}|$ and $|r|$, respectively (h is a characteristic mesh dimension, e.g., the size of a finite element):

$$g_h(\mathbf{w}, \xi) \equiv \left[(1 + 2\xi) \mathbf{w}^T \mathbf{w} + \frac{\xi^2}{h^2} \right]^{1/2} - \frac{\xi}{h};$$

$$j(r, \xi) \equiv \left[(1 + 2\xi) r^2 + \xi^2 \right]^{1/2} - \xi$$

The smoothing, based on the parameter ξ, circumvents numerical problems associated with the non-differentiability of the absolute value operators appearing in the expression for the total variation. Note that $\lim_{\xi \to 0} g_h(\nabla \rho, \xi) = |\nabla \rho|$ and $\lim_{\xi \to 0} j(\langle \rho \rangle, \xi) = |\langle \rho \rangle|$. Further, the approximations are exact in the limit of a discrete approximation to an integer design, even for $\xi > 0$. That is for all $\xi \geq 0$, $g_h(\nabla\rho,\xi) \to |\nabla\rho|$ as $|\nabla\rho| \to 0$ or $\frac{1}{h}$, and $j(\langle\rho\rangle,\xi) \to |\langle\rho\rangle|$ as $|\langle\rho\rangle| \to 0$ or 1

The constitutive model given by $\mathbf{C} = \chi\overline{\mathbf{C}}$ in problem (2) must be replaced by a continuous model consistent with the interpolation parameter ρ. The choice of the continuous constitutive model is not unique, and the model need not correspond to any specific physical system. It is only required that the model define a smooth interpolation between the elastic properties of the solid isotropic material and void. For example, one could use the simple relationship $\mathbf{C} = \rho^p\overline{\mathbf{C}}$; $p > 0$ as used by many authors. If we select $p = 1$, then ρ can be interpreted as a variable-thickness parameter in two-dimensional problems. Alternatively, we can view ρ as an artificial density parameter in three-dimensional problems. We emphasize that these physical interpretations are extraneous to the proposed method. The interpolation models introduced here are solely seen as continuous approximations to the integer problem (2), where there is no need for a physical interpretation for the conditions between solid material and void.

The optimized rank-2 microstructure model from the un-penalized homogenization method also provides a suitable interpolation for the constitutive model. Again, the details of the microstructural model and the fact that rank-2 microstructures are optimal for compliance design are not relevant in the perimeter-control method. Here we are only interested in the fact that the effective properties of the rank-2 model define a continuous interpolation between the properties of solid material and void (one can effectively use the effective properties of the *optimized* rank-2 model [7]). One could also use the effective properties derived from a partial relaxation based on microstructures consisting of for example square holes in square cells.

We must ensure that intermediate values of ρ are suppressed in solutions to the continuous approximation of the integer design problem. This can be accomplished either by introducing an explicit penalty on intermediate values of ρ or by manipulating the material model to obtain an implicit penalty. For example, we can specify $p > 1$ in $\mathbf{C} = \rho^p\overline{\mathbf{C}}$ or use the effective properties from a partial relaxation to generate an implicit penalty. Alternatively, we can append a term to the compliance objective function that explicitly penalizes intermediate values of ρ. We follow this approach here.

In our current implementation, we use an optimality criterion method (see, e.g., [7]) to solve the design problem. This method is best suited to problems with a single equality constraint. Accordingly, we treat the constraint $P \leq P^*$ with an interior penalty method and enforce the equality constraint on the volume with a Lagrange multiplier method. We hereby obtain the modified problem,

$$\sup_{\rho\in V_\rho}\left\{\alpha S_1 + \beta S_2 - \lambda\left[V(\rho) - V^*\right] + \inf_{u\in V_u}\Pi\right\} \tag{3}$$

in which α and β are positive scalars, S_2 is an interior penalty function for the constraint $P \leq P^*$, S_1 penalizes intermediate values of ρ and λ is the Lagrange multiplier associated with the volume constraint. A variety of choices are possible for the penalty functions. In the work reported here, they are given by (with $\overline{\rho} = V^*|\Omega|^{-1}$)

$$S_1 = \int_\Omega f \, d\Omega; \quad f(\rho) = \begin{cases} \dfrac{(\rho - \bar{\rho})^2}{2\bar{\rho}} & \text{for } 0 \le \rho \le \bar{\rho} \\ \dfrac{(\rho - \bar{\rho})^2}{2(1 - \bar{\rho})} & \text{for } \bar{\rho} < \rho \le 1 \end{cases}; \quad S_2 = P^* \log[P^* - P]$$

3. Discrete Formulation, Solution Algorithm and Results

Problem (2) comprise a mixed variational problem for the optimal solution $\rho, \mathbf{u}$. To solve this problem we use a mixed finite element method where the finite element basis functions for the interpolation parameter admits jumps across the inter-element boundaries. We note that the mixed variational form of the problem is significant and care must be exercised in the selection of the discrete function spaces to avoid "checkerboard" patterns and other grid-scale anomalies in the solution.

We employ an optimality criterion type algorithm to solve the discretized version of problem (3), i.e. a nested, iterative solution strategy based on the first-order Kuhn-Tucker. Each outer iteration includes a stiffness analysis and a design update. In our implementation, the penalty parameter α is held constant while β is reduced as in each outer iteration to improve the accuracy of the method. As with any penalty method, optimal performance requires some experimenting and tuning problems are difficult to avoid. We show below some example structures obtained by the method.

4. References

1. Allaire, G.; Kohn, R. V. 1993: Topology design and optimal shape design using homogenization. In: Bendsøe, M. P.; Mota Soares, C. A. (eds.) *Topology design of structures*, 207-218. Dordrecht: Kluwer
2. Ambrosio, L.; Buttazzo, G. 1993: An Optimal Design Problem with Perimeter Penalization. *Calc. Var.* 1, 55-69
3. Bendsøe, M. P.; Kikuchi, N. 1988: Generating optimal topologies in structural design using a homogenization method. *Comp. Meth. Appl. Mech. Engng.* 71, 197-224
4. Eschenauer, H.A.; Kobelev, V.; Schumacher, A. 1994: Bubble Method of Topology and Shape Optimization of Structures. *Struct. Optim.* 8, 42-51.
5. Evans, L. C.; Gariepy, R. F. 1992: *Measure theory and fine properties of functions*, Boca Ratan: CRC Press
6. Haber, R. B.; Jog, C. S.; Bendsøe, M. P. 1994: Variable-topology shape optimization with a control on perimeter. In: Gilmore, B. J.; Hoeltzel, D. A.; Dutta, D.; Eschenauer, H. A. (eds.) *Advances in Design Automation*, ASME DE-Vol. 69-2, 261-272.
7. Jog, C. S.; Haber, R. B.; Bendsøe, M. P. 1994: Topology design with optimized, self-adaptive materials. *Int. J. Num. Methods Engng.* 37, 1323-1350
8. Rodrigues, H. C.; Fernandes, P. 1994: A material based model for topology optimization of thermoelastic structures. *Int. J. Num. Methods Engng.* 37 (to appear)
9. Rozvany, G.I.N.; Bendsøe, M.P.; Kirsch, U. (1994): Layout Optimization of Structures. *Appl. Mech. Rev.* 48, 41-118.
10. Sigmund, O. 1995: *Design of Material Structures using Topology Optimization*. DCAMM Special Report no. S69, Techn. Univ. of Denmark, 1995
11. Suzuki, K.; Kikuchi, N. 1991: Shape and topology optimization for generalized layout problems using the homogenization method. *Comp. Meth. Appl. Mechs. Engng.* 93, 291-318
12. Wheedon, R. L.; Zygmund, A. 1977: *Measure and integral: an introduction to real analysis*, Monographs in Pure and Applied Math. 43, New York: Marcel Dekker

Acknowledgements
This work was supported in part by the Danish Technical Research Council, through the Programme of Research on Computer Aided Design (MPB, RBH), the US National Science Foundation (RBH) and the National Center for Supercomputing Applications (RBH, CJ). All computations were performed on the Silicon Graphics Power Challenge System at the National Center for Supercomputing Applications, Ill..

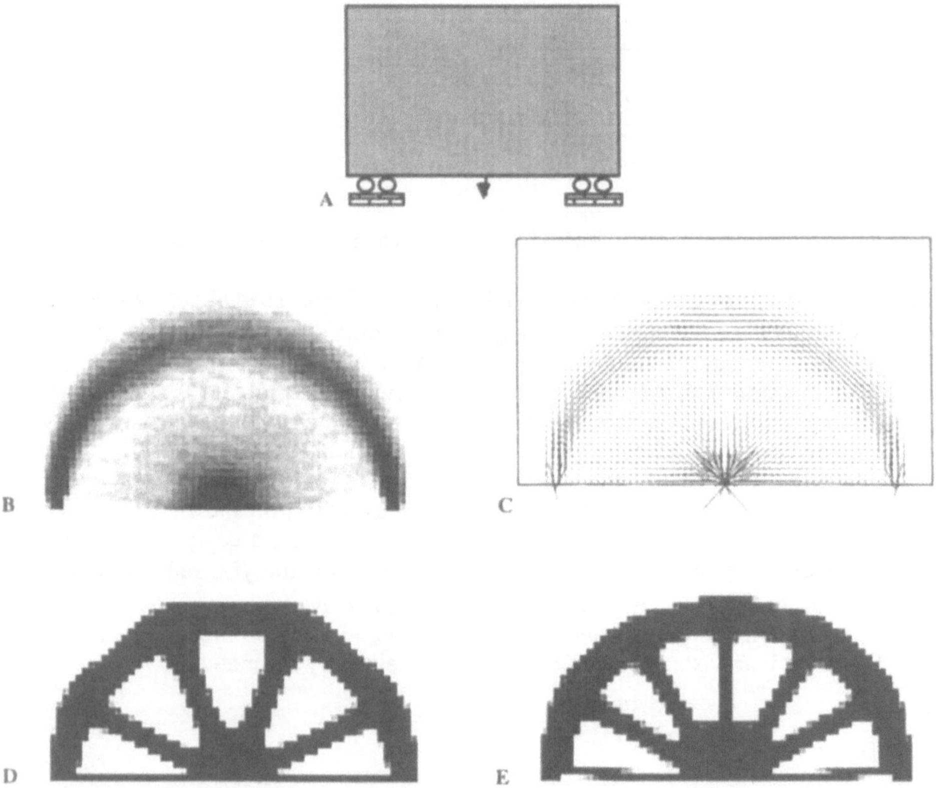

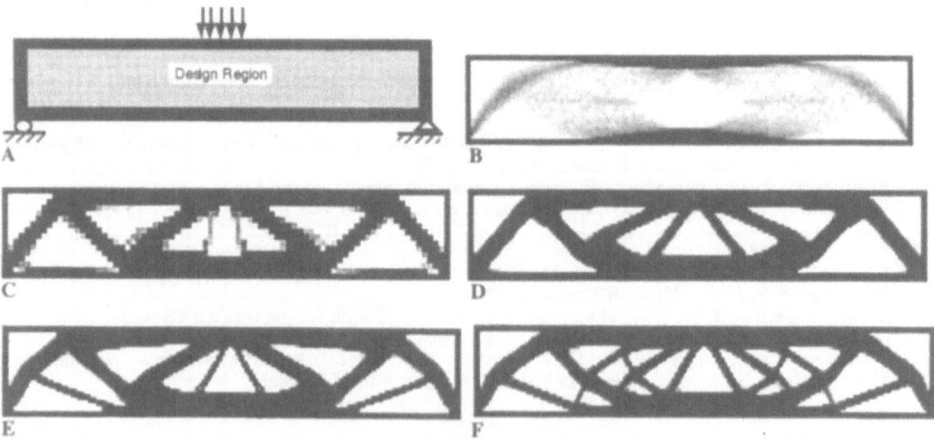

Figure 3. A bicycle wheel designed in for the situation in (A). (B), (C): Optimal design using an optimized rank-2 material strain energy density, showing the density distribution and principal stress distribution. In (D), (E) intermediate densities are penalized and the perimeter is constrained, with increasing perimeter from (D) to (E).

Figure 4. Optimal design using an optimized rank-2 material with penalties on intermediate densities and on the perimeter. The problem is the MBB beam shown in (A). (B) shows the density distribution for the unconstrained case. In (C)-(F) intermediate densities and perimeter are penalized, with (D) being a fine mesh variant of (C).

OPTIMIZATION OF LAMINATE STRENGTH

VELAJA B. HAMMER and PAULI PEDERSEN
Department of Solid Mechanics
Technical University of Denmark
DK-2800 Lyngby, Denmark

Abstract. Laminates subjected to multiple load cases and tested in relation to several strength criteria are optimized using the fiber orientations and the thickness distributions throughout each layer as design variables. The design is parametrized with the use of orthogonal functions in a global description over the laminate domain. Hereby we obtain designs more likely to be manufacturable, and much more stable with respect to the optimization procedure.

1. Introduction

In mechanical systems laminates are used more and more, and extensive studies with the goal of optimizing the stiffness of laminates have been carried out. However, studies that also account for strength constraints are much fewer, see [3]–[8].

We present the results of a rather general optimization problem, which formulated in words is: maximize a common load factor, subjected to multiple strength constraints in multiple layers and with multiple loads. Formulated in this general way, different strength criteria $(F/F_0)_{jk\ell n} \leq 1$ are evaluated, with F/F_0 being the strength ratios for the strength criterion n (such as Tsai–Wu, maximum strain etc.) considered in every element j, layer k, and for the load case ℓ.

The design parameters are the fiber orientations and ply thicknesses defined locally in every finite element. The design parametrization is of vital importance especially for the orientational design, where local optima are an inherent problem. If the optimization is performed allowing the design variables to change from point to point independently of the neighbouring elements, then the resulting designs are often very disordered configurations.

A new global design description is presented in which the tayloring of the laminate is obtained by mapping of mutually orthogonal design functions. Hereby is achieved a reduction of the size of the optimization problem as well as smoother designs, that are more likely to be manufactured.

121

D. Bestle and W. Schielen (eds.), IUTAM Symposium on Optimization of Mechanical Systems, 121–128.
© 1996 *Kluwer Academic Publishers.*

Comparisons with more heuristic approaches like design according to principal stress directions are carried through. Also a hierarchically method is used, which primarily designs for the lamination parameters in each finite element and then secondary for the individual ply parameters. The big advantages of the latter approach are, that the energy expression is linear in the lamination parameters together with the possibility of choosing the ply parameters according to specific needs. It is shown how these designs can be used as good starting designs for the strength optimization. This also puts forward the difference between optimizing for stiffness and optimizing for strength. Many examples will illustrate the essential points of these optimization methods.

2. Problem Formulation, Analysis and Sensitivity Analysis

Let $\{A_\ell\}$ be the load distribution vector corresponding to the load case ℓ, then from the finite element analysis the resulting nodal displacement vector $\{D_\ell\}$ is found from

$$[S]\{D_\ell\} = \lambda\{A_\ell\} \quad \text{for} \quad \ell = 1, 2, ..., L \qquad (2.1)$$

where $[S]$ is the stiffness matrix of the actual design (finally the optimal design) and λ is a load factor common to all the load cases. From $\{D_\ell\}$ strains and stresses in every layer k of every element j, follows directly

$$\{D_\ell\} \Rightarrow \{\varepsilon_{jk\ell}\} \ , \ \{\sigma_{jk\ell}\} \quad \text{for} \quad j = 1, 2, ...J \ ; \ k = 1, 2, ...K \ ; \ \ell = 1, 2, ...L \quad (2.2)$$

Now the load strength F_n being the strength necessary in the given load case and corresponding to a given strength criterion n can be determined for each of these $'jk\ell'$ and compared to the strength limit $(F_0)_n$, reflecting the strength possessed by the laminate at that point. Formulating our problem in relation to first ply (layer) failure (FPF) we thus have **the constraints**

$$\left(F/F_0\right)_{jk\ell n} \leq 1$$

$$(2.3)$$

$$\text{for all} \quad j = 1, 2, ..., J \ ; \ k = 1, 2, ..., K \ ; \ \ell = 1, 2, ..., L \ ; \ n = 1, 2, ..., N$$

The **objective of our optimization** is by laminate design of orientations and thicknesses to

$$\text{Maximize} \ \lambda \ \text{over feasible} \ \theta_{jk} \ ; \ t_{jk} \qquad (2.4)$$

where θ_{jk} is the orientation of layer k, element j covering a domain in the plane. Analogous t_{jk} is the thickness of layer k in element j. In the next section we shall specify how these fields are parametrized.

The **analysis** of our examples is done by the finite element method with the plate discretized into triangular domains with constant strain, i.e. the total stiffness matrix is obtained by the accumulation over element stiffnesses $[S_j]$ which are again accumulated over specific layer stiffnesses $[\tilde{S}_{jk}]$ with layer thicknesses t_{jk} as factors

$$[S] = \sum_j [S_j] = \sum_j \sum_k [\tilde{S}_{jk}] t_{jk} \tag{2.5}$$

With the results from the total analysis we can in addition to displacements, strains and stresses for load case ℓ evaluate work W_ℓ and elastic strain energy $U_\ell = W_\ell/2$

$$W_\ell = \{A\}_\ell^T \{D\}_\ell \quad ; \quad U_\ell = \sum_j a_j \sum_k \frac{1}{2} \{\sigma_{jk\ell}\}^T \{\varepsilon_{jk\ell}\} t_{jk} \tag{2.6}$$

with a_j being the area of element j in the laminate plane. The **sensitivity analysis** has been performed analytically. The derivations are given in Hammer [1] where formulas for all the necessary

$$\frac{\partial(F/F_0)_{jk\ell n}}{\partial \theta_{jk}} \quad \text{and} \quad \frac{\partial(F/F_0)_{jk\ell n}}{\partial t_{jk}} \tag{2.7}$$

are given. The treated failure criteria are the maximum strain criterion and the Tsai–Wu criterion, but extensions to other criteria are straight forward.

3. Design Parametrization

The design parametrization is an issue of vital importance. Many possibilities exist, especially in relation to a build up structure like a laminate. Classifications in local descriptions and global descriptions give some overview, and for laminates it is especially important to keep in mind the two different accumulations: through the thickness (layers) $\sum k = \int dz$ and over the laminate domain (elements) $\sum j = \int\int dx_1 dx_2$.

The **through the thickness** parametrization is in this paper for a given basic material, which means that a local description is specified by

$$\theta = \theta(z) \quad \text{for} \quad -h/2 \le z \le h/2 \quad \text{or by} \quad \theta_k, t_k \quad \text{for} \quad k = 1, 2, ..., K \tag{3.1}$$

with the latter corresponding to layers of constant orientation. From this follows directly the laminate membrane stiffness matrix [A]

$$[A] = \int [C(\theta)] dz = \sum [C_k] t_k \tag{3.2}$$

which with definition of the **lamination parameters**

$$\xi_{1,2,3,4} = \int \cos 2\theta(z), \cos 4\theta(z), \sin 2\theta(z), \sin 4\theta(z) dz \tag{3.3}$$

can be written

$$[A] = h\big([\tilde{C}_0] + [\tilde{C}_1]\xi_1 + [\tilde{C}_2]\xi_2 + [\tilde{C}_3]\xi_3 + [\tilde{C}_4]\xi_4\big) \tag{3.4}$$

where the matrices $[\tilde{C}_i]$ consist of material data and are independent of orientation and thickness. Note that by this parametrization the stiffness expression is linear in the lamination parameters ξ_i, and furthermore we note that only four parameters are necessary to find the laminate displacements and strains, even for say a 20 layer laminate. However, to find the laminate stresses a complete through the thickness description is necessary. This adds to the essential difference between design for stiffness as evaluated by W_ℓ or U_ℓ from (2.6) and design with strength constraints as specified in (2.3).

Now, the **over the laminate domain** parametrization may also be a local description pointwise, i.e. elementwise by

$$\theta_{jk} \, , \, t_{jk} \tag{3.5}$$

constant in layer k of element j (may be extended to Gauss integration points in the element). When the orientations are included as design parameters, the inherent problem of many local optima causes great difficulties. Therefore, a new method for a global parametrization is introduced.

A detailed description of this **parametrization by orthogonal functions** is given in Pedersen & Hammer [2], so here we shall only explain the basic idea.

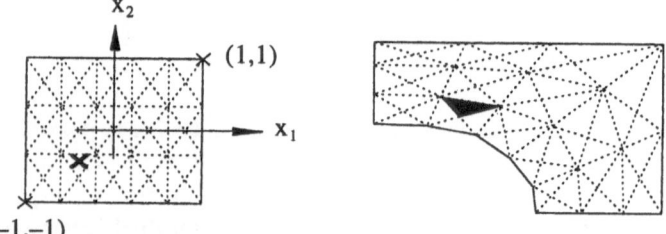

Figure 3.1. The mapping between the reference area and the finite element model.

In Figure 3.1 we show to the left a reference area $-1 \leq x_1 \leq 1$, $-1 \leq x_2 \leq 1$ and indicate that the parameter value at the cross $\times$ determines the orientation or thickness for the black triangle in the actual finite element model to the right. Thus there is a one to one relation between the elements of the two models. In the reference domain a value y_k (orientation θ_k or thickness t_k in layer k) is given by the expansion

$$y_k(x_1, x_2) = \sum_{m,n} (y_{mn})_k \Phi_m(x_1)\Phi_n(x_2) \quad \left(\sum_{m,n} := \sum_m \sum_n \right) \tag{3.6}$$

For our examples we have used the orthogonal functions (from vibration modes of free plates)

$$\Phi_0(x) \equiv 1 \; ; \; \Phi_m(x) = \cosh(\alpha_m x) + \beta_m \cos(\alpha_m x) \quad \text{for} \quad m = 2, 4, \ldots$$

$$\Phi_1(x) = x \; ; \; \Phi_m(x) = \sinh(\alpha_m x) + \beta_m \sin(\alpha_m x) \quad \text{for} \quad m = 3, 5, \ldots \tag{3.7}$$

$$\alpha_m \quad \text{from} \quad \tan(\alpha_m) + (-1)^m \tanh(\alpha_m) = 0 \quad \text{and}$$

β_m from $\cosh(\alpha_m)/\cos(\alpha_m)$ for m even and $\sinh(\alpha_m)/\sin(\alpha_m)$ for m odd

The design variables are no longer the local orientation and thickness, but instead the expansion coefficients $(\theta_{mn})_k$, $(t_{mn})_k$ for each layer k . This means that the design description is now global in the sense that a variation of one design variable leads to a change in the design everywhere in the specific laminate layer and not just in a single element (domain). A major advantage of this description is the drastic reduction of design variables from having two in each element in each layer to having only, say 8 or 18 per layer. The importance of a design description not directly given by the model for analysis is also a well known fact.

Then with y_{mn} as a common notation for orientation parameters θ_{mn} and thickness parameters t_{mn} , our linear programming formulation for **optimal redesign** (sequential optimal design) reads

$$\text{Maximize} \quad \Delta\lambda \quad \text{(max increase in safety factor)}$$

subject to

$$\left(F/F_0\right)_i + \sum_{\text{layer } k} \sum_{m,n} \sum_{\text{elem. } j} \frac{\partial\left(F/F_0\right)_i}{\partial y_{jk}} \Phi_m(x_{1j})\Phi_n(x_{2j})\Delta(y_{mn})_k + \frac{\partial\left(F/F_0\right)_i}{\partial\lambda}\Delta\lambda \leq 1 \quad (3.8)$$

(for all critical strength i (active set strategy))

$$\left.\begin{array}{c} \Delta\lambda_{min} \leq \Delta\lambda \leq \Delta\lambda_{max} \\ (\Delta y_{mn})_{min} \leq \Delta(y_{mn})_k \leq (\Delta y_{mn})_{max} \end{array}\right\} \text{move--limits}$$

From the start only one expansion coefficient for each layer orientation and thickness is allowed to vary until convergence is reached. Then increased to $2 \times 2 = 4$ functions for each layer orientation and thickness, next $3 \times 3 = 9$ functions and so on. All the variables are still allowed to vary but with smaller move limits on the 'old' variables than on the 'new'. By this procedure the design is gradually getting more and more complex and more and more refined.

The designs made using this global description are seen to be smooth with a continuous variation of orientation and thickness from one element to its neighbouring elements throughout each layer. This description also allows for putting control on for instance the slopes and/or the boundary conditions, i.e. on the ability for manufacture.

4. Examples of Strength Optimization

In this section we will show some results of the different optimization techniques presented. The laminates are all subjected to loads in the plane only. All plies consist of graphite/epoxy which is a very orthotropic material with widely different properties in the longitudinal and transverse directions and in tension and compression.

The first example is a simple model of wing. Thinking of the wing as a hollow structure with thin sheets of fiber material on both sides, the bending from an end load can be modelled by in--plane loads acting in each point in the plate away from the body of the airplane. The state is as such pure membrane. A second load--case consists of forces applied perpendicular to the others. The laminate is clamped in one end as is practical for an airplane, see figure 4.1a.

The weighted sum of compliances for these two load cases is minimized using a parametrization of lamination parameters as described in section 3. The resulting configuration of a three–ply laminate is shown in figure 4.1b. In relation to our strength criteria the max. load factor will be $\lambda \approx 2.86$. The hatches illustrate the fiber direction in each finite element. The density of the hatches corresponds to the relative thickness of each ply, i.e. the more dense the thicker. The total thickness is kept constant equal to one.

Below to the left we show the result of using a global parametrization based on the orthogonal functions as expressed in (3.8). Still only orientational design is considered. Figure 4.1c shown the design obtained by use of three functions in each of the two layers, i.e. a total of only 18 design variables for the whole plate. Hereby is achieved a safety factor of $\lambda \approx 3.32$.

This design is used as a starting design for a strength optimization where the fiber orientation is allowed to vary independent of the neighbouring elements. The result is shown in figure 4.1d where the load factor is improved by 20% compared to c , to a value of $\lambda \approx 3.99$.

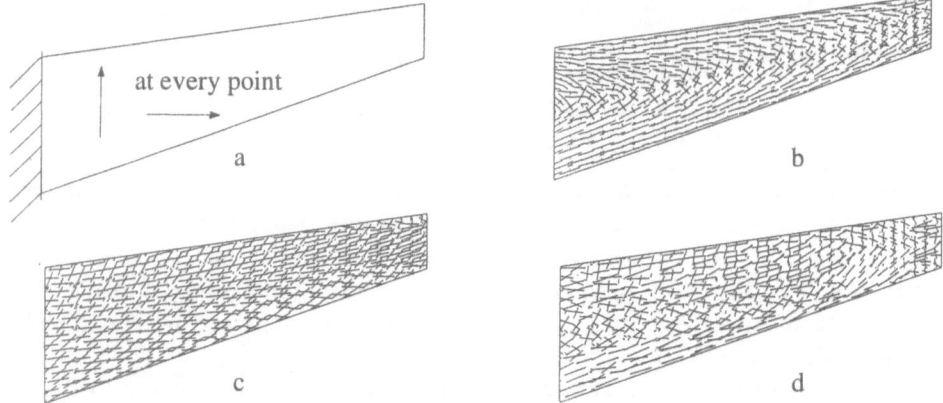

Figure 4.1. a: The model with the two load–cases. Optimization of a wing. b: energy optimized design $\lambda \approx 2.86$. Strength optimized designs with c: orthogonal functions, $\lambda \approx 3.32$, d: local parametrization started out from c, $\lambda \approx 3.99$.

The next example is a plate clamped in both ends with two point loads applied in the middle, working in opposite directions. Using the symmetry the model is as sketched in figure 4.2.

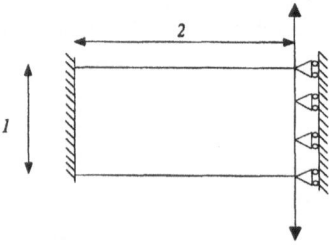

Figure 4.2. Square plate subjected to two independent point loads.

Again the formulation of lamination parameters is applied yielding the designs in figure 4.3 with the laminate consisting of one and three plies, respectively. The total thickness is kept constant but the relative thickness of each ply is optimized in the three–ply laminate. The total weighted energy is equal for the two laminates because they are constructed from the same set of lamination parameters in each finite element, but the resulting safety factor λ is rather different. This demonstrates the difference in the two objectives, maximum stiffness contra maximum strength.

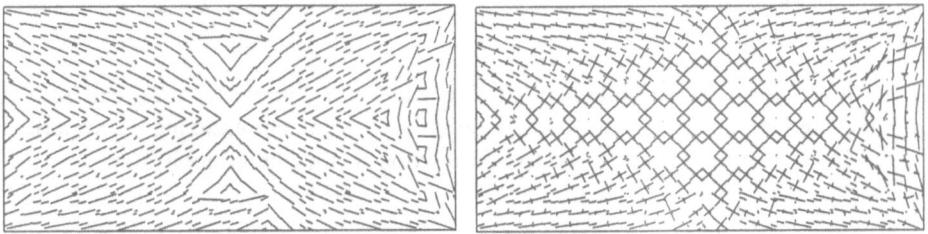

Figure 4.3. Energy optimized designs, a: one ply laminate $\lambda \approx 7.34$. b: three ply laminate $\lambda \approx 9.19$.

The last example is a model of a quarter–part of a square plate with an elliptical hole, see figure 4.4. The laminate is loaded with a uniform distributed load on one of the edges.

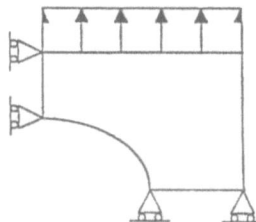

Figure 4.4. Square plate with elliptical hole.

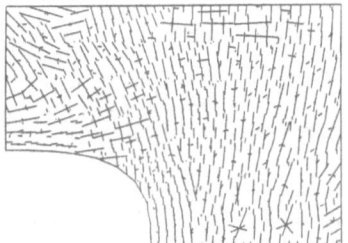

Figure 4.5. Optimal three–ply laminate with respect to the stiffness, $\lambda \approx 1.23$.

A safety factor of $\lambda \approx 1.23$ is achieved with a three ply laminate as in figure 4.5, optimized to maximize the stiffness.

Applying the global formulation based on orthogonal design fields yields the designs shown in figure 4.6. To the left is displayed the result obtained by a variation of the orientational field of the fibers only, to the right both the fiber orientations as well as the two–layer thicknesses are optimized simultaneously. In the latter the total plate thickness is changing from point to point. The load bearing capacities have improved significantly by 74% and 213%, respectively.

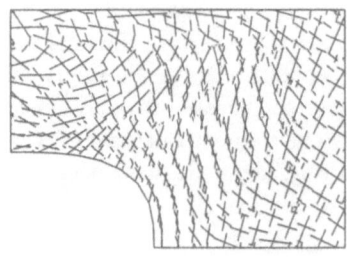 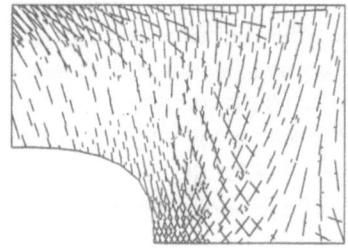

Figure 4.6. Strength optimized designs by varying. a: the orientational field, $\lambda \approx 2.14$ and b: both the orientations and the ply thicknesses, $\lambda \approx 3.86$.

It is seen that the achieved laminate configurations are completely different from the optimal design with respect to the compliance, hereby again emphasizing the importance of choosing the most relevant objective in relation to the actual application.

5. Conclusion

A stable and robust optimization procedure for laminates is presented. The localized strength constraints with multiple load cases and several strength criteria calls for mathematical programming and we have chosen sequential linear programming.

The design parametrization is found to be a crucial point and the introduced description with orthogonal expansion functions works in a very satisfactory way. A large number of numerical examples based on finite element analysis and analytical sensitivity analysis illustrates the optimizations with fiber orientations as well as thickness distributions treated as design variables.

References

[1] Hammer, V.B.: Strength Optimization by Fiber Orientation (to appear).
[2] Pedersen, P. and Hammer V.B.: On Global Description for Orientational Strength Optimization, in Proc. ASME Conf. "Advances in Design Automation", Minneapolis, DE–Vol. 69-2, pp. 221–224, 1994.
[3] Schmit, L.A. Jr. and Farshi, B.: Optimum Laminate Design for Strength and Stiffness, Int. J. Num. Meth. Engng., Vol. 7, pp. 519–536, 1973.
[4] Schmit, L.A. Jr. and Farshi, B.: Optimum Design of Laminated Fibre Composite Plates, Int. J. Num. Meth. Engng., Vol. 11, pp. 623–640, 1977.
[5] Park, W.J.: An Optimal Design of Simple Symmetric Laminates under the First Ply Failure Criterion, J. Composite Materials, Vol. 16, pp. 341–355, 1982.
[6] Katz, Y., Haftka, R.T. and Altus, E.: Optimization of Fiber Directions for Increasing the Failure Load of a Plate with a Hole, in Proc. of the American Society of Composites 4th Technical Conference (ed. K.L. Reifsnide), Tecnomics, Lancaster, PA, 1989.
[7] Fukunaga, H. and Vanderplaats, G.N.: Strength Optimization of Laminated Composites with Respect to Layer Thickness and/or Layer Orientation Angle, Computers and Structures, Vol. 40, No. 6, pp. 1429–1439, 1991.
[8] Fukunaga, H. and Sekine, H.: Optimum Design of Composite Structures for Shape, Layer Angle and Layer Thickness Distributions, J. of Composite Materials, Vol. 27, pp. 1479–1492, 1993.

AN EFFICIENT METHOD FOR SYNTHESIS OF MECHANISMS USING AN OPTIMIZATION METHOD

JOHN M. HANSEN
Department of Solid Mechanics
Technical University of Denmark
DK-2800 Lyngby, Denmark

AND

DANIEL A. TORTORELLI
Department of Mechanical and Industrial Engineering and
Department of Theoretical and Applied Mechanics
University of Illinois at Urbana-Champaign
Urbana, IL 61801, USA

Abstract. The objective here is to design a mechanism so that a tracer point follows a given curve during a portion of the working cycle. To quantify a proposed design, the deviation between the desired curve the tracer point trajectory is measured. A gradient based optimization procedure is used to minimize this deviation. The design sensitivity analyses are efficiently computed using the direct differentiation method. The joint coordinate method is used to analyze the mechanisms. This formulation is advantageous because superfluous variables are eliminated from the planar analysis.

1. Introduction

The task of analyzing and synthesizing mechanisms has been performed for many years. Automated approaches to analyze mechanisms were introduced in [1, 2, 3], and automated synthesis methods appeared in [4, 5, 6, 7, 8]. Most of these methods are developed for a specific topology as in e.g. [9].

In this work the synthesis problem of designing a planar mechanism so that a tracer point on one body follows a given path is resolved. The analysis uses the joint coordinate method which is derived for spatial dynamics in

129

D. Bestle and W. Schielen (eds.), IUTAM Symposium on Optimization of Mechanical Systems, 129–138.
© *1996 Kluwer Academic Publishers.*

[10] and [11] and is specialized here for planar kinematics. Further the joint coordinate method is derived in such a way that all of the position vectors, rotation angles, vectors defining joints, etc., that describe the mechanism are expressed directly in terms of the parameters, i.e. design variables, that are varied during the design process. Consequently, the joint coordinate method offers two advantages over the Cartesian coordinate formulation in that only a minimum number of degrees of freedom and design variables is needed to define the mechanism.

The design problem is formulated as a minimization problem. The cost function is the RMS error between the tracer point trajectory and the desired trajectory. The design sensitivities of the cost function are derived using the direct differentiation method and the chain rule.

2. Design variable identification

To streamline the design process we minimize the number of variables used to define the mechanism. In Figure 1(a) a four-bar mechanism is shown in its most general description, based on the joint coordinate formulation. Notice that e.g. all of the s-vectors are described with two components relative to the local coordinate frames.

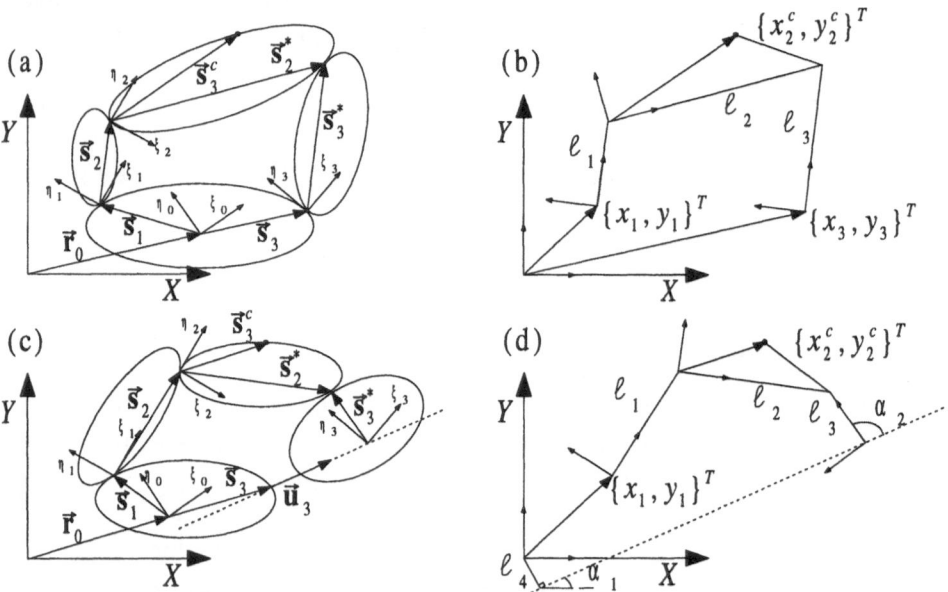

Figure 1. Two examples of four-bar mechanisms. In (a) and (c) the mechanisms are shown in their most general description and in (b) and (d) they are described by a minimum number of design variables

For a designer, however, only one component, as shown in Figure 1(b), is of interest, i.e. there are 9 design variables of interest in this example. A similar observation may be made for the slider-crank mechanism illustrated in Figure 1(c) where the 10 design variables of interest are shown in Figure 1(d). Defining the coordinate systems as shown in Figures 1(b) and (d) eliminates the superfluous variables.

For planar mechanisms which contain only revolute and translational joints, it is possible to directly identify the design variables by placing the local coordinate system as seen in Figure 2. For bodies with two joints, of which no connection is to ground, the proper coordinate system and design variable specification appears in Figure 2(a-c), and for bodies with connections to ground or with more than two connections, the appropriate specifications are depicted in Figure 2(d-e).

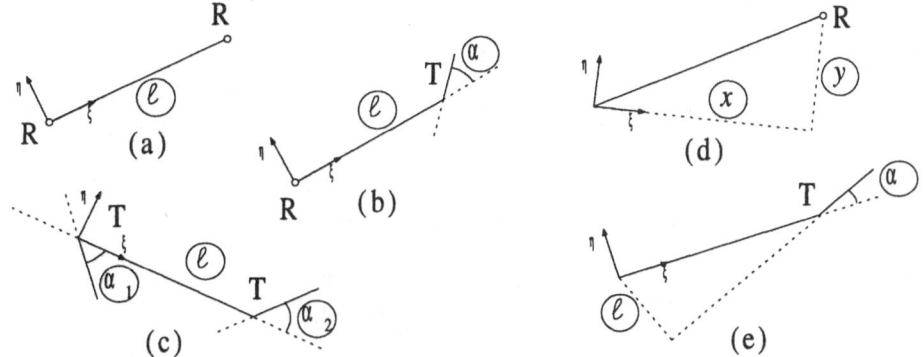

Figure 2.   Identification of design variables and coordinate systems. For connections in binary links that are not connected to ground, use (a)-(c). For bodies with connections to ground or with more than two connections, use (d)-(e).

3. The parameterized joint coordinate method

Here, the joint coordinate formulation is parameterized by the design variables. For a thorough derivation of the joint coordinate method, see [11] and [13]. If the Cartesian coordinates of body $j - 1$, the design variables of body j, and the relative coordinate(s), or joint coordinate(s), between body $j - 1$ and body j are known, then the Cartesian coordinates of body j can be obtained by the "forward updating" process. Two examples of this forward updating process are illustrated in Figure 3; the updating formulas for body j in these two cases becomes

$$\phi_j = \phi_{j-1} + \theta_j$$

$$\mathbf{r}_j = \mathbf{r}_{j-1} + \mathbf{A}_{j-1} \left\{ \begin{array}{c} \ell \\ 0 \end{array} \right\} = \mathbf{r}_{j-1} + \ell \left\{ \begin{array}{c} \cos \phi_{j-1} \\ \sin \phi_{j-1} \end{array} \right\}$$

(1)

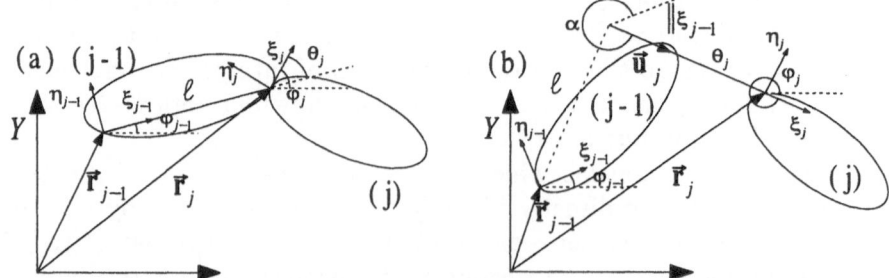

Figure 3. The forward updating process from body $j-1$ to body j, is exemplified for the mechanisms illustrated in Figure 2(a) and (e).

and

$$
\begin{aligned}
\phi_j &= \phi_{j-1} + \alpha \\
\mathbf{r}_j &= \mathbf{r}_{j-1} - \ell \mathbf{A}_{j-1} \hat{\mathbf{u}}'_j + \theta_j \mathbf{A}_{j-1} \mathbf{u}'_j \\
&= \mathbf{r}_{j-1} + \begin{bmatrix} \cos\phi_{j-1} & -\sin\phi_{j-1} \\ \sin\phi_{j-1} & \cos\phi_{j-1} \end{bmatrix} \left(\ell \left\{ \begin{array}{c} -\sin\alpha \\ \cos\alpha \end{array} \right\} + \theta_j \left\{ \begin{array}{c} \cos\alpha \\ \sin\alpha \end{array} \right\} \right)
\end{aligned}
\tag{2}
$$

for the two cases in Figure 3, respectively. In Equations 1 and 2, ϕ_i and $\mathbf{r}_j = \{x_j, y_j\}^T$ are the Cartesian coordinates of body j,

$$
\mathbf{A}_{j-1} = \begin{bmatrix} \cos\phi_{j-1} & -\sin\phi_{j-1} \\ \sin\phi_{j-1} & \cos\phi_{j-1} \end{bmatrix}
\tag{3}
$$

is the rotation matrix of body $j-1$, α and ℓ are design variables, and $\mathbf{u}'_j = \{\cos\alpha, \sin\alpha\}^T$ is the translation axis for the translational joint, defined in the local coordinate system.

The position of any point c on body i, i.e. $\mathbf{r}_i^c$, may be calculated via the updating formula from a summation of the form

$$
\mathbf{r}_i^c = \sum_{m=1}^{i} \mathbf{s}_m + \sum_{\substack{m=1 \\ m \in M_T(i)}}^{i} \theta_m \mathbf{u}_m + \mathbf{s}_i^c
\tag{4}
$$

in which the $M_T(i)$ denote the translational joints found along the path from the base body to body i, and $\mathbf{s}_i^c$ is the vector from the joint reference point in body i to the point of interest, here the tracer point. This expression appears in several similar forms which may be found in [13].

Ultimately, the position of any body in the mechanism may be found assuming that the design variables, joint coordinates, and the base body location are known. If the mechanism has loops, e.g. four-bar mechanisms, some of the joints are cut to generate constraint equations which are enforced to ensure that the mechanism maintains its connectivity. These constraints are assembled in the vector $\boldsymbol{\Psi}(\boldsymbol{\theta}(t), t)$, which is a function of time

t and the joint variables, (assembled in $\boldsymbol{\theta}$), see the similar developments in [11]. Driver constraints are also included to constrain the remaining degrees-of-freedom, this accounts for the explicit time dependence in $\boldsymbol{\Psi}$.

To determine the mechanism's configuration at time t, the nonlinear equation

$$\boldsymbol{\Psi}(\boldsymbol{\theta}(t), t) = \mathbf{0} \tag{5}$$

is solved for $\boldsymbol{\theta}(t)$, using the Newton-Raphson procedure. That is, we solve

$$\mathbf{C}_{m-1}\Delta\boldsymbol{\theta}_{m-1} = -\boldsymbol{\Psi}_{m-1}(\boldsymbol{\theta}_{m-1}(t), t)$$

$$\boldsymbol{\theta}_m(t) = \boldsymbol{\theta}_{m-1}(t) + \Delta\boldsymbol{\theta}_{m-1} \tag{6}$$

iteratively until the constraint equation, cf. Equation 5, is satisfied. Hence, the matrix, $\mathbf{C}_{m-1} \equiv \partial\boldsymbol{\Psi}(\boldsymbol{\theta}_{m-1}(t), t)/\partial\boldsymbol{\theta}$ is required. This matrix is calculated directly from the updating formulas and the constraints equations, as discussed in [13]. When Equation 5 has been solved for joint variables $\boldsymbol{\theta}(t)$ the positions of all bodies may be found by the updating process as described above. Once the position of body j is found, the global position of any point on that body may be found as in e.g. [2].

4. The cost function

The present goal is to establish an automated, yet efficient, procedure that is capable of redesigning a mechanism so that a tracer point follows a given trajectory, C^d. The desired trajectory is defined by an arbitrary number of points, for example see Figure 4 where C^d is described by N_d points. Figure 4 also illustrates a curve, C^c, defined by the N_c points that lie along the trajectory of the current design, i.e. the design with dimensions as specified by the current design variables. The goal is to minimize the

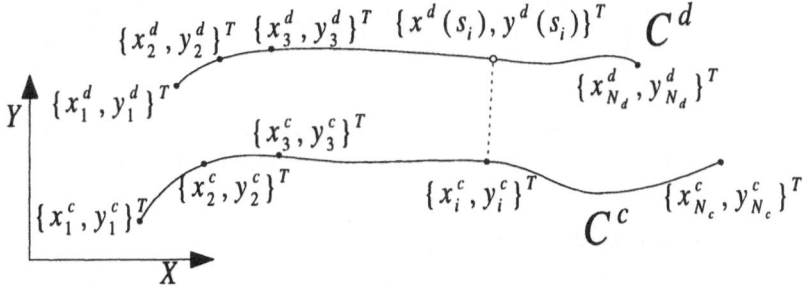

Figure 4. The desired curve C^d, defined by the N_d points $\{x_i^d, y_i^d\}^T$, and the calculated curve C^c, defined by the N_c points $\{x_i^c, y_i^c\}^T$.

distance between these curves, without requiring an equal number of points for each curve and without placing relations between the distinct points on the two curves.

To meet these design requirements, C^d is represented with a cubic spline, [12], and the cost function, ε to minimize is defined as

$$\varepsilon = \frac{1}{2} \sum_{i=1}^{N_c} (\mathbf{r}^d(s_i) - \mathbf{r}_i^c)^T \cdot (\mathbf{r}^d(s_i) - \mathbf{r}_i^c) \tag{7}$$

Here $\mathbf{r}_i^c = \{x_i^c, y_i^c\}^T$ is the i-th calculated point and $\mathbf{r}^d(s_i) = \{x^d(s_i), y^d(s_i)\}^T$ is the point on C^d that is closest to $\mathbf{r}_i^c$ (see Figure 4).

The point $\mathbf{r}^d(s_i)$ in Equation 7 is determined by solving the minimization problem

$$\min_s \frac{1}{2}(\mathbf{r}^d(s) - \mathbf{r}_i^c)^T \cdot (\mathbf{r}^d(s) - \mathbf{r}_i^c) \tag{8}$$

for each of the N_c points $\mathbf{r}_i^c$. The optimality conditions of Equation 8 require

$$\beta(s) = (\mathbf{r}^d(s) - \mathbf{r}_i^c)^T \cdot \frac{d\mathbf{r}^d(s)}{ds} = 0 \tag{9}$$

which is solved by the Newton-Raphson iteration

$$\Delta s = -\beta(s_{m-1})/J(s_{m-1})$$
$$\tag{10}$$
$$s_m = s_{m-1} + \Delta s$$

in which

$$J(s) = \frac{d}{ds}\left((\mathbf{r}^d(s) - \mathbf{r}_i^c)^T \cdot \frac{d\mathbf{r}^d(s)}{ds}\right)$$
$$\tag{11}$$
$$= \left(\left(\frac{d\mathbf{r}^d(s)}{ds}\right)^T \cdot \frac{d\mathbf{r}^d(s)}{ds} + (\mathbf{r}^d(s) - \mathbf{r}_i^c)^T \cdot \frac{d^2\mathbf{r}^d(s)}{ds^2}\right)$$

Summarizing, Equation 9 is solved N_c times, once for each of the calculated points, and then the cost function, i.e. Equation 7, is evaluated.

5. Design sensitivities

The design sensitivities are computed directly from equation 7 as

$$\frac{d\varepsilon(\mathbf{b})}{db_k} = \sum_{i=1}^{N_c} (\mathbf{r}^d(s_i(\mathbf{b})) - \mathbf{r}_i^c(\mathbf{b}))^T \cdot \left(\frac{\partial \mathbf{r}^d(s_i(\mathbf{b}))}{\partial s_i}\frac{ds_i(\mathbf{b})}{db_k} - \frac{d\mathbf{r}_i^c(\mathbf{b})}{db_k}\right) \tag{12}$$

in which we assemble the design parameters into the **b** vector.

The derivative $\frac{\partial r^d(s_i(\mathbf{b}))}{\partial s_i}$ is calculated directly from the cubic spline curve equations. Next consider the derivative $ds_i(\mathbf{b})/db_k$ in Equation 12. Since β must be zero (cf. Equation 9), the derivative with respect to b_k must also be zero, i.e.

$$\frac{d\beta(s(\mathbf{b}),\mathbf{b})}{db_k} = \frac{d\left((\mathbf{r}^d(s(\mathbf{b})) - \mathbf{r}_i^c(\mathbf{b}))^T \cdot \frac{d\mathbf{r}^d(s(\mathbf{b}))}{ds}\right)}{db_k} = 0 \Rightarrow$$

$$\frac{\partial\left((\mathbf{r}^d(s(\mathbf{b})) - \mathbf{r}_i^c(\mathbf{b}))^T \cdot \frac{d\mathbf{r}^d(s(\mathbf{b}))}{ds}\right)}{\partial s} \frac{ds(\mathbf{b})}{db_k} + \tag{13}$$

$$\frac{\partial\left((\mathbf{r}^d(s(\mathbf{b})) - \mathbf{r}_i^c(\mathbf{b}))^T \cdot \frac{d\mathbf{r}^d(s(\mathbf{b}))}{ds}\right)}{\partial b_k} = 0 \Rightarrow$$

$$\frac{ds(\mathbf{b})}{db_k} = -\frac{\frac{\partial}{\partial b_k}\left((\mathbf{r}^d(s(\mathbf{b})) - \mathbf{r}_i^c(\mathbf{b}))^T \cdot \frac{d\mathbf{r}^d(s(\mathbf{b}))}{ds}\right)}{\frac{\partial}{\partial s}\left((\mathbf{r}^d(s(\mathbf{b})) - \mathbf{r}_i^c(\mathbf{b}))^T \cdot \frac{d\mathbf{r}^d(s(\mathbf{b}))}{ds}\right)}$$

Note that the denominator of Equation 13 equals $J(s(\mathbf{b}),\mathbf{b})$ which we have already calculated (cf. Equation 11). Also note that $\mathbf{r}^d$ does not depend explicitly on b_k. Finally we obtain

$$\frac{ds(\mathbf{b})}{db_k} = \left(\frac{d\mathbf{r}_i^c(\mathbf{b})}{db_k}\right)^T \cdot \frac{d\mathbf{r}^d(s(\mathbf{b}))}{ds} / J(s(\mathbf{b}),\mathbf{b}) \tag{14}$$

All that remains in Equations 12 and 14 is to evaluate the derivative $\frac{d\mathbf{r}_i^c(\mathbf{b})}{db_k}$. Differentiation of Equation 4 yields the desired derivative expression

$$\frac{d\mathbf{r}_i^c}{db_k} = \sum_{m=1}^{i}\left(\frac{\partial\mathbf{s}_m}{\partial\boldsymbol{\theta}} \cdot \frac{d\boldsymbol{\theta}}{db_k} + \frac{\partial\mathbf{s}_m}{\partial b_k}\right) + \sum_{\substack{m=1 \\ m \in M_T(i)}}^{i} \theta_m\left(\frac{\partial\mathbf{u}_m}{\partial\boldsymbol{\theta}} \cdot \frac{d\boldsymbol{\theta}}{db_k} + \frac{\partial\mathbf{u}_m}{\partial b_k}\right) +$$

$$\sum_{\substack{m=1 \\ m \in M_T(i)}}^{i} \frac{d\theta_m}{db_k} \cdot \mathbf{u}_m + \frac{\partial\mathbf{s}_i^c}{\partial\boldsymbol{\theta}} \cdot \frac{d\boldsymbol{\theta}}{db_k} + \frac{\partial\mathbf{s}_i^c}{\partial b_k} \tag{15}$$

Similar expressions may be derived for the other connections.

To determine $d\boldsymbol{\theta}/db_k$ in equation 15 we differentiate the constraint equation 5

$$\boldsymbol{\Psi}(\boldsymbol{\theta}(t,\mathbf{b}),t,\mathbf{b}) = \mathbf{0} \Rightarrow$$

$$\frac{\partial\boldsymbol{\Psi}}{\partial\boldsymbol{\theta}} \cdot \frac{d\boldsymbol{\theta}}{db_k} + \frac{\partial\boldsymbol{\Psi}}{\partial b_k} = \mathbf{0} \Rightarrow \tag{16}$$

$$\mathbf{C}\frac{d\boldsymbol{\theta}}{db_k} = -\frac{\partial\boldsymbol{\Psi}}{\partial b_k}$$

Notice above that $\mathbf{C}$ already has been assembled and decomposed during the position analysis, so it is inexpensive to compute $d\boldsymbol{\theta}/db_k$. Finally we have all of the expressions required to evaluate sensitivities of the cost function.

6. Optimization

The optimization is performed using the Broyden-Fletcher-Goldfart-Shanno method as described in [12]. To assist the optimizer a first approximation to the desired design is determined in which the first and last points on the calculated curve are coincident with the first and last desired points, i.e. the initial design of the final optimization problem is the solution to

$$\min_{\mathbf{b}} \frac{1}{2}(\mathbf{r}_1^d - \mathbf{r}_1^c)^T \cdot (\mathbf{r}_1^d - \mathbf{r}_1^c) + \frac{1}{2}(\mathbf{r}_{N_d}^d - \mathbf{r}_{N_c}^c)^T \cdot (\mathbf{r}_{N_d}^d - \mathbf{r}_{N_c}^c) \qquad (17)$$

7. Examples

Two examples are presented to demonstrate the method's design capability and it's generality. In the first example a four-bar slider-crank mechanism is considered, and in the second six-bar linkage is designed.

7.1. THE FOUR-BAR SLIDER-CRANK

In this example a four-bar slider-crank is designed to trace a straight line. The initial design and its trajectory are shown in Figures 5 (a) and (c). The desired trajectory is also shown in this figure. This problem uses 10 design variables. The optimized design appear in Figure 5 (b) and (c), respectively. Note the agreement between the desired and optimal trajectories.

7.2. THE SIX-BAR LINKAGE

The above problem is now repeated for the six-bar linkage shown in Figure 6 (a). This problem uses 15 design variables. The optimal design appear in Figure 6 (b) and (c), respectively. Again, note the agreement between the desired and optimal trajectories.

8. Conclusion

An efficient method is developed to optimize mechanisms so that a point on the mechanism follows a prescribed trajectory. The method uses analytical sensitivities which are computed via the direct differentiation method. The sensitivities are computationally efficient because they utilize the existing

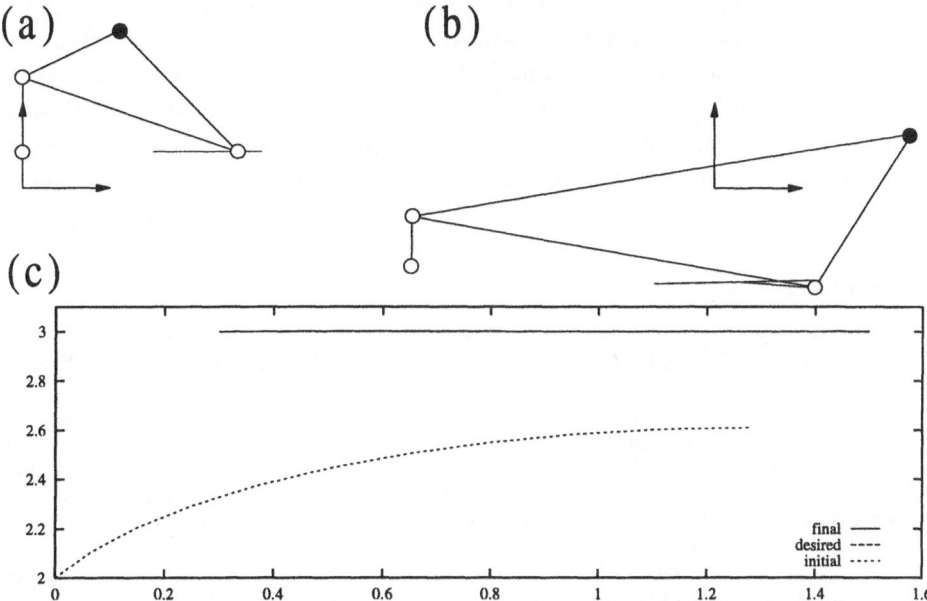

Figure 5. The initial (a) and final (b) design for the four-bar slider-crank and the initial, desired, and optimal trajectories (c). The final and desired curve appear coincident on the figure

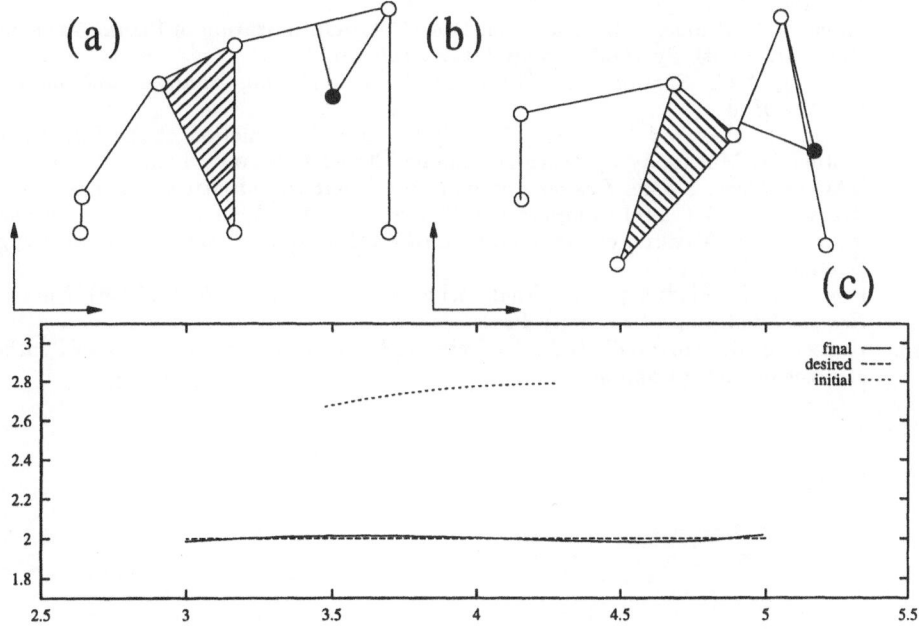

Figure 6. The initial and optimal designs for the six-bar mechanism and the initial, desired, and optimal trajectories

decomposed Jacobian matrices. In addition, the mechanism is defined directly in terms of the design variables, so that the mechanism analysis and optimization data are easily identified. Finally the method is general in that first the underlying analysis tool is a general purpose method that works equally well for all topologies (within the limits of the defined joints) and that second the desired tracer curve is defined by an arbitrary number of points.

References

1. Wittenburg, J. (1991) *Dynamics of Systems of Rigid Bodies* Teubner
2. Nikravesh, P.E. (1988) *Computer Aided Analysis of Mechanical Systems*, Prentice-Hall
3. Haug, E.J. (1989) *Computer-Aided Kinematics and Dynamics of Mechanical Systems*, Allyn and Bacon
4. Haug, E.J., Sohoni, V.N. (1984) Design Sensitivity Analysis and Optimization of Kinematically Driven Systems. In E.J. Haug, editor, *Computer Aided Analysis and Optimization of Mechanical System Dynamics*, volume F9 of NATO ASI, pages 499-554, Springer-Verlag
5. Bruns, T. (1992) *Design of Planar, Kinematic, Rigid Body Mechanisms*, Master's thesis, University of Illinois at Urbana-Champaign, Urbana, IL.
6. Hansen, M.R. (1992) A General Procedure for Dimensional Synthesis of Mechanisms. *Mechanism Design and Synthesis*, vol 46, pp. 67-71
7. Hansen, J.M. (1993) Synthesis of Spatial Mechanisms Using Optimization and Continuation Methods. In M.S. Pereira and J.A.C. Ambrósio, editors, *Computer Aided Analysis of Rigid and Flexible Mechanical Systems*, vol. 2 of NATO ASI, pp. 423-439, Troia, Portugal
8. Pesch, V.J., Hinkle, C.L., Tortorelli, D.A (1995) Optimization of Planar Mechanism Kinematics with Symbolic Computation. *submitted*
9. Erdman, A.G., Sandor, G.N (1991) *Mechanism Design, Analysis and Synthesis*, Prentice-Hall
10. Nikravesh, P.E. and Gim, G. (1989) Systematic Construction of the Equations of Motion for Multibody Systems Containing Closed Kinematic Loops. Paper no. 89-DAC-58, Proc. *ASME Design Automation Conference*, Montreal, Canada
11. Nikravesh, P.E. (1991) *Computational Methods in Multibody Systems*, Lecture notes from COMETT course on *Computer Aided Analysis of Mechanical Systems*, Lyngby, Denmark, May, 1991
12. Press, W.H., Flannery, B.P, Teukosisky, S.A, Vetterling, W.T (1986) *Numerical Recipes* Cambridge University Press.
13. Hansen, J.M., Tortorelli, D.A. An Efficient Method for the Analysis and Synthesis of Mechanisms, *to appear*

SYNTHESIS OF MECHANISMS INCLUDING THE SHAPE OF BODIES AS DESIGN VARIABLES

MICHAEL R. HANSEN
Institute of Mechanical Engineering
Aalborg University
DK-9220 Aalborg East, Denmark

ABSTRACT.   This paper deals with the possibilities in rigid body mechanism design when the shape of the bodies are included as design variables. Typical criteria upon which mechanism design is based are listed and their dependencies on the actual body presentation is discussed, together with an introduction of the advantages and new possibilities obtained by adding shape to the bodies. A general way of describing the shape of a body is put forward. Special emphasis is put on one of the new possibilities, namely the inclusion of higher pair joints. The necessary mathematical background for defining a contact surface is shown and some examples based on cam-mechanisms with rotating roller-followers are presented.

1. Introduction

In recent years a lot of work has been devoted to the synthesis of rigid body mechanisms containing lower pair joints, [1], [2], [3]. These types of synthesis are always based on kinematic criteria such as path-, motion-, and function-generation [4]. A practical problem also often involves territorial criteria as well [5], that define areas into which the entire mechanism or part of it is not allowed to enter. Finally, dynamic criteria such as the minimization of reactive forces or the power consumption are often added.

The bodies of planar mechanisms are classically represented as skeleton diagrams. Naturally, real bodies look different, and this would affect the evaluation of the abovementioned criteria in certain ways.

Kinematic criteria, however, are not affected since the motion solely depends on the position and orientation of the joints relative to each other.

D. Bestle and W. Schielen (eds.), IUTAM Symposium on Optimization of Mechanical Systems, 139–146.
© 1996 Kluwer Academic Publishers.

Territorial criteria can be treated without having the exact shape [5], but
without being absolutely sure that no constraints are violated. For high
speed mechanisms the inertia data are of significance to the dynamic criteria. In cases where the inertia data are thought of as dependent variables
to be derived from the body, the shape is required.

Thus, inclusion of shape is motivated by the desire to properly handle
territorial and dynamic criteria. It is, however, even more motivated by the
new types of criteria and mechanisms that may be taken into consideration. Having the actual shape expressed, it is possible to address criteria
involving, e.g., deflections, stresses and eigenfrequencies. At the same time
the possibility to include higher pair joints appears, because the contact
surface may be modeled.

2. Segment library

In order to be able to define any kind of shape of a body its contour must
be made up of segments, hence, a segment library is required for a general
purpose approach. Another consideration involves the higher pair joints. In
general it will be necessary to be able to prescribe an n'th order continuity
between adjoining contour segments. This may be accomplished by means
of bezier curves.

These considerations boil down to a segment library consisting of the
basic segment types shown in Fig. 1.

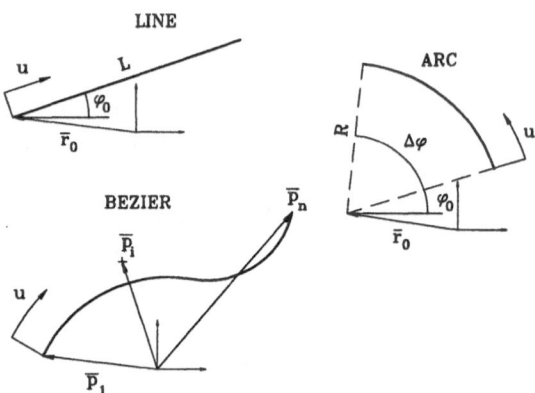

Figure 1. Segment library with design variables.

The potential design variables are shown in Fig. (1). They are all measured relative to a body fixed coordinate system and for the 3 segment
types they comprise:

$$\vec{x}_{line} = \vec{r}_0, \ \phi_0, \ L \tag{1}$$

$$\vec{x}_{arc} = \vec{r}_0, \ \phi_0, \ \Delta\phi, \ R \tag{2}$$

$$\vec{x}_{bezier} = \vec{p}_1, \ ..., \ \vec{p}_n, \tag{3}$$

All of these variables will not necessarily be independent. Especially in the case of defining a contact surface for a higher pair joint a bezier segment will be restricted by the surrounding segments in order to obtain a satisfactorily degree of continuity.

3. Higher pair joint

In the following emphasis will be put on the higher pair joints that may be modeled via the introduction of the shape of the bodies.

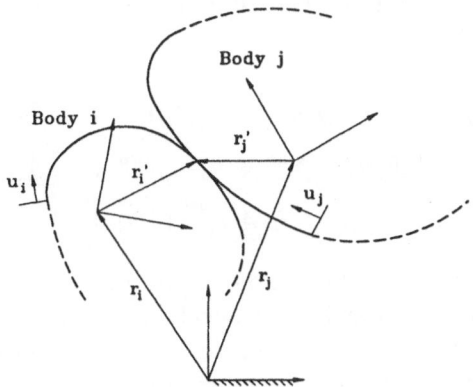

Figure 2.   General higher pair contact joint.

A general mathematical formulation of a simple point contact joint, see Fig. 2, is set up as a combination of a closed vector loop equation and an alignement of the contact surface tangents at the point of contact:

$$\vec{r}_i + \vec{r}_i\,' - \vec{r}_j - \vec{r}_j\,' = \vec{0}$$
$$\Downarrow$$
$$\vec{r}_i + \underline{\underline{A}}_i\vec{\rho}_i - \vec{r}_j + \underline{\underline{A}}_j\vec{\rho}_j = \vec{0} \tag{4}$$

$$\underline{\underline{A}}_i\widehat{\vec{\rho}}_i^{\,u} \ \underline{\underline{A}}_j\vec{\rho}_j^{\,u} = 0 \tag{5}$$

In the above equations $\underline{\underline{A}}_i$ is the transformation matrix of the i'th body and $\vec{\rho}_i$ is the position of the contact point measured in the body fixed coordinate system of the i'th body. The vector $\vec{\rho}_i$ is a function of u_i which is a surface parameter of the i'th body. The velocity constraints are obtained by differentiating (4) and (5) with respect to time:

$$\vec{v}_i + \dot{\theta}_i \underline{\underline{B}}_i \vec{\rho}_i + \dot{u}_i \underline{\underline{A}}_i \vec{\rho}_i^u - \vec{v}_j - \dot{\theta}_j \underline{\underline{B}}_j \vec{\rho}_j - \dot{u}_j \underline{\underline{A}}_j \vec{\rho}_j^u = \vec{0} \tag{6}$$

$$[\, \dot{\theta}_i \underline{\underline{B}}_i \hat{\vec{\rho}}_i^u + \dot{u}_i \underline{\underline{A}}_i \hat{\vec{\rho}}_i^{uu} \,] \underline{\underline{A}}_j \vec{\rho}_j^u + [\, \dot{\theta}_j \underline{\underline{B}}_j \vec{\rho}_j^u + \dot{u}_j \underline{\underline{A}}_j \vec{\rho}_j^{uu} \,] \underline{\underline{A}}_i \hat{\vec{\rho}}_i^u = 0 \tag{7}$$

The matrix $\underline{\underline{B}}_i$ is $\underline{\underline{A}}_i$ differentiated with respect to θ_i. The acceleration constraints are obtained by differentiating (6) and (7) with respect to time:

$$\vec{a}_i + \ddot{\theta}_i \underline{\underline{B}}_i \vec{\rho}_i - \dot{\theta}_i^2 \underline{\underline{A}}_i \vec{\rho}_i^u + 2\dot{\theta}_i \dot{u}_i \underline{\underline{B}}_i \vec{\rho}_i^u + \ddot{u}_i \underline{\underline{A}}_i \vec{\rho}_i^u + \dot{u}_i^2 \underline{\underline{A}}_i \vec{\rho}_i^{uu} - \vec{a}_j + \ddot{\theta}_j \underline{\underline{B}}_j \vec{\rho}_j$$
$$-\dot{\theta}_j^2 \underline{\underline{A}}_j \vec{\rho}_j^u + 2\dot{\theta}_j \dot{u}_j \underline{\underline{B}}_j \vec{\rho}_j^u + \ddot{u}_j \underline{\underline{A}}_j \vec{\rho}_j^u + \dot{u}_j^2 \underline{\underline{A}}_j \vec{\rho}_j^{uu} = \vec{0} \tag{8}$$

$$[\, \ddot{\theta}_i \underline{\underline{B}}_i \hat{\vec{\rho}}_i^u - \dot{\theta}_i^2 \underline{\underline{A}}_i \hat{\vec{\rho}}_i^u + 2\dot{\theta}_i \dot{u}_i \underline{\underline{B}}_i \hat{\vec{\rho}}_i^{uu} + \ddot{u}_i \underline{\underline{A}}_i \hat{\vec{\rho}}_i^{uu} + \dot{u}_i^2 \underline{\underline{A}}_i \hat{\vec{\rho}}_i^{uuu} \,] \underline{\underline{A}}_j \vec{\rho}_j^u +$$
$$[\, \ddot{\theta}_j \underline{\underline{B}}_j \vec{\rho}_j^u - \dot{\theta}_j^2 \underline{\underline{A}}_j \vec{\rho}_j^u + 2\dot{\theta}_j \dot{u}_j \underline{\underline{B}}_j \vec{\rho}_j^{uu} + \ddot{u}_j \underline{\underline{A}}_j \vec{\rho}_j^{uu} + \dot{u}_j^2 \underline{\underline{A}}_j \vec{\rho}_j^{uuu} \,] \underline{\underline{A}}_i \hat{\vec{\rho}}_i^u +$$
$$2[\, \dot{\theta}_i \underline{\underline{B}}_i \hat{\vec{\rho}}_i^u + \dot{u}_i \underline{\underline{A}}_i \hat{\vec{\rho}}_i^{uu} \,][\, \dot{\theta}_j \underline{\underline{B}}_j \vec{\rho}_j^u + \dot{u}_j \underline{\underline{A}}_j \vec{\rho}_j^{uu} \,] = 0 \tag{9}$$

As may be seen from (9) the 3rd derivative of the surface contour of both bodies must be finite in order for the accelerations to be finite. Accordingly the 3rd derivative of the surface contour must be continuous in order for the accelerations of the system to be continuous. The latter requirement is the basic rule of cam design, [6], and states that the time derivative of the accelerations, namely the jerk, should be finite at any time, in order to avoid shock loads and wear.

It is obvious that arcs and lines in any combination must be separated by bezier curves in order for the contact surface to fulfil the requirement for the continuous accelerations. Whenever a bezier segment is connected to the end of any segment type the first 4 points are dependent variables that may be determined from the variables of the previous segment. Similarly, the last 4 points depends on the next segment unless this is another bezier curve.

4. Examples

In this section a number of examples is presented to show some of the possibilities of the previously mentioned way of defining the body shapes and the point contact joint. The examples are carried out on a well-known type of cam mechanism with a rotating roller follower, see Fig. 3.

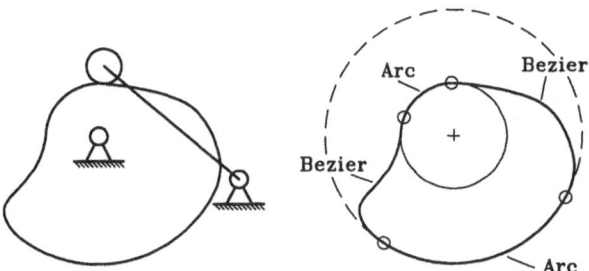

Figure 3. Cam mechanism with rotating roller follower. The subdivision of the cam into contour segments is shown.

Actually, the contact between the roller follower and the cam should be a rolling contact joint and not a point contact joint. However, if the friction force between the cam and the roller is eliminated then the roller and the follower arm may be treated as one body and the contact as a point contact.

The cam is modeled as a closed contour made up by 4 segments: arc - bezier - arc - bezier. The body fixed coordinate system of the cam is positioned in the center of rotation and is free to move relative to the center of rotation of the follower. The position of the center of the 2 arcs relative to the body fixed coordinate system of the cam are fixed to (0,0). Furthermore, as explained in the previous section, the first 4 points and the last 4 points of both bezier curves are determined directly by the arcs. The bezier curves are modeled by means of 12 control points, thus, leaving 4 points on each curve as independent variables.

The follower is modeled as a single arc with a 360° sweep angle. All its physical dimensions are fixed.

A couple of side constraints are constantly imposed on the system. Firstly, the minimum concave radius of curvature of the cam must be larger than the radius of the roller follower. Secondly, the cam must not move into the center of rotation of the follower. The latter constraint is imposed with a view to save bearing costs.

The basic kinematic criteria that the cam should fulfil is a standard motion sequence for the follower: rise - dwell - fall - dwell. Each period lasts a 90° rotation of the cam and the difference in rotation of the follower in the two dwell positions is 30°.

A dynamic criteria is also imposed, namely to minimize the maximum angular acceleration of the follower. This is, however, not activated until the kinematic criteria is exactly fulfiled. By minimizing the maximum angular acceleration of the follower the classical approach in cam design is followed.

In Fig. (4) the initial cam (left) and the resulting cam (right) are shown.

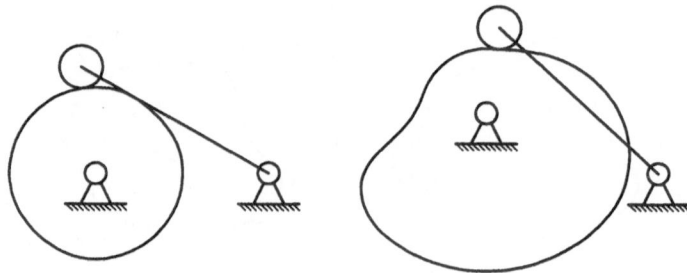

Figure 4.   Initial cam mechanism (left) and cam mechanism with desired kinematic criteria and minimized maximum angular acceleration of follower (right).

In Fig. 5 the angular acceleration of follower of the resulting cam mechanism is shown. As may be seen an acceleration history not much unlike the acceleration laws normally put forward in textbooks [6] on cams as desirable (sinusoidal and trapezoidal) has been obtained.

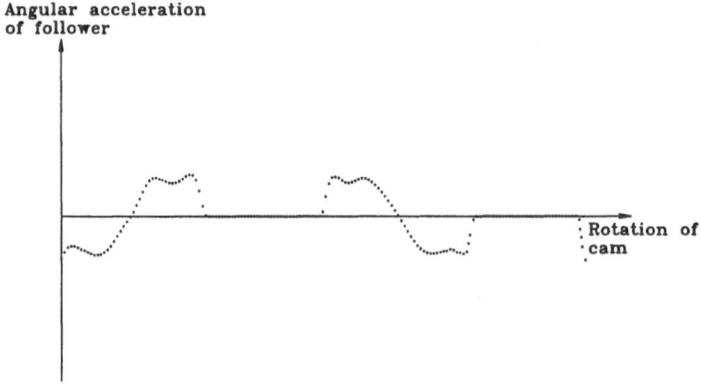

Figure 5.   Variation of angular acceleration of follower of the cam mechanism shown in Fig. 4.

From a dimensioning point of view, however, the size of the angular acceleration of the follower is not as interesting as the normal force between the cam and the follower, or the hertz pressure at the contact point. In the following example the normal force is considered.

It is assumed that the payload as well as the spring force holding the follower and cam together are approximately constant. In that case the normal force may be written as:

$$N = C_0 \, \frac{C_1 - \alpha_f}{sin\beta} \tag{10}$$

In (10) C_0 and C_1 are constants and β is the angle between the common normal at the contact point and a line along the length of the follower. The angular acceleration of the follower, α_f is taken to be positive in the counterclockwise direction.

If the maximum value of N is minimized the cam shown to the right in Fig. (6) appears. The one to the left is the initial mechanism that corresponds to the one with minimized maximum angular acceleration of the follower, hence, it is the same as the one shown in Fig. (4) to the right.

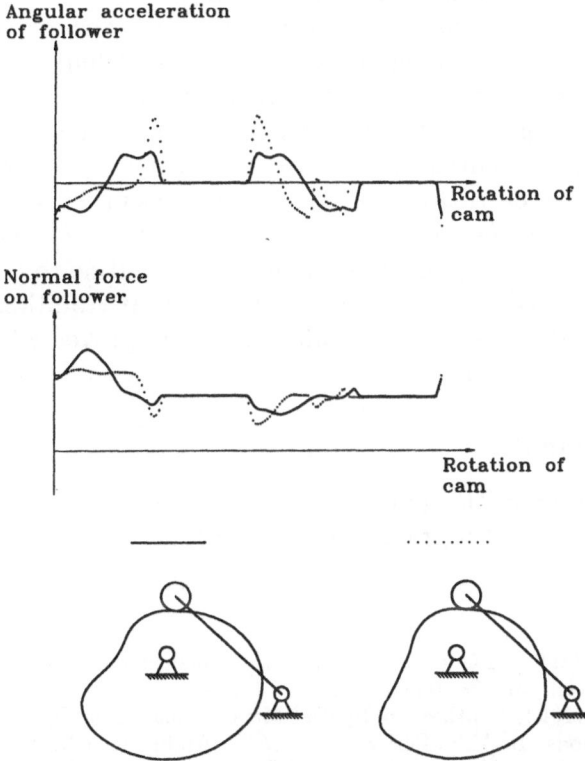

Figure 6. Initial mechanism (left) and mechanism with minimized maximum normal force between follower and cam (right). A graphical presentation of the angular acceleration of the follower as well as the normal force between follower and cam are shown for the 2 mechanisms.

In Fig. 6 the angular acceleration of the follower is shown for the two cam mechanisms, and, as may be seen, a substantial change in accelerational behavior is present and a much higher peak angular acceleration is accepted for the mechanism to the right because it happens in a position with good transmission properties.

Also the variation of the normal forces are shown. It is clear from the graphical presentation that the maximum normal force has been substantially minimized.

5. Conclusions

There are several strong motivations to introduce the shape of bodies as design variables when synthesizing mechanisms. Firstly, territorial and dynamic criteria may be evaluated more correctly, and, secondly, new types of criteria as well as mechanisms may be considered.

A segment library consisting of line, arc and bezier curve was put forward as a suitable set of building stones when modeling a body contour.

Especially the bezier curves are imperative when considering mechanisms with higher pair joints, in which case a 3rd order continuity of the contact surfaces is required in order to get continuous accelerations.

Some examples carried out on a cam mechanism with a rotating roller follower seems to suggest that a contour made up by a segment sequence of: arc - bezier - arc - bezier, allows a minimization routine to find satisfactory cam shapes. The examples also clearly showed that traditional cam design based on acceleration laws can be substantially improved when focusing on the variation of the actual loads.

Acknowledgement

The work presented in this paper received support from the Danish Technical Research Council, Programme of Research on Computer Aided Design.

References

1. Hansen, M.R. (1992) A General Procedure for Dimensional Synthesis of Mechanisms, *Mechanism Design and Synthesis* **46** pp. 67-71.
2. Hansen, J.M. (1993) Synthesis of Spatial Mechanisms Using Optimization and Continuation Methods, In M.S. Pereira and J.A.C. Ambrosio, editors, *Computer Aided Analysis of Rigid and Flexible Mechanical Systems* NATO ASI **2**, pp. 423-439.
3. Pesch, V.J., Hinkle, C.L., Tortorelli, D.A. (1995) Optimization of Planar Mechanism Kinematics with Symbolic Computation, *submitted* , .
4. Erdman, A.G., Sandor, G.N. (1991) *Mechanism Design, Analysis and Synthesis*, Prentice Hall , .
5. Hansen, M.R. (1995) A Multi Level Approach to Synthesis of Planar Mechanisms, *Journal of Nonlinear Dynamics* To appear, .
6. Norton, R.L. (1992) *Design of Machinery*, McGraw-Hill , .

OPTIMIZATION OF A SPATIAL TRANSMISSION
LINKAGE FOR HEAVY MANIPULATORS

S. HARTMANN
IMECH GmbH.
Bergwerkstraße
47445 Moers
Germany

AND

T. KRUPP AND M. HILLER
Gerhard-Mercator-Universität – GH Duisburg
Lotharstraße 1
47057 Duisburg
Germany

1. Introduction

The problem described in this paper is part of the development of a hydraulically driven heavy manipulator at the IMECH GmbH. The manipulator is intended for handling payloads up to 500 kilograms in a constrained environment. Therefore it is necessary to equip the manipulator with redundant degrees of freedom – which means more than six in the spatial case. On the other hand, in contrast to existing large manipulators (Wanner, 1990), the size of the manipulator is limited. Consequently it is difficult to provide as many joints as there are degrees of freedom, which leads to the idea of using spherical joints to replace three rotational joints. A spatial transmission linkage is necessary to control the movement of the spherical joint, since the swivel motion covers an angle up to 205°. The complete manipulator is built up of modular components, called ball-joint-units consisting of a central spherical joint and a transmission linkage. A brief description of a ball-joint-unit is given in section 2.

The aim now is to determine the system's parameters so that the (static) joint forces reach a minimum while maintaining the nominal workspace

D. Bestle and W. Schielen (eds.), IUTAM Symposium on Optimization of Mechanical Systems, 147–154.
© 1996 *Kluwer Academic Publishers.*

and avoiding collisions within the transmission linkage and to provide a parameter configuration which is suitable for the later-on realization of the manipulator. This leads to an optimization problem which was solved with an SQP-algorithm (Gill *et al.*, 1981). Section 3 presents some details of the optimization problem, including the avoidance of kinematic singularities and section 4 shows results of an optimization run.

2. System Description and Modelling

The ball-joint-unit described in Fig. 1 consists of a manipulator arm supported by a spherical joint S_1, which is connected to a transmission linkage and the support joint S_2. The central joint S_1 and the supporting joint S_2 define a rotation axis around which the manipulator arm can be swiveled when moving piston 1. Motion of piston 2 and piston 3 controls the position of S_2, so that the orientation of the rotation axis u can be changed. The transmission linkage, working over the swivel plate which is connected to the base by the rotational joint R (angle β), ensures a large nominal workspace up to 205° around the main axis for the swivel angle θ_1. (For the example described here only a workspace up to 180° must be achieved.) In contrast to θ_1 the angles θ_2 and θ_3, which describe the orientation of the rotation axis u relative to a reference orientation u_0, (Fig. 1) will be confined to ±30°.

When modelling the kinematical structure of the system the main problem is to compute the angles θ_1, θ_2 and θ_3 for the known piston strokes s_1 (piston 1), s_2 (piston 2) and s_3 (piston 3). The relation between the piston strokes and the joint angles can be described by a set of nonlinear constraint equations of the form

$$\left.\begin{array}{ccccc} g_1 & = & (r_5(\beta) - r_6)^2 & = & s_1^2 \\ g_2 & = & (r_2(\theta_1, \theta_2, \theta_3) - r_7)^2 & = & s_2^2 \\ g_3 & = & (r_2(\theta_1, \theta_2, \theta_3) - r_8)^2 & = & s_3^2 \\ g_4 & = & (r_3(\theta_1, \theta_2, \theta_3) - r_4(\beta))^2 & = & d^2 \end{array}\right\} , \qquad (1)$$

which can be solved explicitly or numerically by using a *Newton-Raphson-*scheme.

A kinetostatic model of the ball-joint unit was developed using the object-oriented tool-set for multibody systems M⊃BILE (Kecskeméthy *et al.*, 1993), which is based on the concept of *kinetostatic transmission elements*. This allows computation of kinematics and statics and simulation of the dynamics of a system. For the current problem the position kinematics is calculated as well as the statics, where the loads applied to all bodies and joints of the system are computed. Therefore a single load of 800 kg

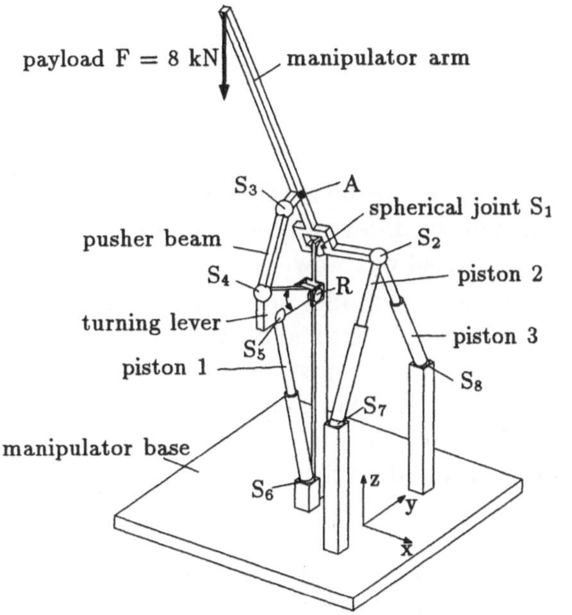

geometric parameters

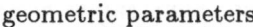

$$
\begin{aligned}
r_1 &= \overrightarrow{S_1R} & r_2 &= \overrightarrow{S_1S_2} \\
r_3 &= \overrightarrow{S_1S_3} & r_4 &= \overrightarrow{S_1S_4} \\
r_5 &= \overrightarrow{S_1S_5} & r_6 &= \overrightarrow{S_1S_6} \\
r_7 &= \overrightarrow{S_1S_7} & r_8 &= \overrightarrow{S_1S_8} \\
d &= \overline{S_3S_4} & h_1 &= \overline{RS_5} \\
h_2 &= \overline{RS_4} & a_1 &= \overline{S_1A} \\
a_2 &= \overline{AS_3} \\
\alpha &= \angle(h_1, h_2)
\end{aligned}
$$

payload F = 8 kN

manipulator arm

S_3

A

spherical joint S_1

pusher beam

S_2

S_4

R

piston 2

turning lever

piston 3

S_5

piston 1

S_8

S_7

manipulator base

S_6

z

y

x

θ_1

u

θ_3

θ_2

u_0

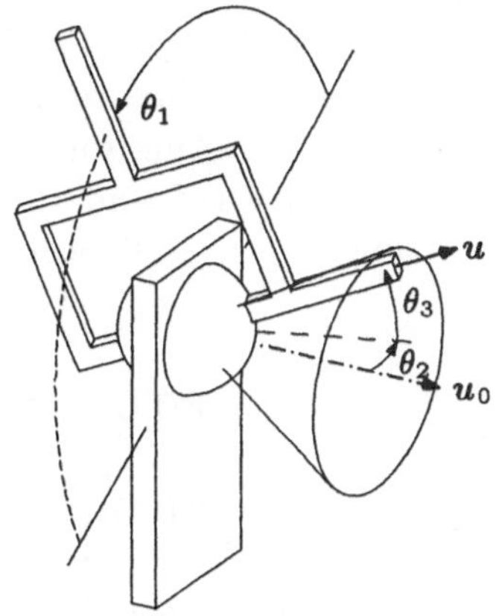

Figure 1.   Spherical Joint and Transmission Linkage

was applied at the top of the manipulator arm, including the payload of 500 kg, the weight of the manipulator arm itself and the gripper.

3. Optimization

As mentioned above, the design problem is to minimize the forces acting on the links within the ball-joint-unit. While formulating the optimization criterions the following two subtasks arise: (A) the determination of forces as a function of given parameters and (B) the determination of optimal parameters which result in minimal forces. Subtask (A) can be done quickly with the help of the object-oriented tool-set M⟨⟩BILE, where a kinetostatic model of the ball-joint-unit is used as described in section 2. For Subtask (B) an optimization problem has to be solved. The set of $n = 10$ design variables $x = [r_{1_y}, r_{1_z}, r_{6_y}, r_{6_z}, a_1, a_2, d, h_1, h_2, \alpha]^T$ can be found in Fig. 1. Lower and upper bounds of x (l_B and u_B) arise from mechanical limitation. With nonlinear constraints $c(x)$ due to the restrictions of the nominal workspace the following optimization problem is formulated:

$$x : f(x) \overset{!}{=} \text{Min} \ , \quad x = [x_1, \ldots, x_n]^T \\ l_B \leq x \leq u_B \\ c(x) = 0 \ \Big\} . \tag{2}$$

SQP-algorithms are well suited for such problems if the objective function $f(x)$ is smooth, i.e. twice-continously differentiable. The definition of an appropriate objective function is carried out now.

Simulation results from the kinetostatic model show that the crucial point of the ball-joint-unit is the central spherical joint S_1. Therefore the main objective function of the design process must be the minimization of the forces acting on this joint. The relationship between the magnitude of the joint force and the angle θ_1 is shown in Fig. 2, where two critical positions $\theta_{1_a} = -90°$ and $\theta_{1_b} = 90°$ can be identified.

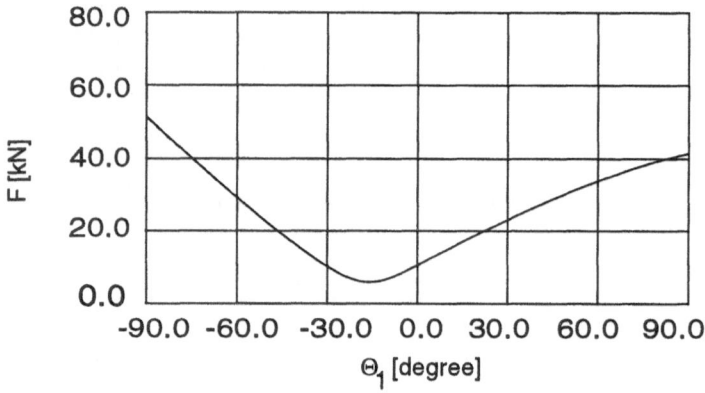

Figure 2. Magnitude of joint force as a function of the joint angle θ_1

The variation of θ_2 and θ_3 for these two positions leads to the joint forces shown in Fig. 3. The first idea was to use the average of several discrete force computations over the nominal workspace as objective function, but as one can see from Fig. 3b a small average does not avoid extrem values. Taking the maximum values of all computed forces as objective function may guarantee lower extremal values, but it does not fullfill the criterion of smoothness. Therefore a mixed criterion is chosen, taking into account the averages for the critical positions θ_{1_a} and θ_{1_b} on the one hand, and the deviation of these averages on the other hand:

$$f = \alpha \, (\bar{F}_a^2 + \bar{F}_b^2) + \sum_{i=1}^{9}(F_{a_i} - \bar{F}_a)^2 + \sum_{i=1}^{9}(F_{b_i} - \bar{F}_b)^2 \; . \qquad (3)$$

Here α denotes a weighting factor and $\bar{F}$ denotes the averages, which are computed out of 9 discrete force values (Fig. 3) for θ_{1_a} and θ_{1_b}.

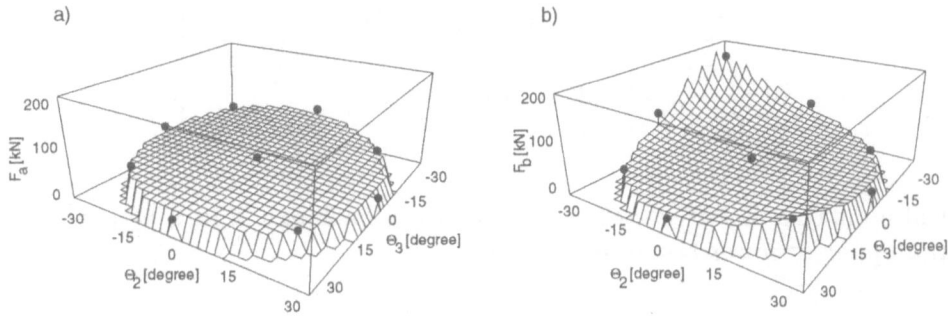

Figure 3. Joint force as a function of θ_2 and θ_3 for a) $\theta_1 = \theta_{1_a}$ and b) $\theta_1 = \theta_{1_b}$

Equation (3) guarantees a smooth objective function if there are no kinematic singularities, which often occur in such complex systems (Fig. 4). The condition number of the (10×10)-HESSIAN-matrix $H_{opt} = \partial^2 f / \partial x^2$ may be used to detect singular positions, but its calculation is too expensive.

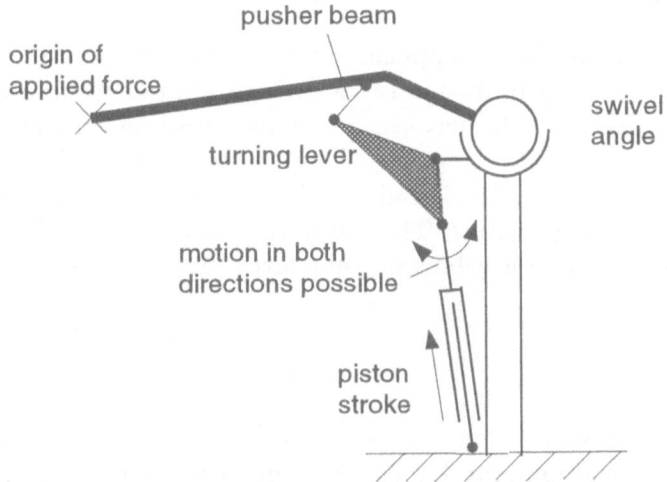

Figure 4. Singular position for the joint kinematics

While the influence of the kinematics on the condition number increases with the value of the condition number, one can detect singular positions by controlling the condition number κ of the JACOBIAN-matrix $J_\beta = \partial g / \partial \beta$ ($g = [g_1, g_2, g_3, g_4]^T$ in Eq. 1, $\beta = [\theta_1, \theta_2, \theta_3, \beta]^T$), which is only a ($4 \times 4$) dimensional one. If κ exceeds a prescribed limit, a correction of the parameter bounds can be estimated (Kecskeméthy et al., 1993). Therefore a direction i has to be determined for which the sensitivity

$$\left. \begin{aligned} \frac{\triangle \kappa^{(i)}}{\epsilon} &= \frac{\kappa(J_\beta(x + \triangle \hat{x}^{(i)})) - \kappa(J_\beta(x))}{\epsilon} \\ \triangle \hat{x}_j^{(i)} &= \begin{cases} \epsilon & for \quad j = i \\ 0 & for \quad j \neq i \end{cases} \end{aligned} \right\} \tag{4}$$

will be maximal. Afterwards κ can be forced back to a limit $\bar{\kappa}$ by correcting the upper or lower bound of x_i with

$$\triangle x_i = \frac{\bar{\kappa} - \kappa(J_\beta(x))}{\triangle \kappa^{(i)} / \epsilon} \tag{5}$$

as follows:

$$\left. \begin{aligned} u_{B_i} &= x_i + \triangle x_i \quad if \quad \triangle x_i < 0 \\ l_{B_i} &= x_i + \triangle x_i \quad if \quad \triangle x_i > 0 \end{aligned} \right\} . \tag{6}$$

4. Example

The described algorithm is applied successfully to the system described in section 2. The first optimization run starting from the given initial position displayed in Fig. 5a fails because a singular position appears within the parameter bounds of h_1 (Fig. 5b). The correction procedure (Eq. 4 - 6) sets the lower bound automatically to $\hat{h}_1$. As a result further optimization runs are free of singularities. The optimum configuration is shown in Fig. 5c, where the maximum values of joint forces are reduced to 50% of their initial values.

5. Conclusions

The presented design problem is part of the development process of a new type of heavy manipulator for use in constrained environment. The manipulator consists of modular components, called ball-joint-units, which combine three degrees of freedom to reduce the size of the manipulator without restricting its dexterity. Each ball-joint-unit is driven by hydraulic pistons. A kinetostatic model of the ball-joint-unit was generated using the object-oriented tool-box MⓄBILE. The model was used for minimization

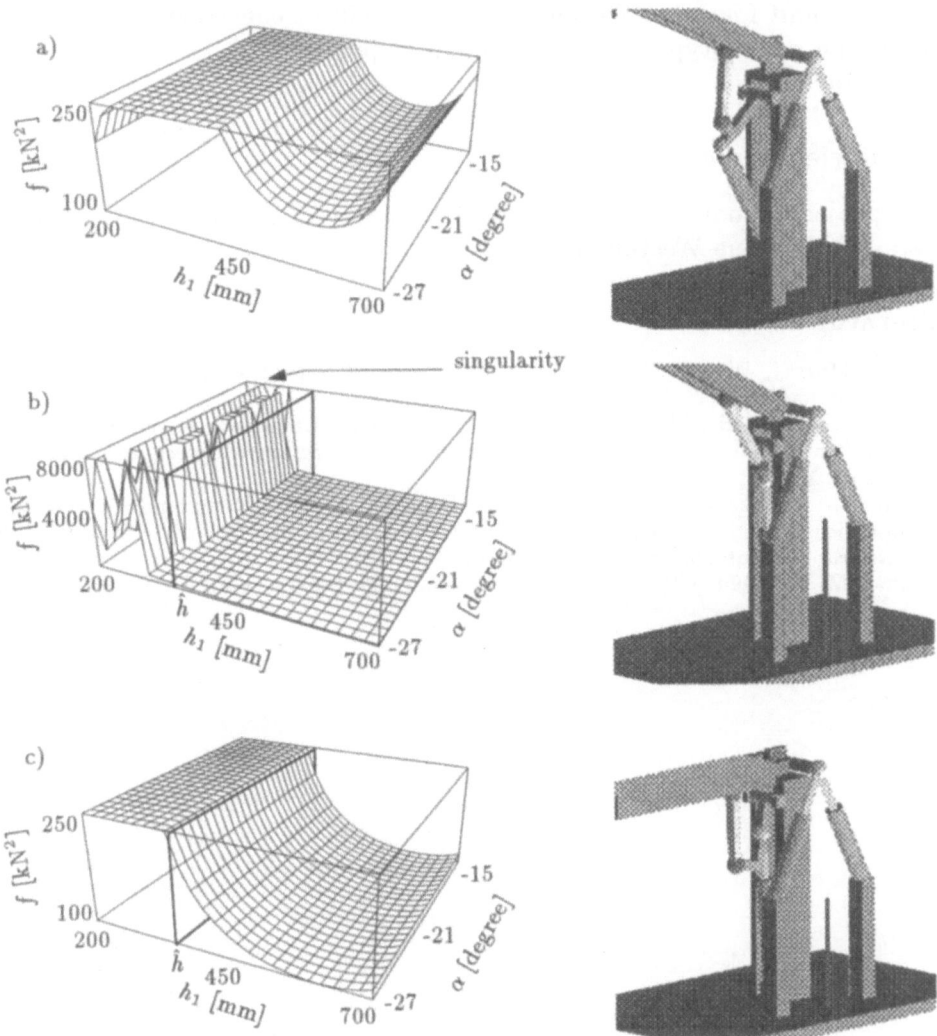

Figure 5. Example of an optimization run: a) start, b) singular, c) optimal configuration

of crucial forces within the system by using an SQP-Algorithm. The definition of an objective function and a method to avoid kinematic singularities were described.

Different types of ball-joint-units are currently under investigation at the IMECH GmbH. In addition to kinematic and static computations dynamical models, including the hydraulic components, are generated. Therefore models for the hydraulic componentes are generated for simulations.

A testing unit for a complete ball-joint-unit will be constructed in the near future and a prototype of the complete manipulator will be built before the end of 1996.

Acknowlegements

This project is mainly financed by the Ministery of Economy of the federal state North-Rhine-Westphalia.

Literatur

Gill, P., Murray, W. and Wright, M. (1981) *Practical Optimization*, Acacdemic Press, London and New York.

Kecskeméthy, A. and Hiller, M. (1993) An object-oriented approach for an effective formulation of multibody dynamics, *Computer methods in applied mechanics and engineering* **115**, pp. 287-314

Krupp, T. , Kecskeméthy, A. and Hiller, M. (1995) Modellbildung und Optimierung räumlich anstellbarer Kugelgelenke für den Einsatz in der Robotik, *Zeitschrift für angewandte Mathematik und Mechanik* **75**, pp. 127–128

Wanner, M. C. (1990) HfH Hochflexible Handhabungssysteme, KFK-PFT Bericht 153

ROBUST INTEGRATED DESIGN OF CONTROLLED MULTIBODY SYSTEMS

P.K.KIRIAZOV

Institute of Mechanics
Bulgarian Academy of Sciences
"Acad. G. Bonchev" Str., bl.4, BG-1113 Sofia, Bulgaria

E-mail: kiriazov@bgearn.bitnet

Abstract: A simple approach for robust integrated design of MBS with decentralized control structure is proposed. The main design objective is to find optimal trade-off between the given bounds of disturbances, the required system output accuracy, and the control force limits. The proposed criterion makes it possible a decomposition of the overall design problem into a hierarchical sequence of design problems for the subsystems: mechanics, actuators, sensors, and controls.

1. Introduction

The paper is, in general, addressed to integrated control/structure design problems of various-type multibody systems (MBS) like vehicles, robots, mechanical systems with active vibration control. They can be equipped with various-type actuators: electric motors, hydraulic or pneumatic cylinders, as well as jet, inertial, piezoelectric, or electromagnetic actuators. MBS are assigned to perform shape or position/orientation change, and, due to elasticity, they may undergo undesirable vibration motion.

Along with the basic design requirement for strength/load capacity additional design criteria for controlled MBS are needed to meet the continuously increasing demands for improved precision, higher speed, and reduced energy consumption. The fundamental property of such systems that reflects these performance demands is the controllability studied in the system theory for more than 30 years. This property is firstly discussed in the paper of Kalman *et al.* [1] where the so-called standard controllability matrix is introduced to give answer to the question of whether or not a dynamical system is completely controllable.

D. Bestle and W. Schielen (eds.), IUTAM Symposium on Optimization of Mechanical Systems, 155–162.
© 1996 *Kluwer Academic Publishers.*

Several control problems for satellites [2-5] motivate the need of further development of the concept of controllability towards optimization of this property. In [6], a set of possible scalar measures quantifying the complete controllability of linear dynamical systems are axiomatically defined and physically interpreted. A recent application of these concepts has been done in [7] where the degree of controllability is used for the actuator placement problem of a shallow spherical shell.

Some attempts to study the complete controllability of a class of nonlinear dynamic systems under minimum time/energy requirements are given in [8,9]. Employing an appropriate set of simple control functions, a weak condition is shown to be sufficient for the existence of such a property.

Along with the investigations on the complete controllability, there is extensive research [10-12] on feedback control design of multi-input multi-output (MIMO) systems. For complex MIMO problems where reliability and fast response are required, decentralized control structures appear simpler, cheaper, and easier to implement than alternative structures [12-14]. A common feature of the design methods for linear MIMO systems with decentralized control is that stability in the face of parameter uncertainties can be ensured if the feedback transfer matrix is generalized diagonal dominant. With this condition fulfilled, an interaction index (e.g., the Perron root) can be defined using the non-negative matrix theory and Niquist-like control design methods can be applied.

MBS are highly nonlinear and difficult to model, identify, and control dynamical systems due to dynamic couplings, friction, back-lash, elasticity, actuator limits, and load change. For the quality of their feedback stabilization, it is very important to have estimated parameters like mass and moments of inertia of the indivudual bodies, mass centers' locations and all other geometrical data, actuator/ sensor placements, and control gain coefficients. All of these parameters enter the feedback matrix that transfers the control inputs into accelerations of the bodies. The control input matrix plays the central role in the integrated design optimization of MBS as shown in [15-17] where classical control design methods are applied.

A simple deterministic approach for worst-case identification of the coefficients of MBS control input matrices is proposed in [18]. Thus for robust control design purposes, we can have the identi-fication errors estimated with given bounds of model uncertainties.

Recently [19-22], an explicit condition on the control input matrix of MBS has been proven to be necessary and sufficient for a decentralized control system to be robust to arbitrary, but otherwise bounded, disturbances. And what is more, this condition gives optimal trade-off relations between the bounds of disturbances, the system output accuracy, and the control force limits. The proposed criterion makes it possible a decomposition of the overall design problem into a hierarchical sequence of design problems for the mechanical, actuator, sensor, and control subsystems. The basic structural design requirement for strength/load capacity is included as a constraint in the optimization scheme.

2. General Design/Performance Requirements of Controlled MBS

A controlled MBS presents a functionally directed assembly of four mutually influencing MIMO subsystems: control, actuator, mechanical, and sensor subsystems. The conflicting demands for better accuracy, higher speed and reduced energy-loss of the compound system and the dynamic complexities with it impose stringent requirements to the design of its components. The subsystems should have, in general, the following features:

o the control subsystem has as small as possible sampling frequency; the feedforward control laws are time/energy optimized under control and state constraints [8,9]; and the feedback controllers are optimally robust in the face of bounded disturbances.

o the actuators are as light and powerful as possible; their placements are most appropriate for both the mechanical and control subsystems;

o the mechanical subsystem is as light and inflexible as possible; it may be equipped with springs and dampers in order to reduce undesirable vibration motion [23]; the inertia couplings between the joints are as weak as possible.

o the sensors give the required system output information with maximum accuracy and their locations are most appropriate as regards the mechanical, actuator, and control subsystems.

The requirements stated above are rather contradictory and we should search for optimal trade-offs between them. To do that, we need some quantitative relations between the design parameters and an overall design criterion which optimization leads to the best possible dynamic performance of the controlled MBS. Further, as the design of the subsystems of controlled MBS is usually performed by different design teams, our main guideline in setting and solving the overall design problem will be such that this problem can be decomposed into a series of design problems for the sub-systems.

Before setting the problem of robust integrated design of a controlled MBS, we have to define the structure of the input-output relations between its MIMO subsystems. That depends on the type of the control subsystem: centralized or decentralized. The decentralization is in the sense that, during the motion, each controller refers only to the corresponding measured output. We should prefer this manner of control for its main advantages to the centralized one: fast response, higher reliability, and minimum computation effort.

Thus with decentralized controllers, the sampling frequency can be the highest one which means fastest response for the controlled MBS. To meet the other dynamic performance requirements for accuracy and control effort, in the face of internal and external disturbances, we have to deal with the following design problem: *With known bounds of disturbances and desired system output accuracies, design a MBS with decentralized controllers that need minimum actuator forces to robustly stabilize the motion.*

3. Mathematical Background

The mechanical system, though actually flexible, can be approximated by a composition of rigid bodies connected by joints, springs, dampers, and actuator forces [24]. For simplicity, we consider MBS with collocated actuators/sensors that can be modelled by the following system of differential equations

$$\ddot{q}_i(t) = \sum_{i=1}^{n} A_{ij}(q,t) f_j(t) + a_i(q,\dot{q},t) + d_i(t) , \quad i=1,...,n \qquad (1)$$

where A is the inverse inertia matrix, f and q are the input force and controlled output, respectively. We assume that a can be feedforward compensated to some extent, and, for the purpose of feedback stabilization, the vector d will further stand for the uncompensated terms as well as the model imprecision, measurement and environment noises. Therefore, for the feedback design considerations, we shall use the following error model

$$\ddot{e}_i(t) = -\sum_{i=1}^{n} A_{ij}(q,t) f_j(t) + d_i(t) , \quad e_i = q_i^{ref} - q_i, \quad i=1,...,n \qquad (2)$$

As a measure of the tracking precision, we take the absolute value of $s_i = e_i + \lambda \dot{e}_i$, where $\lambda \geq 0$.

Definition: A system of decentralized feedback controllers $f_i = g_i s_i$ is robust if it gets the local subsystem state $(q_i, \dot{q}_i)$ at each joint i to track the desired state $(q_i^{ref}, \dot{q}_i^{ref})$ with maximum allowable values δ_i of errors $|s_i|$, in the presence of random disturbances d_i with known upper bounds d_i^+, $i=1,...,n$.

Introduce matrix B: $B_{ij} = -|A_{ij}|$ if $i \neq j$, and $B_{ii} = A_{ii}$. Then

Theorem.1 [12,19,20]: Matrix A is generalized diagonal dominant GDD if and only if B is positive definite.

Theorem.2 [21,22]: A necessary and sufficient condition for a dynamic system (1) to be robustly controlled is that the comparison matrix B be positive definite.

When B is positive definite, there exist positive g_i satisfying

$$\sum_{j=1}^{n} B_{ij} g_j \delta_j = d_i^+ + \mu_i, \quad \mu_i > 0, \quad i=1,\ldots,n \tag{3}$$

These relations present optimal trade-off between the tracking precision, the range of disturbances, and the amount of feedback control effort. With given δ and d^+, Eqs. (3) give the minimum values of the control gains that are necessary for the robust stability of MBS with decentralized control structure. The sufficiency of these conditions fas been verified using sliding mode controllers and applying Lyapunov theory.

The system of linear equations (3) shows that the greater the determinant of matrix B the less control forces are required to overcome the disturbances. In other words, *detB* quantifies the capability of MBS to be robustly controlled in decentralized manner, and for that purpose it can be used as a design index for those subsystems whose parameters enter the inverse inertia matrix. Therefore, we have the opportunity to decompose the design problem of the entire system into design problems for its subsystems.

4. Integrated Design Optimization Scheme

In practice, the order of designing the subsystems of a controlled MBS is the following: (1) mechanics, (2) actuators, (3) sensors, and (4) controls. This order will correspond to the hierarchy in a multi-level optimization procedure in which a series of design problems for these subsystems are to be solved.

Along with our desire to improve the decentralized controllability of MBS, we have to observe, in general, several structural design requirements, e.g., strength/load capacity, eigen-structure assignment, and so on. Due to the presence of several criteria the design issue can be considered as a Pareto-optimization problem and the fast convergent SQP algorithm can be applied [25] to solving the resulting nonlinear programming problems.

In this study, the main design objective is to provide MBS with maximum degree of robust decentralized controllability, and the other design requirements will be considered as constraints. For simplicity, only a structural constraint on the mechanical strength will be taken into account. The mathematical considerations given in the preceding section, enable us to propose the following multi-level optimization procedure for robust integrated design of controlled MBS:

1. Mechanics
-design parameters: all inertial/geometrical data of the bodies;
-design constraints: strength and GDD conditions;
-design objective: maximize $det(B_{ij})$;

2. Actuators
-design parameters: actuator masses and positions;
-design constraints: strength and GDD conditions;
-design objective: maximize $det(B_{ij})$;

3. Sensors
-design parameters: sensor locations;
-design constraints: GDD condition;
-design objective: maximize $det(B_{ij})$;

4. Controls
-design parameters: control gains g_i;

-design constraints: robust stability conditions (3);

-design objective: find $g_i^* = \max_q (\min g_i)$, $i=1,...,n$;

There is a set of conditions to be checked in order to stop this overall, iterative in general, design procedure. Denote by f_i^+ the actuators' maximum output forces, the feedack control forces needed to robustly stabilize the motion in the whole working space must satisfy the following conditions

$$g_i^* \delta_i \leq f_i , \quad i=1,...,n \tag{4}$$

If this is not the case we have to go back to level 2. choosing larger sizes for the actuators. Conditions (4) become more restrictive when there are feedforward control forces and their maximum values should be added to the left part of these inequalities.

5. Conclusion

A simple approach for robust integrated design of MBS with decentralized control structure has been proposed. The main design objective is to find optimal trade-off between the given bounds of disturbances, the required system output accuracy, and the control force limits.

The definition of the design criterion is based on the GDD of the control input matrix which condition is necessary and sufficient for robust decentralized controllability of MBS. This criterion makes it possible a decomposition of the overall design problem into a hierarchical sequence of design problems for the sub-systems.

The criterion for robust integrated design of controlled MBS does not conflict with the basic design requirement for strength/load capacity. The viability of the proposed design concepts has been verified on several examples of MBS with two, three, and six degrees of freedom.

Acknowledgement: The financial support from the National Science Foundation of Bulgaria (Project MM-426/94) as well as the German Research Foundation (DFG) is gratefully acknowledged.

References

1. Kalman,R.E., Ho,Y.C., and Narenda,K.S.: Controllability of linear dynamical systems, *Contributions to Differential Equations*, 1 (1961), 182-213.
2. Sagirow,P.: *Satelliten-Dynamik*, Bibliographisches Institut, Band 719/719a, Manheim, 1970.
3. Schiehlen,W. and Kolbe,O.: Gravitationsstabilisierung von Satelliten auf elliptischen Bahnen, *Ing.-Arch.*, 38 (1969), 389-399.
4. Weber,H.I. and Schiehlen,W.: Attitude controllability of a satellite with flywheels, *Journal Spacecr. & Rockets*, 7 (1970), 501-2.
5. Mueller,P.C.: Schnelligkeitsoptimales Ausrichten von Traegheitsplattformen. *Ing.-Arch.*, 40 (1971), 248-265.
6. Mueller,P.C. and Weber,H.I.: Analysis and optimization of certain qualities of controllability and observability for linear dynamic systems, *Automatica*, 8 (1972), 237-246.
7. Xing Guangqian and Bainum, Peter M.: Actuator placement using degree of controllability for discrete-time systems, *ASME Journal of Dynamic Systems, Measurement, and Control,* 114 (1992), 508-16.
8. Marinov,P. and Kiriazov,P.: Point-to-point motion of robotic manipulators: dynamics, control synthesis and optimization, *IFAC Symposia Series, Robot Control 1991, Eds. I.Troch & K.Desoyer*, 1992, 149-152.
9. Kiriazov,P.: Controllability of a class of dynamic systems, *ZAMM*, 75 (1995) SI, 85-86.
10. Singh,M.G.: *Decentralized Control*, North-Holland, New York, 1981.
11. Siljak,D.D.: *Decentralized Control of Complex Systems*, Academic Press, San Diego, CA, 1991.
12. Lunze,J.: *Feedback Control of Large-Scale Systems*, Prentice Hall, U.K., 1992.

13. Nwokah,O.D.I. and Yau,C.-H.: Quantitative feedback design of decentralized control systems, *ASME Journal of Dynamic Systems, Measurement and Control*, **115** (1993), 452-466.

14. Perez,R.A. and Lou,K.-N.: Decentralised multivariable control and stability of gas turbine engine, *IEE-Proc.-Control Theory Appl.*, **141** (1994), 357-366.

15. Lim,K.B., and Junkins, J.L.: Robust optimization of structural and controller parameters, *J. Guidance, Control and Dynamics*, **12** (1989), 89-96.

16. Park,J.H., and Asada,H.: Integrated structure/control design of a two-link non-rigid robot arm for high speed positioning, *Proc. IEEE Conf. on Robotics and Automation, Nice, France*, 1992.

17. Khot,N.S. and Heise,S.A.: Consideration of plant uncertainties in the optimum structural-control design, *AIAA Journal*, **32** (1994), 610-615.

18. Kiriazov,P.: On the identification of the inertia coupling co-efficients of robotic manipulators, *Identification of Nonlinear Mechanical Systems from Dynamic Tests, Eds. L. Jezequel & C. Lamarque, Balkema, Rotterdam*, 1992, 195-196.

19. Kiriazov,P. and Marinov,P.: Independent joint controllability of manipulator system, *Advanced Robotics: 1989, Ed. K.J.Waldron*, 1990, 604-611.

20. Kiriazov,P.: A decentralized controllability measure for robotic manipulators, *Proc. IEEE Conf. on Robotics and Automation, Nice, France*, 1992, 2141-2145.

21. Kiriazov,P.: A necessary and sufficient condition for robust decentralized controllability of robot manipulators, *Proc. American Control Conf. , MD, Baltimore*, 1994, ASME Paper 159.

22. Kiriazov,P.: On the optimal design of active vibration control systems, *Inter. Symp. MV2: Active Control in Mechanical Engineering, Ed. L.Jezequel, Hermes, Paris*, 1995, 295-302.

23. Guergoese,M. and Mueller,P.C., Optimal positioning of dampers in multi-body systems, *J. Sound and Vibration*, **158** (1992), 517-525.

24. Sciehlen,W. (ed.): *Multibody Systems Handbook*. Berlin: Springer, 1990.

25. Bestle,D. and Eberhard,P.: Analyzing and optimizing multibody systems, *Mech. Struct. and Mach.*, **20** (1992), 67-92.

OPTIMAL ROBOT PATH PLANING

DIETER KRAFT
Fachhochschule München
Dachauerstr. 98b, D-80335 München
email: kraft@maschinenbau.fh-muenchen.d400.de

1. Introduction

To balance ecology and economy it is necessary to raise the productivity of automation processes, e.g. assembling or welding by robotic manipulators. Industrial robots are now widely used in various fields of application. But the conventional control of their movements is by no means optimal [2]. To raise the productivity of their operation optimal control calculations have been proposed recently [9].

In this contribution a combined modeling and optimization approach is described which is suitable for application in small and medium sized industrial companies where large and costly special purpose modeling software is not available. Rather we propose commonly used universal packages like Axiom, Maple, Mathematica or Dymola [7] for modeling purposes together with AutoCAD as design tool and the optimal control calculator TOMP, which is in the public domain [4].

To be specific, a six-degree-of-freedom industrial robot of the Puma or Manutec type is modeled symbolically and simulated numerically as a worksheet in Maple. But Maple, as an interpreting rather than compiling software, is too slow to be used as an optimizing tool. Therefore, the symbolic model is transfered to the optimal control package TOMP. The first computational results show an encouraging gain of productivity compared with conventional controls.

This approach is by no means limited to robotic manipulators but can be applied to more general and complex mechatronic multibody systems like mechanisms, motor cars, railway wheelsets, etc.. Recent overviews are given by Schiehlen [8] from the side of the modeling process and by Bulirsch [1] from the aspects of optimal control calculations.

D. Bestle and W. Schielen (eds.), IUTAM Symposium on Optimization of Mechanical Systems, 163–170.
© *1996 Kluwer Academic Publishers.*

2. The Physical Model

It is well known that the algorithm based on the Newton-Euler equations is one of the most efficient for formulating the differential equations (de) of mechanical multibody systems [2]:

Algorithm 1: Newton-Euler-Algorithm

1. Calculate velocities and accelerations from base to tip:
   repeat steps (a) – (f) for $i = 1, \ldots, n$ joints

   (a) angular velocity

       i. rotational joint

   $$\,_i^i\Omega = \,_{i-1}^i R\,_{i-1}^{i-1}\Omega + \dot{\theta}_i k_i,$$

       ii. prismatic joint

   $$\,_i^i\Omega = \,_{i-1}^i R\,_{i-1}^{i-1}\Omega,$$

   (b) angular acceleration

       i. rotational joint

   $$\,_i^i\dot{\Omega} = \,_{i-1}^i R(\,_{i-1}^{i-1}\dot{\Omega} + \,_{i-1}^{i-1}\Omega \times \dot{\theta}_i k_i) + \ddot{\theta}_i k_i,$$

       ii. prismatic joint

   $$\,_i^i\dot{\Omega} = \,_{i-1}^i R\,_{i-1}^{i-1}\dot{\Omega},$$

   (c) linear acceleration

       i. rotational joint

   $$\,_i^i\dot{v} = \,_{i-1}^i R(\,_{i-1}^{i-1}\dot{v} + \,_{i-1}^{i-1}\dot{\Omega} \times \,_i^{i-1}p + \,_{i-1}^{i-1}\Omega \times (\,_{i-1}^{i-1}\Omega \times \,_i^{i-1}p)),$$

       ii. prismatic joint

   $$\,_i^i\dot{v} = \,_{i-1}^i R(\,_{i-1}^{i-1}\dot{v} + \,_{i-1}^{i-1}\dot{\Omega} \times \,_i^{i-1}p + \,_{i-1}^{i-1}\Omega \times (\,_{i-1}^{i-1}\Omega \times \,_i^{i-1}p))$$

   $$+ 2\,_i^i\Omega \times \dot{d}_i k_i + \ddot{d}_i k_i,$$

   (d) linear acceleration of center of gravity

   $$\,_i^i\dot{v}_m = \,_i^i\dot{v} + \,_i^i\dot{\Omega} \times \,_i^i p_m + \,_i^i\Omega \times (\,_i^i\Omega \times \,_i^i p_m),$$

   (e) force at center of gravity

   $$\,_i^i F = m_i \,_i^i\dot{v}_m,$$

   (f) moment at center of gravity

   $$\,_i^i N = I_i \,_i^i\dot{\Omega} + \,_i^i\Omega \times I_i \,_i^i\Omega.$$

2. Calculate forces and moments from tip to base: reapeat steps (a) – (c) for $i = n, \ldots, 1$ joints

   (a) forces

   $$_{i-1}^{i-1}\mathbf{f} = {}_{i}^{i-1}\mathbf{R}\,_{i}^{i}\mathbf{f} + {}_{i-1}^{i-1}\mathbf{F},$$

   (b) moments

   $$_{i-1}^{i-1}\mathbf{n} = {}_{i}^{i-1}\mathbf{R}\,_{i}^{i}\mathbf{n} + {}_{i}^{i-1}\mathbf{p} \times {}_{i}^{i-1}\mathbf{R}\,_{i}^{i}\mathbf{f} + {}_{i-1}^{i-1}\mathbf{P_m} \times {}_{i-1}^{i-1}\mathbf{F} + {}_{i-1}^{i-1}\mathbf{N},$$

   (c) controlling forces and moments

   i. rotational joint

   $$u_{i-1} = {}_{i-1}^{i-1}\mathbf{n}^{\mathbf{T}}\mathbf{k}_{i-1},$$

   ii. prismatic joint

   $$u_{i-1} = {}_{i-1}^{i-1}\mathbf{f}^{\mathbf{T}}\mathbf{k}_{i-1}.$$

3. Set initial values:

   (a) base angles
   if base is not rotating

   $$_{0}^{0}\mathbf{\Omega} = {}_{0}^{0}\dot{\mathbf{\Omega}} = 0,$$

   else

   $$_{0}^{0}\mathbf{\Omega} = \mathbf{\Omega_0} \qquad _{0}^{0}\dot{\mathbf{\Omega}} = \dot{\mathbf{\Omega}}_0,$$

   (b) influence of gravitational acceleration g:

   $$_{0}^{0}\dot{\mathbf{v}} = (0,\ 0,\ g)^{\mathbf{T}},$$

   (c) forces and moments acting on the end effector
   if the robot hand is moving free in space
   then

   $$_{n}^{n}\mathbf{f} = {}_{n}^{n}\mathbf{n} = 0,$$

   else

   $$_{n}^{n}\mathbf{f} = \mathbf{f_n} \qquad _{n}^{n}\mathbf{n} = \mathbf{n_n}.$$

   □

In algorithm 1, $\mathbf{R}$ are transformation matrices for neighboring link coordinate frames, θ and d are free Denavit-Hartenberg parameters for rotational and prismatic joints, respectively, $\mathbf{p}$ are coordinate system origins of neighboring frames, $\mathbf{f}$ and $\mathbf{n}$ are internal forces and moments, and $\mathbf{k}$ is the unit vector of a link coordinate system pointing along the joint axis. Step (1) and (2) of algorithm 1 are implemented as MAPLE functions **outward** and **inward**, respectively, and the initial values of step (3) are used to start the iterations. The resulting equations are combined as matrix-vector de's of the form

$$M(q)\ddot{q} + V(q, \dot{q}) + G(q) = u,$$

with the $n \times n$ positive-definite mass matrix M, the $n \times 1$ vector V of the centripedal and Coriolis terms and the $n \times 1$ vector G of the gravity components. This set of n second-order de's is transformed to first-order de's by introducing $2n$ state variables $z = (q^T, \dot{q}^T)^T$. The robot is moved from given initial conditions

$$z(t_0) = z_0 \quad \text{to given final conditions} \quad z(t_1) = z_1,$$

where this point-to-point movement is constrained by state constraints

$$z_{min} \leq z(t) \leq z_{max}, \quad \forall t \in [t_0, t_1],$$

and control constraints

$$u_{min} \leq u(t) \leq u_{max}, \quad \forall t \in [t_0, t_1],$$

and it is to be performed in such a way as to minimize a certain optimality criterion $J(u(t), t_1)$, e.g.

$$J = t_1,$$

that means time-minimal motion.

3. The Mathematical Model

The mathematical representation of the physical model of the previous section is formulated as the following optimal control problem (OCP):

$$\textbf{(OCP):} \quad \min_{\substack{u \in U \\ \pi \in R^{q_\pi}}} \int_a^b \phi_0(t, z(t), u(t), \pi) dt + \psi_0(a, z(a), b, z(b)),$$

subject to differential constraints

$$\dot{z}_i(t) - \phi_i(t, z(t), u(t), \pi) = 0, \quad i = 1, \ldots, p, \tag{1}$$

boundary conditions, with $a < \tau < b$,

$$\psi_i(a, z(a), \tau, z(\tau), b, z(b)) = 0, \quad i = 1, \ldots, r, \tag{2}$$

and (algebraic) state or control constraints

$$\xi_i(t, z(t), u(t), \pi) \geq 0, \quad i = 1, \ldots, s. \tag{3}$$

In problem (OCP) functions are considered depending on one independent variable $t \in [a, b] = [t_0, t_1] =: I$. The state $z \in R^p$ is assumed to be absolutely continuous, the control $u \in U \subset R^q$ is assumed to be bounded and measurable, which practically means piecewise continuous. The initial and final time a and b may be fixed or free. The trajectory $z(t)$ is controlled by a vector of time functions $u_i(t)$, $i = 1, \ldots, q$, and a finite dimensional vector of design parameters π_i, $i = 1, \ldots, q_\pi$.

In the software package TOMP[1] problem (OCP) is converted into a nonlinear programming problem (NLP):

$$\textbf{(NLP):} \quad \min_{x \in R^n} f(x)$$

subject to general equality and inequality constraints

$$g_j(x) = 0, \quad j = 1, \ldots, m_e,$$

$$g_j(x) \geq 0, \quad j = m_e + 1, \ldots, m,$$

and to lower and upper bounds on the variables

$$l_i \leq x_i \leq u_i, \quad i = 1, \ldots, n.$$

3.1. CONTROL PARAMETERIZATION

To use an (NLP) solver for problem (OCP), the infinite dimensional control functions $u(t)$ have to be represented by a finite set of control parameters x_u. This is accomplished as follows:

1. a number of q time grids Δ_i are defined

$$\Delta_i := \{a = t_{i_1} < t_{i_2} < \cdots < t_{i_{n_i-1}} < t_{i_{n_i}} = b\}, \quad i = 1, \ldots, q,$$

as partitions of the interval I, one for each control function $u_i(t)$,
2. with each breakpoint or knot t_{i_j} a control parameter $x_{u_{i_j}}$ is associated

$$x_{u_{i_j}} = u_i(t_{i_j}), \quad j = 1, \ldots, n_i, \quad i = 1, \ldots, q,$$

which represent the control functions $u_i(t)$ at the breakpoints,
3. within each subinterval

$$I_{i_j} = [t_{i_j}, t_{i_{j+1}}], \quad j = 1, \ldots, n_{i-1}, \quad i = 1, \ldots, q,$$

of the partitions Δ_i, each control function $u_i(t)$ is approximated by an interpolating piecewise polynomial (pp) function $v_{\Delta_i}(t)$ of a certain order

$$u_i(t) \approx v_{\Delta_i}(t), \quad i = 1, \ldots, q. \tag{4}$$

[1]To be acquired as algorithm 733.Z by anonymous ftp from research.att.com in /netlib/toms.

Thus the vector $u(t)$ is completely and uniquely defined by the finite vector

$$x_u := \{x_{1_1}, x_{1_2}, \ldots, x_{q_{n_q}-1}, x_{q_{n_q}}\}.$$

The entire vector of controlling parameters x is composed of the control parameters x_u and the design parameters π

$$x = (x_u^T, \pi^T)^T = \{x_{1_1}, x_{1_2}, \ldots, x_{q_{n_q}-1}, x_{q_{n_q}}\} \cup \{\pi_1, \pi_2, \ldots, \pi_{q_\pi}\}.$$

The total number of controlling parameters sums up to $n = \dim(x) = n_1 + n_2 + \cdots + n_q + q_\pi$. It should be noted that a and b are elements of the set of controlling parameters if they are free parameters.

3.2. PROBLEM APPROXIMATION

With $u(t)$ approximated by (4) which in turn is represented by a finite number of parameters x_u, the differential constraints (1)

$$\dot{\zeta}_i(t) - \phi_i(t, \zeta(t), v_\Delta(t), \pi) = 0, \quad i = 1, \ldots, p, \tag{5}$$

are solved with initial values

$$\psi_{a_i}(a, \zeta(a)) = 0, \quad i = 1, \ldots, r_a, \tag{6}$$

where $\zeta(t)$ is an approximation to $z(t)$, and $\zeta(a) = z(a)$. For non-stiff or mildly stiff problems one-step discretization methods of the Runge-Kutta-Fehlberg class are preferred as solvers for (5,6).

For given x the trajectory ζ is uniquely defined, and it only depends on x. Thus the cost function

$$\int_a^b \phi_0(t, \zeta(t), v_\Delta(t), \pi)dt + \psi_0(a, \zeta(a), b, \zeta(b)) =: f(x), \tag{7}$$

and the boundary conditions

$$\psi_{b_i}(b, \zeta(b)) =: g_i(x) = 0, \quad i = 1, \ldots, r_b, \tag{8}$$

$$\psi_{\tau_i}(\tau, \zeta(\tau)) =: g_i(x) = 0, \quad i = r_b + 1, \ldots, r_b + r_\tau, \tag{9}$$

can be evaluated with $m_e = r_\tau + r_b$. To calculate the state constraints

$$\xi_i(t, \zeta(t), v_\Delta(t), \pi) \geq 0, \quad i = 1, \ldots, s, \tag{10}$$

a pointwise approach is applied. Equation (10) is evaluated on a communication grid Δ_c

$$\Delta_c := \{a = t_{c_1} < t_{c_2} < \ldots < t_{c_{t-1}} < t_{c_t} = b\}$$

pointwise at certain communication knots t_{c_i}

$$\xi_i(t_{c_j}, \zeta(t_{c_j}), v_\Delta(t_{c_j}), \pi) =: g_{m_e+k}(x) \geq 0,$$

$$i = 1, \ldots, s, \quad j = 1, \ldots, t, \quad k = 1, \ldots, s \times t, \tag{11}$$

and the number of inequality constraints $m_i = m - m_e$ is $m_i = s \times t$. Now the equations (7,8,9,11) approximate problem (OCP) as problem (NLP).

4. Example and Results

As a realistic example the industrial robot Manutec r3 [6] is chosen for modeling and optimization. We confine the model to 3 degrees of freedom which are essentially responsible for the robot position, while the movement of the end effector affects its orientation and has no essential influence on optimal trajectories.

All geometric and mass parameters as well as the admissable states and controls are taken from [6] and can be found there. As reference we have chosen a trajectory from [3] which represents a wide overhead movement:

$$\mathbf{q_0} = (13.1°, 17.8°, 100.5°)^\mathbf{T} \quad \text{and} \quad \mathbf{q_1} = (-8.4°, 41.9°, -104.8°)^\mathbf{T}, \tag{12}$$

and a 'quarter turn' from [9]:

$$\mathbf{q_0} = (0°, -85.94°, 0°)^\mathbf{T} \quad \text{and} \quad \mathbf{q_1} = (57.3°, -111.73°, 57.3°)^\mathbf{T}. \tag{13}$$

In [3] suboptimal results are compared with industrial controls (RCM). We summarize the results in column 2–4 of Table 1 for condition (12)

TABLE 1. Times for different robot controls

configuration	RCM[3]	Kempkens[3]	TOMP	vStryk[9]	TOMP
constrained control	—	1.125	0.949	0.445	0.273
constrained state	1.664	1.280	1.009	0.495	0.445

and in column 5–6 for condition (13). We observe a considerable gain in productivity by optimization. The trajectories calculated by TOMP are shown in Figure 1 for the second example and the state-constrained case. The lower right graphs in addition give the angular velocities for the control-constrained case.

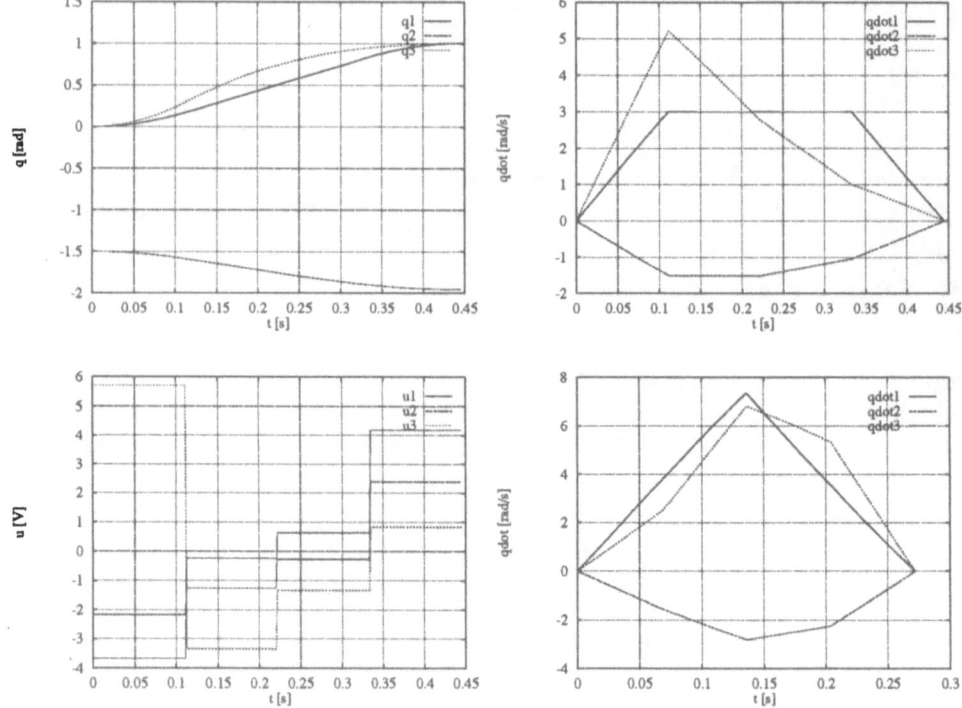

Figure 1.   Optimal trajectories for the Manutec r3 robot

References

1. Bulirsch, R., D. Kraft (eds.): *Computational Optimal Control*. Proceedings of 9th IFAC Workshop on 'Control Applications of Optimization'. Birkhäuser, Basel, 1994.
2. Craig, J.J.: *Robotics – Mechanics and Control*. Addison-Wesley, Reading, 1989.
3. Kempkens, K.: Optimierung der Verfahrparameter von Industrieroboter-Steuerungen. Robotersysteme Vol. 6, pp. 145–529, 1990.
4. Kraft, D: TOMP — Fortran Modules for Optimal Control Calculations. ACM Trans. Math. Softw., Vol. 20, No. 3, pp. 262–281, 1994.
5. Kraft, D: Modeling and Simulation of Robots in MAPLE. MapleTech, Vol. 1, No. 2, pp. 39–49, 1994.
6. Otter, M., S. Türk: The DVFLR Models 1 and 2 of the Manutec r3 Robot. DFVLR-Mitt. 88-13, Köln, 1988.
7. Otter, M.: Objektorientierte Modellierung mechatronischer Systeme am Beispiel geregelter Roboter. Fortschritt-Berichte VDI Reihe 20 Nr. 147, VDI-Verlag, Düsseldorf, 1995.
8. Schiehlen, W. (ed.): *Advanced Multibody System Dynamics*. Simulation and Software Tools. Kluwer, Dordrecht, 1993.
9. von Stryk, O.: Numerische Lösung optimaler Steuerungsprobleme: Diskretisierung, Parameteroptimierung und Berechnung der adjungierten Variablen. Fortschritt-Berichte VDI Reihe 8 Nr. 441, VDI-Verlag, Düsseldorf, 1995.

TOPOLOGY OPTIMIZATION OF INTEGRAL RIB REINFORCEMENT OF PLATE AND SHELL STRUCTURES WITH WEIGHTED-SUM AND MAX-MIN OBJECTIVES

LARS A. KROG AND NIELS OLHOFF
Institute of Mechanical Engineering
Aalborg University
DK-9220 Aalborg East, Denmark

ABSTRACT.   This paper deals with the problem of determining the optimum integral rib reinforcement of statically loaded or freely vibrating plate and shell structures. The problem is treated as one of topology optimization where the rib reinforcement is considered as an orthotropic microstructure with continuously varying orientation and concentration of a two-way system of reinforcing ribs attached symmetrically with respect to the mid-surface of the base plate or shell structure. Multiobjective weighted-sum and max-min formulations are considered for problems of stiffness optimization of plates and shells subjected to several static load cases and problems of optimization of eigenfrequencies of free vibrations. Special emphasis is devoted to the occurrence of multiple eigenfrequencies in max-min formulations of vibration problems. Illustrative examples of optimum rib reinforcement are presented for the problems considered.

The present paper is based on a recent study by the authors, and the reader is referred to [1] for details.

1. Introduction

Topology optimization of continuum structures was introduced in the literature by the landmark paper [2] by Bendsoe & Kikuchi by way of compliance minimization of planar structures subjected to a single system of in-plane static loads. Recent extensions of the method include handling of multiple loads [3], bi-material structures [4], plate and shell bending problems [5], [6] and [7], eigenfrequency optimization [8], and buckling eigenvalue opti-

D. Bestle and W. Schielen (eds.), IUTAM Symposium on Optimization of Mechanical Systems, 171–179.
© 1996 Kluwer Academic Publishers.

mization problems [9]. Prior related work on layout optimization of discrete structures can be seen in [10], and the reader is referred to the excellent recent review paper [11] for an exhaustive overview of the very new and rapidly expanding field of structural topology or layout optimization.

Section 2 of this paper presents the basic concepts for the plate and shell rib reinforcement topology optimization problems considered herein, and Section 3 deals with multiobjective problems where the optimization is carried out with respect to several cases of static loads. These problems encompass minimization of a weighted sum of the compliances associated with the different load cases, and minimization of the maximum of the weighted compliances.

The principal aim of this paper, however, is to report on a recent extension [1] of the method of topology optimization to encounter max-min optimization of eigenfrequencies of vibration of continuum plate and shell structures, in which the possibility of multiple optimum eigenfrequencies is properly accounted for in the mathematical formulation and the solution procedure. This development is discussed in Section 4 where multiple optimum eigenvalue solutions to topology optimization problems will be presented for the first time.

As is well-known (see, e.g. the recent paper [12]), special attention must be devoted to the fact that multiple eigenvalues are not differentiable in the usual sense with respect to design variables. In our current study (see [1] for details), we have been able to meet this difficulty by imposing some restrictions on the design changes, and to develop a new general method of solution of max-min problems which can efficiently handle optimization of multiple eigenfrequencies. The method is particularly attractive as it only requires ordinary procedures for sensitivity analysis and mathematical programming.

The fact that rib reinforced plate and shell structures often have a very dense spectrum of eigenvalues, and that multiple eigenfrequencies of vibration or buckling loads are very often found, has been the incentive for the present work as it was obvious that this feature of rib reinforced plates and shells would be considerably accentuated by max-min optimization of the eigenvalues. The fact that possible multiplicity of optimum eigenvalues must necessarily be taken into account in the problem formulation in order to obtain correct results, was already established in [13].

2. Concepts of Topology Optimization

In the topology optimization problems for integral rib reinforcement considered in this paper, the shape of the boundary, the boundary conditions, and the form of the mid-surface is assumed to be given for a base plate or shell of given, uniform thickness of material. For the static problems, also the different sets of loads are given. Now, assuming that the ribs are to be made of the same kind of material as the base plate or shell, and that the total amount of rib material is given, the problem consists in distributing this material optimally into a system of thin integral ribs on the upper and lower surfaces of the given base plate or shell.

Fig. 1 depicts that the base plate or shell, in any small subdomain, is assumed to be equipped with an orthotropic microstructure of mutually orthogonal ribs. The model is termed a rank-2 microstructure. The dark and grey areas in Fig. 1 represent structural material and very soft material (void), respectively. The concentrations of the ribs in the two orthogonal directions are governed by the variables ρ_1 and ρ_2, while the thickness h_1 of the base plate and the height h_2 of the reinforced plate are fixed. The limiting case of $\rho_1 = \rho_2 = 0$ represents local vanishing of the ribs, and the cases of $\rho_1 = 1$ and/or $\rho_2 = 1$ represents reinforcement of the plate or shell by solid, isotropic material.

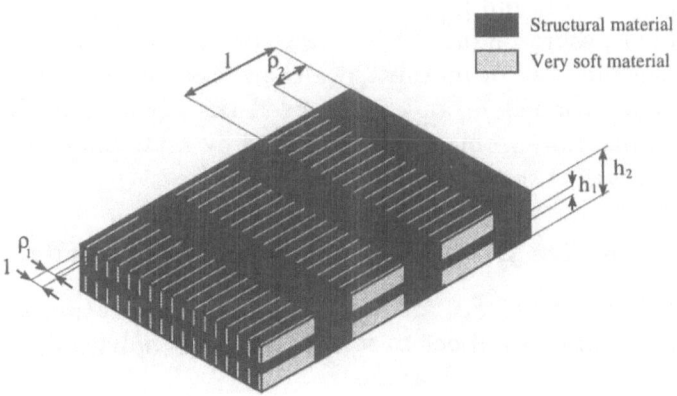

Figure 1.   Rank-2 microstructure model for rib-reinforced plate or shell.

In [1] the effective stiffness properties of the orthotropic microstructural plate or shell model shown in Fig. 1 are determined analytically by a homogenization procedure as developed by [6]. Mindlin theory is used, and the constitutive properties of the plate model are established in terms of constitutive matrices for the membrane, bending, and shear action which depend analytically on the rib concentration variables ρ_1 and ρ_2. The dependence of the matrices on rotation can be derived by standard formulas. The finite element method is adopted for analyses, in which a fixed mesh is used, and the rib reinforcement is assumed to be constant within each (Mindlin) finite element.

With this discretization we have three design variables for each finite element, namely the values of the rib concentrations ρ_1 and ρ_2, and an angle θ that governs the local orientation of the system of mutually orthogonal ribs. Hence, the topology optimization problem consists in determining the optimum values of these three design parameters for each of the finite elements in the structure.

Work concerning introduction of a more general microstructural model [7] as a basis for the current multiobjective topology optimization problems is in progress.

In [1] the optimization problems are formulated as minimization or maximization of a given (scalarized) multicriterion objective function subject to given total amount of structural material and side constraints for the design variables. The optimization is performed by iterations, in each of which a two-level scheme of redesign is used:

In the first step of redesign, the optimum orientations θ of the rib microstructures are sought for fixed, current values of the rib concentration variables ρ_1 and ρ_2. In [1] this problem is solved analytically via optimality criteria (see [14] and [3]) for weighted-sum objective problems, and by mathematical programming for min-max or max-min objective problems.

In the second step of redesign, we then seek optimum values of the rib concentration variables ρ_1 and ρ_2 for fixed rib orientations. This is a constrained optimization problem, and is solved by mathematical programming based on analytical sensitivity analysis.

3. Multiobjective Optimization under Static Loads

Topology design problems associated with maximization of the integral plate or shell stiffness subject to several mutually independent sets of static loads shall now be briefly discussed.

3.1. WEIGHTED-SUM FORMULATION

In the very few problems with multiple objectives treated up to now, see, e.g., [3] and [7], authors have formed and minimized an objective function taken to be a weighted sum of the compliances associated with each load case. Such an objective function will be differentiable in general, and the problem can be solved using ordinary sensitivity analysis and standard optimization algorithms. Also, the weighting factors may be chosen so as to obtain a pseudo-type of max-min problem for the integral structural stiffness.

3.2. EXAMPLE

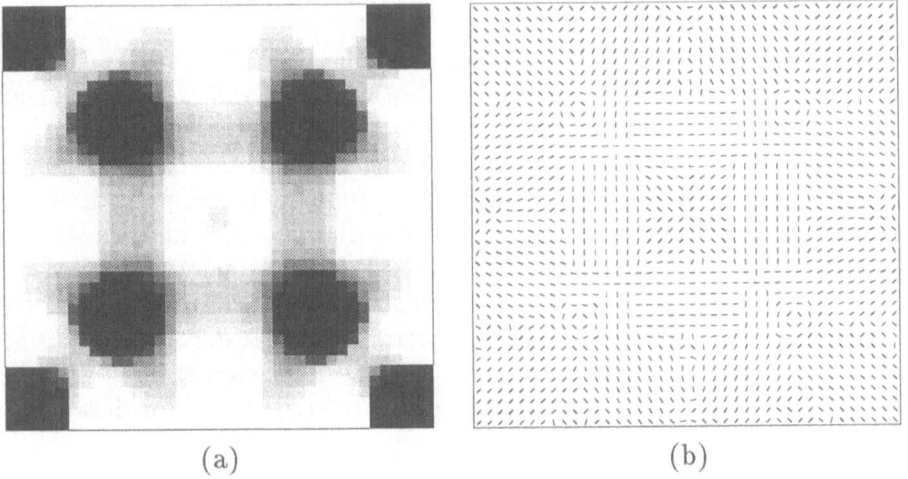

(a)   (b)

Figure 2.   (a) Rib density and (b) Rib orientations

Fig. 2 shows as an example the optimum topology result (see [1]) from
a weighted-sum formulation (with unit weighting factors) of maximizing
the integral stiffness of a square, simply supported plate subjected to four
point loads, each acting independently at a plate diagonal quarter point as
a separate load case. The plate thickness ratio is $h_2/h_1 = 5$ (see Fig. 1),
and it is prescribed that ribbed plate sections of the height h_2 may occupy
30 pct. of the plate surface.

3.3. MIN-MAX FORMULATION

The min-max formulation covers minimization of the maximum of a set of
weighted compliances associated with different sets of given loads. In [1],
this initially non-differentiable problem is cast in differentiable form via a
bound formulation [15], whereby the problem can be solved by standard
methods of linear programming.

We believe that, from the point of view of practical engineering design,
the min-max approach to the multiobjective optimization problem is more
interesting than the weighted-sum approach discussed in Section 3.1, since
it directly represents "design for the worst case".

4. Max-min of Eigenfrequencies

Multiple optimum eigenvalues are prone to manifest themselves in optimization of rib stiffened plates and shells, see, e.g., [12]. Multiplicity of eigen-

values implies loss of the usual property of differentiability with respect to design changes because such eigenvalues are only directionally differentiable, see [16], [17] and [18].

In [1], rib reinforcement topology optimization of plates and shells is conducted with attention devoted to both simple and multiple eigenfrequencies of free vibrations, and it is shown that, despite the lack of usual differentiability properties of multiple eigenfrequencies, the problems may be treated like differentiable optimization problems if some restrictions are imposed on the direction of the vector of design changes at each iteration. Along these lines, we have been able to develop a new, general method of solution of max-min problems which can also handle optimization of multiple eigenfrequencies. This method is very attractive because it only requires ordinary procedures for sensitivity analysis and mathematical programming for constrained optimization. The reader is referred to [1] for details.

4.1. EXAMPLES

Two illustrative examples of optimal rib reinforcement on plate and shell structures with respect to different eigenfrequency objectives shall be finally presented. The same data for the rib thickness ratio and amount of available rib material were assumed as in the example of Section 3.2.

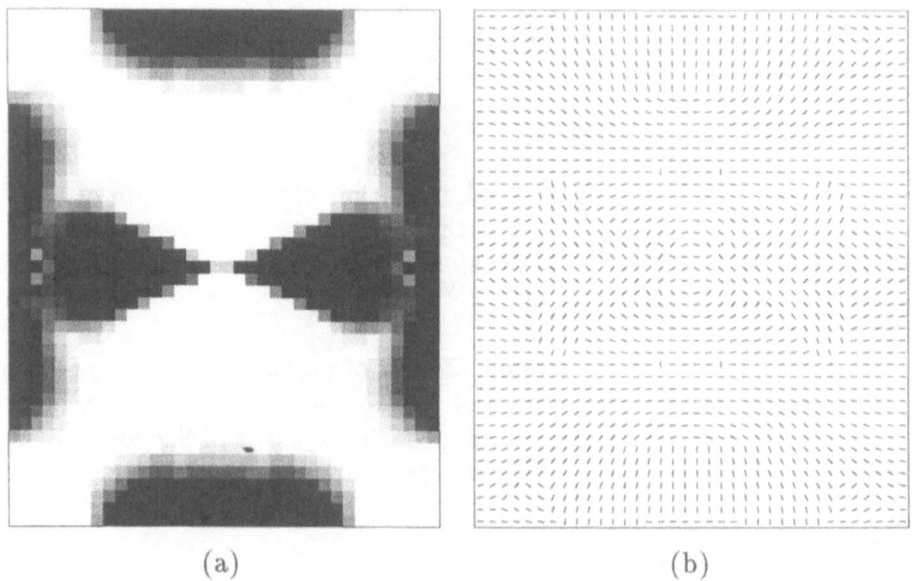

(a)   (b)

Figure 3.   (a) Rib density and (b) Rib orientations

In the first example, the problem was posed as maximization of the lowest eigenfrequency from among the second, third and fourth eigenfrequencies of a clamped rectangular plate with the aspect ratio 1.2. This resulted in coalescence of the second and third eigenfrequency into an optimum double second eigenfrequency, while the first and fourth eigenfrequencies remained simple.

Fig. 3 shows the optimum rib reinforcement topology of the corresponding plate in terms of rib density distribution and orientation.

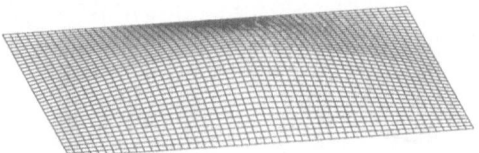

Figure 4.   Shell mid surface.

In the second example, a parabolic shell with a clamped rectangular boundary of aspect ratio 1.2 was considered, see Fig. 4. Here, the optimization problem was formulated in the more usual way, namely as maximization of the smallest eigenfrequency, and the ten lowest frequencies (and associated modes) were considered candidates in the solution procedure. It should be mentioned that in the solution of this problem, optimization of the rib orientations was not carried out.

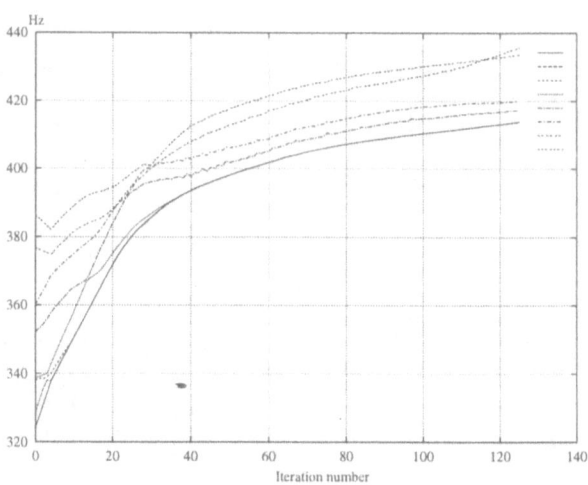

Figure 5.   Iteration history.

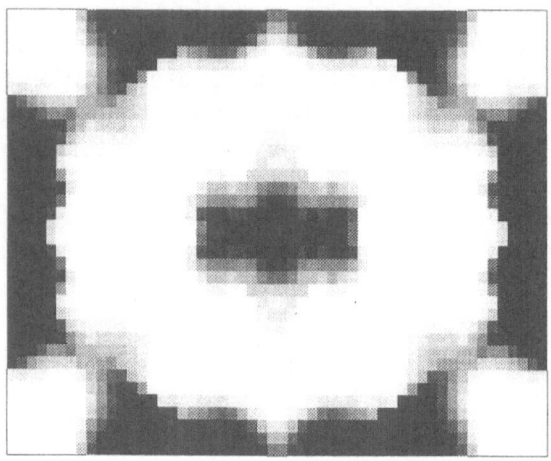

Figure 6. Rib density.

This problem resulted in a no less than four-fold multiple fundamental eigenfrequency of vibrations! The iteration history is depicted in Fig. 5, and Fig. 6 illustrates the optimum rib reinforcement distribution of the shell.

Finally, Fig. 7 shows four linearly independent vibration modes associated with the fundamental eigenfrequency of the shell. One should bear in mind that any linear combination of these modes will be also an eigenmode since the corresponding eigenfrequency is quadruple.

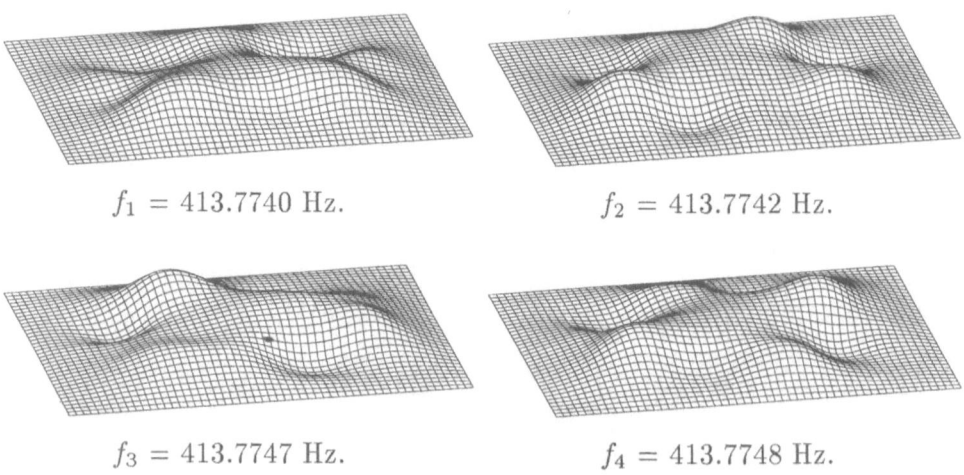

$f_1 = 413.7740$ Hz. $f_2 = 413.7742$ Hz.

$f_3 = 413.7747$ Hz. $f_4 = 413.7748$ Hz.

Figure 7. Eigenmodes.

Acknowledgement

The work presented in this paper received support from the Danish Technical Research Council, Programme of Research on Computer Aided Design.

References

1. Krog, L.A. & Olhoff, N. (1995) Topology Optimization of Integral Rib Reinforcement of Plate and Shell Structures with Respect to Multiple Objectives, Preprint, *Inst. Mech. Engrg., Aalborg University.*
2. Bendsoe, M.P. & Kikuchi, N. (1988) Generating Optimal Topologies in Structural Design using a Homogenization Method, *Comp. Meth. in Appl. Mech. and Engrg.* **71**, pp. 197-224.
3. Diaz, A.R. & Bendsoe, M.P. (1992) Shape Optimization of Structures for Multiple Loading Conditions using a Homogenization Method, *Structural Optimization* **4**, pp. 17-22.
4. Olhoff, N., Thomsen, J. & Rasmussen, J. (1993) Topology Optimization of Bimaterial Structures, in P. Pedersen (ed.), *Optimal Design with Advanced Materials, Elsevier, Amsterdam* , pp. 191-206.
5. Suzuki, K. & Kikuchi, N. (1991) Generalized Layout Optimization of Shape and Topology in Three-dimensional Shell Structures, in V. Komkov (ed.), *Geometric Aspects of Industrial Design, Siam, Philadelphia.*
6. Soto, C. & Diaz, A.R. (1993) On the Modelling of Ribbed Plates for Shape Optimization, *Structural Optimization* **6**, pp. 175-188.
7. Diaz, A.R., Lipton, R. & Soto, C.A. (1994) A New Formulation of the Problem of Optimum Reinforcement of Reissner-Mind- lin Plates, Preprint, Department of Mechanical Engineering, Michigan State University, East Lansing, MI, USA.
8. Diaz, A.R. & Kikuchi, N. (1992) Solutions to Shape and Topology Eigenvalue Optimization Problems using a Homogenization Method, *Int. J. Num. Meth. Engrg.* **35**, pp. 1487-1502.
9. Neves, M.N., Rodrigues, R. & Guedes, J.M. (1994) Generalized Topology Design of Structures with a Buckling Load Criterion, Preprint, Mechanical Engineering Department, Technical University of Lisbon, Portugal.
10. Rozvany, G.I.N. (1989) *Structural Design via Optimality Criteria*, Kluwer Academic Publishers, Dordrecht.
11. Rozvany, G.I.N., Bendsoe, M.P. & Kirch, U. (1995) Layout Optimization of Structures, *Applied Mechanics Reviews* bf 48, pp. 41-119.
12. Seyranian, A.P., Lund, E. & Olhoff, N. (1994) Multiple Eigenvalues in Structural Optimization Problems, *Structural Optimization* **8** 4, pp. 207-227.
13. Olhoff, N. & Rasmussen, S.H. (1977) On Single and Bimodal Optimum Buckling Loads of Clamped Columns, *Int. J. Solids Structures* **13**, pp. 605-614.
14. Pedersen, P. (1993) On Optimal Orientation of Orthotropic Materials, *Structural Optimization* **1**, pp. 101-106.
15. Bendsoe, M.P., Olhoff, N. & Taylor, J.E. (1983) A Variational Formulation for Multicriteria Structural Optimization, *J. Struct. Mech.* **11**, pp. 523-544.
16. Masur & Mróz (1979) Non-stationarity Optimality Conditions in Structural Design, *Int. J. Solids Structures* 15, pp. 503-512.
17. Masur & Mróz (1980) Singular solutions in Structural Optimization Problems, in S. Nemat-Nasser (ed.), *Variational Methods in Mechanics of Solids, Evanston Illinois, Pergamon Press, New York* , pp. 337-343.
18. Haug, E.J. & Rousselet, B. (1980) Design Sensitivity Analysis in Structural Mechanics. II: Eigenvalue Variations, *J. Struct. Mech.* **8**, pp. 161-186.

SENSITIVITY ANALYSIS AND OPTIMIZATION
OF FINITE ELEMENT DISCRETIZED STRUCTURES

ERIK LUND
Institute of Mechanical Engineering
Aalborg University
DK-9220 Aalborg East, Denmark

ABSTRACT.   The aim of this paper is to present some of the results obtained in a Ph.D. project that the author finished in June 1994, see Lund (1994a). A general description of the structural optimization system ODESSY and its facilities for structural analysis and design sensitivity analysis is given. Different approaches of deriving expressions of design sensitivity analysis are discussed and the direct differentiation approach is chosen. An improved method for semi-analytical sensitivity analysis based on "exact" numerical differentiation of element matrices using first order forward finite differences has been developed and will be outlined. Finally, a general and flexible way of formulating problems of mathematical programming in structural design optimization systems is presented.

1. Introduction

In 1991 the development of a general purpose, fully three-dimensional computer aided environment for interactive structural design, analysis, design sensitivity analysis, synthesis, and engineering design optimization was initiated at the Institute of Mechanical Engineering, Aalborg University, Denmark. This system is named "ODESSY" (Optimum DESign SYstem) and is based on the experiences obtained by developing the prototype optimization system CAOS, see Rasmussen *et al.* (1993). ODESSY is being developed by several researchers and is integrated with the commercial CAD systems AutoCAD and Pro/ENGINEER, thereby setting rational design facilities directly at the disposal of the designer.

We have decided to write our own analysis code. It might be advantageous to use a commercial finite element system for the analysis due to the generality of the analysis facilities but we wanted to develop a kind

D. Bestle and W. Schielen (eds.), IUTAM Symposium on Optimization of Mechanical Systems, 181–188.
© 1996 *Kluwer Academic Publishers.*

of computer engineering design laboratory where experiments concerning, for example, structural analysis, design sensitivity analysis, mathematical programming, and links between CAD environments and structural optimization systems can be carried out. Furthermore, the tools available for software developers have reached a high standard, making it easier to develop large and complex software systems. It should be noted that interfaces from ODESSY to commercial finite element systems are available so the designer can verify the results obtained by ODESSY using his favourite finite element code.

The objective of my Ph.D. project has been to develop, implement, and integrate methods for structural analysis, design sensitivity analysis, and optimization into the general purpose computer aided design system ODESSY, and some of the basic concepts of this work will be presented in the following.

2. Design Sensitivity Analysis

In a general purpose computer aided environment for interactive structural design and optimization, *design sensitivity analysis is the basic enabling tool* and has therefore been given the main attention in the project.

The analysis facilities in ODESSY include static stress analysis, natural frequency analysis, steady state thermal analysis, thermo-elastic analysis, eigenfrequency analysis with initial stress stiffening effects due to mechanical or thermal loads, and linear buckling analysis with the possibility of including thermo-elastic effects. About 25 different finite elements have been implemented.

The design variables of the structural design problem in ODESSY can be either geometrical design variables like sizing or shape variables, material design variables like constitutive parameters of materials, support design variables like the position of supports for the structure, and loading design variables like the position of external loads applied to the structure. These design variables can be categorized as

- Generalized shape design variables
- Sizing design variables
- Material design variables

These design variables are denoted by a_i, $i = 1, \ldots, I$, and it is now the aim to establish expressions for design sensitivities of various criteria with respect to these design variables in an accurate and efficient way.

The simplest approach to obtain design sensitivities is the overall finite difference (OFD) method, but as this method is computationally inefficient, another approach has to be used. The derivatives of structural response in principle can be calculated at three stages.

We can (I) *differentiate the continuum equations defining the response of the system*, using the material derivative concept of continuum mechanics, then discretize the problem and solve it by using an adjoint variable technique.

We can (II) use the *direct differentiation approach* which is based on *differentiation of the equations obtained when the continuum equations have been discretized* (in this case by the finite element method). This approach, which is also known as the *implicit differentiation approach*, thus has a reversed order of discretization and differentiation compared to the continuum approach described above. The direct approach is, in both the discrete and adjoint version, very popular due to its ease of implementation compared to the continuum approach.

Finally, we can (III) *differentiate directly the computer program* used to solve the structural response, i.e., use the approach of *automatic differentiation* of a function defined by its program (in Fortran, C, etc.).

The conclusive argument for selection of method for design sensitivity analysis has been that the *continuum approach* takes a lot of analytical work in order to develop the expressions for design sensitivities. I have wanted to implement many different kinds of finite element types and at the same time be able to handle many different kinds of design variables. For this purpose the *direct differentiation approach* seems much easier to implement and to be just as applicable to solve the problem as the continuum approach.

So, the discrete version of the direct differentiation approach, where design sensitivities of criterion functions are computed with respect to each design variable, has been chosen due to its ease of implementation.

In the following a short presentation of the direct differentiation approach for static stress analysis is given, and using this approach it is necessary to determine derivatives of various element matrices with respect to design variables. This problem is adressed in Section 3.

2.1. EXAMPLE: STATIC STRESS ANALYSIS

As an example, the direct differentiation approach is applied for a static stress analysis problem. The global equilibrium equation of a finite element discretized structural design problem with linearly elastic response is given by

$$\mathbf{KD} = \mathbf{F} \tag{1}$$

where $\mathbf{K}$ is the global stiffness matrix, $\mathbf{D}$ is the nodal displacement vector, and $\mathbf{F}$ is the consistent nodal force vector.

The *direct differentiation approach* to obtain design sensitivities of the displacement field *is based on implicit differentiation of the global equilibrium equation*. If Eq. 1 is differentiated with respect to a design variable a_i and the terms are rearranged, the following discrete version of the direct differentiation approach for the displacement sensitivities is obtained

$$K(a)\frac{\partial D}{\partial a_i} = -\frac{\partial K(a)}{\partial a_i}D + \frac{\partial F}{\partial a_i}$$

$$= \sum_{n_e^a}\left(-\frac{\partial k^a}{\partial a_i}d^a + \frac{\partial f^a}{\partial a_i}\right), \quad i = 1,\ldots,I \qquad (2)$$

Here, n_e^a is the number of perturbed (active) finite elements, k^a is the element stiffness matrix, d^a is the element displacement vector, and f^a is the element load vector for a perturbed finite element.

Eq. 2 is of the same form as Eq. 1, so the factorized stiffness matrix K can be reused, and only the new right hand side which is termed the *pseudo load vector* needs to be calculated before the sensitivities $\partial D/\partial a_i$ for each design variable a_i can be found by forward and back substitution. This approach is therefore very efficient.

Having computed the displacement sensitivities $\partial D/\partial a_i$, sensitivities of, e.g., stresses and compliance, are easily computed.

In general, the direct differentiation approach outlined here can be used to derive expressions for design sensitivities for all the different analysis problems described in the beginning of this section. For further details, see Lund (1994a, 1994b).

3. Computing Derivatives of Element Matrices and Vectors

It is seen from the sensitivity expressions derived with the direct differentiation approach, e.g., Eqs. 2, that it is necessary to compute derivatives of various element matrices and vectors. In this section different approaches of computing these element derivatives are described and discussed.

3.1. ANALYTICAL APPROACH

If the derivatives of element matrices and vectors are determined analytically before their numerical evaluation, the approach is called *analytical design sensitivity analysis*. The approach of analytical design sensitivity analysis is easy to apply in case of sizing or material design variables but can be very difficult to implement in a general purpose shape design system if many different kinds of shape design variables and finite element types are available. Thus, a large amount of analytical work and programming will be required in order to develop analytic expressions for derivatives of various element matrices and vectors with respect to possible shape design variables, if possible.

In ODESSY, the generalized shape design variables may be translation, rotation, or scaling of the positions of master nodes that control the shape of boundaries and surfaces, see Rasmussen *et al.* (1993) and Lund (1994a). Using the analytical approach in case of a generalized shape design variable, it is necessary to have explicit relations between the position of nodal coordinates of the finite element and the shape design variable, and such relations may not be available. For example, if unstructured mesh generation is used to generate a 3D surface mesh, smoothing processes are normally involved in this process, and explicit relations between the generalized shape design variables controlling the shape of the surface and the coordinates of finite element nodal points on the surface are not available. In case of using such mesh generation facilities, the analytical approach cannot be implemented for generalized shape design variables.

3.2. SEMI-ANALYTICAL APPROACH

Instead of using the analytical approach to obtain derivatives of element matrices and vectors, it might be advantageous to determine these derivatives by numerical differentiation. This approach is called *semi-analytical design sensitivity analysis.* That is, in semi-analytical design sensitivity analysis the derivatives of the element matrices are approximated by first order forward finite differences (or another finite difference scheme), e.g., the derivative $\partial k/\partial a_i$ of the element stiffness matrix is computed as

$$
\begin{aligned}
\frac{\partial \mathbf{k}(\mathbf{a})}{\partial a_i} &\simeq \frac{\Delta \mathbf{k}(a_1, \ldots, a_I)}{\Delta a_i} \\
&= \frac{\mathbf{k}(a_i + \Delta a_i) - \mathbf{k}(a_i)}{\Delta a_i}
\end{aligned}
\tag{3}
$$

If the stiffness matrix $\mathbf{k}$ depends linearly on the design variable a_i, the first order finite difference expression in Eq. 3 yields the "exact" numerical derivative. Here and in the following, derivatives obtained by numerical differentiation will be termed "exact derivatives" if they have no truncation error due to neglection of higher order terms in their Taylor series expansion and are exact except for computational round-off errors.

The method of semi-analytical design sensitivity analysis is very attractive in a shape design system, as it is easy to implement for many different kinds of shape design variables and finite element types, because simple and computationally inexpensive first order finite differences are used. Therefore, the method of S-A sensitivity analysis is very popular and, *in most cases, this method is very efficient and reliable.*

However, as a finite difference approximation is involved in this method, both truncation and conditions errors may occur.

3.3. SEMI-ANALYTICAL APPROACH BASED ON "EXACT" NUMERICAL DIFFERENTIATION OF ELEMENT MATRICES

This subsection is devoted to the problem of obtaining "exact" numerical derivatives of various finite element matrices as *the accuracy of the first order finite difference approximations in the semi-analytical (S-A) method,* see Eq. 3, *is strongly dependent on the chosen size of perturbation Δa_i of a shape design variable a_i*. This dependency arises as the element matrices generally depend non-linearly on shape design variables, and in some cases, so small perturbations are needed that computational round-off errors become the problem. The inaccuracy problem associated with the S-A method in connection with shape design variables was first discovered by Barthelemy *et al* (1988), and later understood and resolved for different type of problems in Pedersen *et al* (1989), Olhoff & Rasmussen (1991), Cheng & Olhoff (1993), Olhoff *et al* (1993), Lund & Olhoff (1994), and Lund (1994a).

The inaccuracy problem associated with the traditional semi-analytical method may occur for design sensitivities with respect to structural *shape design variables* in problems where *the displacement field is characterized by rigid body rotations which are large relative to actual deformations of the finite elements*, i.e., for example in problems involving linearly elastic bending of long-span, beam-like structures, and of plate and shell structures. However, it should be noted that the semi-analytical method works excellently for most problems and the inaccuracy problem is only encountered for the above-mentioned type of design sensitivity analysis problems.

In order to avoid dependence on the chosen perturbation Δa_i, *a method for "exact" numerical differentiation of element matrices based on computationally inexpensive first order finite differences has been developed.* This goal may seem unattainable, but a closer study of the functions that form the element matrices reveals that the same mathematical forms are common for large groups of finite elements, see Olhoff *et al* (1993). For instance, the element matrices for all isoparametric elements with translational degrees of freedom and isoparametric Mindlin plate and shell elements depend on the same class of functions. Similarly, the element matrices of a large class of finite elements comprising Bernoulli-Euler beam and Kirchhoff plate and shell elements have a similar mathematical structure. The members of these classes of matrices in general depend non-linearly on the design variables, but are defined within a special mathematical form. This mathematical form implies that their *approximate numerical derivatives*, computed by a usual first order finite difference scheme, *can be upgraded to "exact" derivatives* by simple multiplication by appropriate correction factors. The values of these correction factors can be very easily pre-computed and be used throughout the procedure of design sensitivity analysis. The results thereby become totally independent of the magnitude of the perturbation.

The basic idea of this approach to shape design sensitivity analysis is to relate element derivatives w.r.t. a generalized shape design variable A_m to element derivatives w.r.t. nodal coordinates $a_j, j = 1, \ldots, J$, as in the analytical approach. For example, the derivative of the element stiffness matrix $\mathbf{k}$ w.r.t. a generalized shape design variable A_m is found by application of the chain rule

$$\frac{\partial \mathbf{k}}{\partial A_m} = \sum_{j=1}^{J} \frac{\partial \mathbf{k}}{\partial a_j} \frac{\partial a_j}{\partial A_m}, \quad m = 1, \ldots, M \tag{4}$$

The derivatives $\partial \mathbf{k} / \partial a_j, j = 1, \ldots, J$, where a_j represents nodal coordinates, can be computed by means of first order finite differences using the method of "exact" numerical differentiation of element matrices.

As an example, in case of isoparametric finite elements the element functions used as basis for forming finite element matrices and vectors can be written as linear functions of the nodal coordinates $a_j, j = 1, \ldots, J$, and application of first order finite differences, see Eq. 3, directly yields the "exact" numerical derivatives of these element functions.

The mesh sensitivities $\partial a_j / \partial A_m$, i.e., the design velocity field, are computed using first order finite differences. If the generalized shape design variable A_m represents translation or scaling of a master node controlling the shape of a boundary surface, these finite differences yield the "exact" numerical derivatives, whereas truncation errors occur in case of a rotation shape design variable. Further discussion concerning these mesh sensitivities can be found in Lund (1994a).

This improved method of semi-analytical shape design sensitivity analysis *completely eliminates the inaccuracy problem that has been observed for the traditional semi-analytical method.* Detailed information about this method can be found in Olhoff *et al* (1993), Lund & Olhoff (1994), and Lund (1994a), where "exact" numerical derivatives of many different types of finite elements are derived and the accuracy is documented by several numerical benchmarks.

4.   A General and Flexible Method of Defining Problems of Mathematical Programming

Another objective of the Ph.D. project has been to be able to formulate design optimization problems in terms of objective and constraint functions in a general way, and a very flexible method of formulating problems of mathematical programming in structural optimization systems has been developed and implemented. This method makes it possible to evaluate user defined design criteria and their derivatives w.r.t. specific design parameters, and the method enables the formulation and solution of problems involving local, integral, min/max, max/min, and possibly non-differentiable user defined functions in any conceivable mix. The mathematical formulation is based on the so-called bound formulation, see Bendsøe *et al* (1983), and the implementation involves a parser capable of interpreting and performing symbolic differentiation of the user defined objective and constraint functions. This database module can also be used for graphical visualization of user defined mathematical expressions for design criteria.

The development of this database module has made it possible to solve very complicated formulations of design optimization problems without having to reprogram anything in the computer code, and the module is very easy to expand as new analysis facilities or new criterion functions are added to the system. A description of this approach can be found in Lund & Rasmussen (1994) and examples can be found in Lund (1994a).

Acknowledgement

The work presented in this paper received support from the Danish Technical Research Council, Programme of Research on Computer Aided Design.

References

Barthelemy, B., Chon, C.T. & Haftka, R.T. (1988) Sensitivity Approximation of Static Structural Response, Finite Elements in Analysis and Design 4, pp. 249-265.

Bendsøe, M.P., Olhoff, N. & Taylor, J.E. (1983) A Variational Formulation for Multicriteria Structural Optimization, *Journal of Structural Mechanics* 11, pp. 523-544.

Cheng, G. & Olhoff, N. (1993) Rigid Body Motion Test Against Error in Semi-Analytical Sensitivity Analysis, *Computers & Structures* 46 3, pp. 515-527.

Lund, E. (1994a) Finite Element Based Design Analysis and Optimization, Ph.D. Thesis, *Special Report No. 23, Institute of Mechanical Engineering, Aalborg University, Denmark* , 240 pp.

Lund, E. (1994b) Design Sensitivity Analysis and Optimization of Finite Element Discretized Structures, in , *Proc. 5th AIAA/USAF/NASA/ISSMO Symposium on Multidisciplinary Analysis and Optimization* 1, ISBN 1-56347-097-7, pp. 641-660.

Lund, E. & Olhoff, N. (1994) Shape Design Sensitivity Analysis of Eigenvalues Using "Exact" Numerical Differentiation of Finite Element Matrices, *Structural Optimization* 8 1, pp. 52-59.

Lund, E. & Rasmussen, J. (1994) A General and Flexible Method of Problem Definition in Structural Optimization Systems, *Structural Optimization* 8 2, pp. 86-92.

Olhoff, N. & Rasmussen, J. (1991) Study of Inaccuracy in Semi-Analytical Sensitivity Analysis - A Model Problem, *Structural Optimization* 3, pp. 203-213.

Olhoff, N., Rasmussen, J. & Lund, E. (1993) A Method of "Exact" Numerical Differentiation for Error Elimination in Finite Element Based Semi-Analytical Shape Sensitivity Analysis, *Mechanics of Structures and Machines* 21, pp. 1-66.

Pedersen, P., Cheng, G. & Rasmussen, J. (1989) On Accuracy Problems for Semi-Analytical Sensitivity Analysis, *Mechanics of Structures and Machines* 17 3, pp. 373-384.

Rasmussen, J., Lund, E., Birker, T. & Olhoff, N. (1993) The CAOS System, *International Series of Numerical Mathematics* 110, Birkhuser Verlag, Basel, pp. 75-96.

OPTIMIZATION OF THE DYNAMIC RESPONSE OF LINEAR MECHANICAL SYSTEMS USING A MULTIPOINT APPROXIMATION TECHNIQUE

V.L. MARKINE, P. MEIJERS, J.P. MEIJAARD
Laboratory for Engineering Mechanics, Delft University of Technology, Delft, Netherlands. email: V.Markine@wbmt.tudelft.nl

V.V. TOROPOV
** *Department of Civil Engineering, University of Bradford, Bradford, United Kingdom. email: V.V.Toropov@bradford.ac.uk*

ABSTRACT

The optimization problem of the ride characteristics of a travelling truck is considered. Its dynamic behaviour is approximated by linear FE models (both 2-D and 3-D). The road surface profile is presented as a random function with known power spectral density. The design variables comprise geometry as well as spring and damper properties. Limitations are imposed on maximum values of the relative displacements of suspensions, dynamic wheel load to axles, acceleration of the cargo.

The above problem is solved using a multipoint approximation method. To reduce the computational cost a two-level optimization procedure for the truck optimization problem is proposed.

It is demonstrated that the method used is efficient for optimizing the dynamic behaviour of complex structures and it is also promising for geometrically non-linear dynamic problems. Moreover it can easily be coupled with a general-purpose finite-element software package.

1. Introduction

Approximation concepts are now very popular in practical optimization [1]. In most cases the optimization process involves many calculations of objective and constraint functions and/or their derivatives, this often implies use of some numerical response analysis technique. Considering the design of a large engineering system such as a ground vehicle, the response analysis can be time consuming. The computational effort is considerably reduced by introducing approximation concepts, the original functions are then replaced by simplified and explicit ones.

Sometimes, optimization is difficult even if one of the approximation methods is employed because of a large number of design variables and constraints. A most common solution is to break the problem into several smaller subproblems and a coordination problem. Several procedures for decomposing the large structural optimization problem into subproblems have been proposed in the literature (e.g. [2]).

For a problem which combines sizing (stiffness and damping properties of elements) and geometry (nodal coordinates) optimization, a typical decomposition is to consider the geometrical variables as the upper-level (global) variables and the sizing variables as the

D. Bestle and W. Schielen (eds.), IUTAM Symposium on Optimization of Mechanical Systems, 189–196.
© *1996 Kluwer Academic Publishers.*

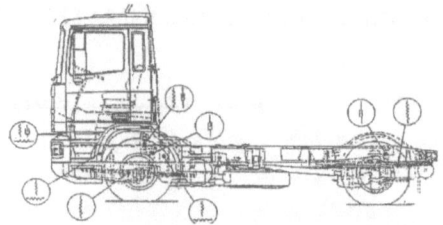

Figure 1.   Suspension system of a truck (I.Besselink and F.van Asperen, [3])

lower-level (local) ones [7]. This is motivated by the fact that the geometrical and sizing design variables are of fundamentally different nature which can lead to numerical difficulties if they are treated together in a single-level optimization [8].

In the present paper a problem of optimization of ride characteristics of a truck is considered. The ride characteristics are related to the vibration of the vehicle due to road irregularities and its effects on a driver and goods. The dynamic behaviour of the truck is described by a linear finite element model and a road surface profile is presented as a random function with known power spectral density. The comfort of the driver is estimated by means of a ride index calculated as a weighted mean square acceleration at the point of the driver's seat. To minimize the ride index, stiffness and damping coefficients of suspension elements and coordinates of several nodes are varied with limitations imposed on the RMS value of the responses such as the relative displacements of the suspensions, dynamic wheel load of the axles, acceleration of the cargo. This problem is solved using a multipoint approximation (MPA) technique [11].

To reduce computational efforts a two-level procedure was developed. Three sizing sub-optimization problems related to the axles, cabin and engine subsystems are considered at the lower level, and the optimization of the geometry at the top level. The results of the optimization of the 2-D and 3-D finite element model of the truck are presented and discussed.

2. Design Problem

2.1. CALCULATION OF DYNAMIC RESPONSE

A truck moving on the road represents a complex vibratory system. It contains the truck and trailer masses, connected to a chassis by suspensions, and resting on axles and tires, while the road represents the excitation input applied to the tires (Figure 1). The dynamic response of such a system depends on the characteristics of the vehicle elements as well as on the road quality.

Assuming only relatively small displacements, the dynamic behaviour of a vehicle can be described by a linear model. The simplified 2-D finite element model of the truck is shown in Figure 2. The cabin and engine are modelled as rigid bodies, beam elements are used for the chassis, whereas linear spring and damping elements describe the cabin and axle suspensions and the engine mountings.

In a simple case, the excitation inputs caused by the road roughness can be described by periodic functions. But, in order to obtain more realistic results the road surface profile should be regarded as a random function with known power spectral density (PSD) which is different for the various types of roads ranging from an unprepared terrain to a highway. In the present study the dynamic response of the truck was evaluated for a "standard" road profile with PSD approximated by the function $S_p(f) = 10^{-6} V/f^2$ $[m^2/Hz]$, where V is the speed of the truck $[m/s]$, f is the frequency $[Hz]$.

For a linear mechanical system, a direct linear relationship between inputs and output exists [10]. If a system has R excitation inputs it reads

$$S_q(f) = \sum_{r=1}^{R} \sum_{s=1}^{R} H_{p_r}^*(f) H_{p_s}(f) S_{p_r p_s}(f)$$ (1)

where S_q - response PSD; H - complex frequency response function; H^* - complex conjugate of H; $S_{p_r p_r} = S_{p_r}$ - PSD of input $p_r(t)$; $S_{p_r p_s}$ - power cross-spectral density of inputs $p_r(t)$ and $p_s(t)$. The frequency response function H_{p_r} is calculated as a response of a system to the excitation input p_r assumed to be a unit harmonic excitation, i.e. $p_r(t) = e^{2\pi i f t}$. The harmonic response analyses were fulfilled using the ANSYS finite element package [12]. When the response PSD has been determined, the mean square response can be calculated directly from

$$E[q^2] = \int_0^\infty S_q(f)\,df$$ (2)

The mean square of the velocity and acceleration can be easily expressed through the PSD of the displacement.

The 2-D model of the truck is affected by two excitation inputs applied at the road contact points of the front and rear wheels with the same PSD $S_{p_1} = S_{p_2} = S_p$. Moreover, since the rear wheels encounter the same road irregularities as the frontal ones (after a time delay T), i.e. $p_1(t) = p_2(t + T)$, the corresponding inputs can be considered as perfectly correlated, so that their power cross-spectral densities are $S_{p_1 p_2} = S_p e^{-2\pi i f T}$ and $S_{p_2 p_1} = S_p e^{2\pi i f T}$. The time delay $T = L/V$ depends on the length of the wheelbase L and the speed V of the truck.

To assess the driver's comfort, a specific kind of mean square response - ride index F_0 - is calculated. It is defined as a weighted mean square acceleration and can be written as

$$F_0 = \left(\int_0^\infty [W^l(f)]^2 S_{q''}^l(f)\,df + \int_0^\infty [W^v(f)]^2 S_{q''}^v(f)\,df + \int_0^\infty [W^{lt}(f)]^2 S_{q''}^{lt}(f)\,df \right)^{1/2}$$ (3)

where $S_{q''}^l$, $S_{q''}^v$ and $S_{q''}^{lt}$ are the PSD of the acceleration in the longitudinal, vertical and lateral direction respectively. The weight functions (W^l, W^v, W^{lt}) reflecting the human

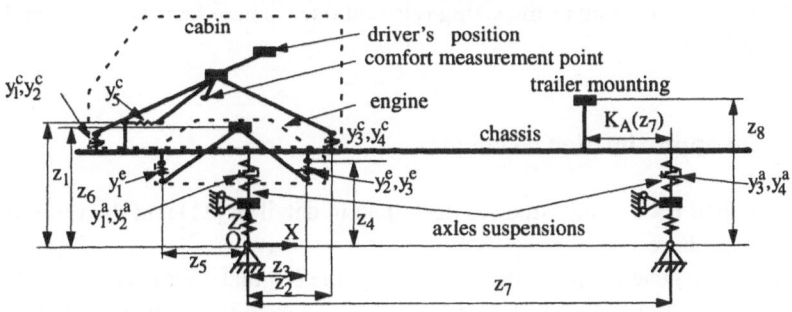

Figure 2. The 2-D FE model of the truck and the design variables

sensitivity to vibrations in a particular direction are taken from ISO standard 2631 [5].

In the problem under consideration all the responses were restricted to the frequency range from 0 to 20 Hz while the speed of the truck $V = 22.2 m/s$ ($80 km/h$).

2.2. FORMULATION OF OPTIMIZATION PROBLEM

As it is mentioned above, the road surface irregularities result in vibrations which affect the driver and goods. They can significantly deteriorate the comfort of the former and simply damage the latter. That is why reduction of the undesirable vibrations is of great importance in the design of such a vehicle. The optimization problem was stated as:

Minimize the ride index $F_0(x)$ at a comfort measurement point (Figure 2) for a given road profile.

The stiffness and damping of the suspension elements (Figure 1) and the coordinates of several nodes were chosen as components of the vector of design variables x. The design variables $x = [y, z]$ shown in Figure 2 can be classified as follows:

$y = [y^a, y^c, y^e]$ is the subvector of sizing design variables;

z is the subvector of geometrical design variables.

The subvectors y^a, y^c and y^e are related to the axle, cabin and engine subsystems respectively. The lower and upper bounds of the design variables are collected in Table 1.

Written in the dimensionless form the constraints can be classified as follows:

$$F_j(y^a, z, [y^c, y^e]) \leq 1, \quad j = 1, 2 \tag{4a}$$

$$F_j(y^c, z, [y^a, y^e]) \leq 1, \quad j = 3, 4, 5 \tag{4b}$$

$$F_j(y^e, z, [y^a, y^c]) \leq 1, \quad j = 6, 7, 8 \tag{4c}$$

$$F_j(y^a, z, [y^c, y^e]) \leq 1, \quad j = 9, 10 \tag{4d}$$

$$F_{11}(z, [y^a, y^c, y^e]) \leq 1 \tag{4e}$$

Here (4a)-(4c) are the limitations on the RMS value of the relative displacements of the axles and cabin suspensions and the engine mountings respectively, (4d) constraint the maximum value of the dynamic wheel load of the axles and (4e) refers to the RMS acceleration of the cargo. The displacements due to static loads were also taken into account. A more detailed information on the constraints is given in [3, 9].

It should be noted, that the constraint functions depend only weakly on the subvectors of the design variables in brackets. The geometrical design variables affect all the constraints though not so strong as the sizing relevant ones. These facts will be used later for the decomposition.

3. Multipoint approximation technique

The MPA method used in this study is described in detail in [11] and therefore it is only briefly presented here.

According to a general approximation concept the original optimization problem is replaced by a succession of the simpler ones, namely in each iteration step k the following problem is to be solved:

Find x_*^k such that $\quad \tilde{F}_0^k(x, a_0) \rightarrow min, \qquad x \in R^N$ (5)

$$\tilde{F}_j^k(x, a_j) \le 1, \, a_m = \begin{bmatrix} a_1^m a_2^m ... a_L^m \end{bmatrix}^T, \quad j = 1, ..., M, \qquad m = 0, ..., M$$ (6)

$$A_i^k \le x_i \le B_i^k, \qquad A_i^k \ge A_i, \qquad B_i^k \le B_i, \qquad i = 1, ..., N$$ (7)

Each functions $\tilde{F}(x, a)$ (the indices j and k are suppressed to simplify the notations) is an explicit approximation of the original function $F(x)$. A vector of tuning parameters a is determined on the basis of the information about the original function (and possibly its derivatives [11]) at several points of the design space. The evaluation of the components of the vector a is formulated as a weighted least-squares optimization problem:

Find the vector a that minimizes

$$G(a) = \sum_{p=1}^{P} \{ w_p^{(0)} [F(x_p) - \tilde{F}(x_p, a)]^2 \}$$ (8)

where $w_p^{(0)}$ is a weight coefficient that characterizes the relative contribution of the information about F at the point x_p [11]. In [4] it is shown that a most accurate approximation can be attained using *a multiplicative* function

$$\tilde{F}(x, a) = a_0 \prod_{l=1}^{N} x_l^{a_l}$$ (9)

The minimum number of the points P needed to determine the coefficients a is equal to $N + 1$. They are obtained by taking steps from the starting point (optimum from the previous iteration) in each coordinate direction of the design space. The quality of the approximation can be improved without additional effort by taking the points from the previous steps (belonging to the current subregion and its neighbourhood) into account.

The strategy of changing of the move limits can be summarized as follows. After each iteration step k the search subregion defined by the move limits A^k and B^k is reduced if in the obtained point x_*^k: 1) the approximation is not adequate at least for one active constraint or 2) none of the move limits is active (the obtained point is internal). If the above conditions are both satisfied the same move limits are used in the next iteration step [11].

The optimization process is terminated when
- the approximations are adequate for all active constraint in the obtained point
- the obtained point is internal
- the subregion has reached a prescribed small size: $max \left[\dfrac{B_i^k - A_i^k}{B_i - A_i} \right] \le 0.01, \, i = 1, ..., N$.

4. Two-level solution

The considered optimization problem can directly be solved using the multipoint approximation method. However, the computational cost considerably increases if a more complex 3-D finite element model is used in the response analysis. Therefore an attempt was made to apply the two-level optimization technique to the same problem decomposing the original one into a number of smaller subproblems.

The sizing design variables y are then optimized for the fixed geometry at the lower level. Thereafter the geometry optimization with all the constraints are fulfilled at the top level coordinating the optimization process. Because of coupling, this procedure must be iteratively repeated until the optimum is reached (Figure 3).

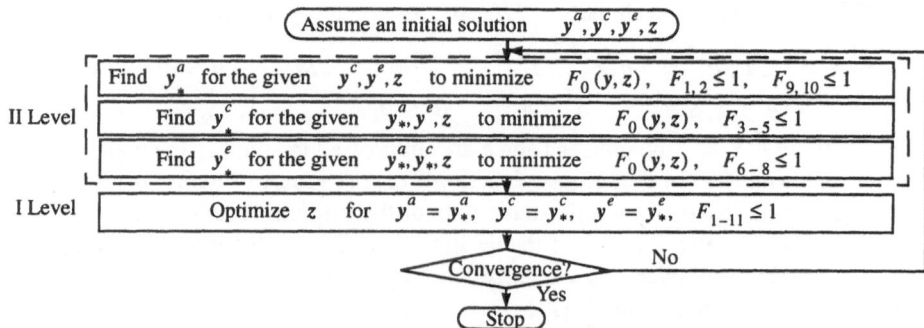

Figure 3. Flowchart for two-level optimization of the truck

We found it was essential that all the constraints should be satisfied (at least not strongly violated) before the geometry optimization, otherwise there might be no feasible solution of the coordination problem. That means that a proper choice of the constraints for the low-level optimization is very important.

5. Results

5.1. THE 2-D MODEL OF THE TRUCK

First the MPA optimizer along with the two-level procedure proposed were tested on a simplified problem with the 2-D model of the truck. All 20 design parameters (Table 1) were varied whereas the number of constraints were reduced to 8 (only limitations on the suspensions travels (6a)-(6c) were taken into account). To estimate the effectiveness of the two-level procedure the straightforward optimization was fulfilled too. The derivatives of the functions with respect to design parameters were not used during the optimization since we had found that taking them into account does not improve the convergence characteristics of the optimization process whereas the evaluation of one partial derivative and one response quantity are approximately of the same computational cost.

The optimization started from an infeasible design (maximum constraint violation of 60%). The two-level optimization was terminated after three cycles when there were no improvements in values of the objective function (Figure 5). Comparing the results (Table 1) it can be noted that the decomposed problem provided almost the same optimal design as the single-level optimization (1.270 and 1.262 m/s^2 respectively) but did not give considerable computational savings though (533 and 547 response analyses). Its usage might be justified for more complex 3-D problem with a large number of design variables and constraints when a straightforward optimization is difficult and even impossible.

5.2. THE 3-D MODEL OF THE TRUCK AND TRAILER COMBINATION

To optimize the 3-D model of the truck and trailer combination (Figure 4) we used all the design variables and constraints (4a)-(4e).The response analysis for such a system was

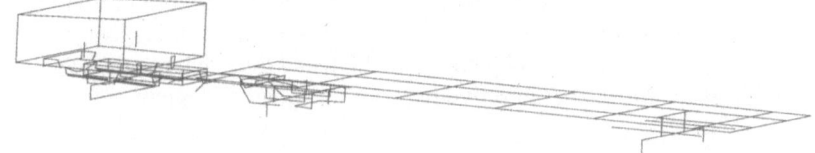

Figure 4. The 3-D finite element model of the truck and trailer combination

more sophisticated and time consuming. The FE model affected by 8 excitation inputs (corresponding to 8 wheels) contained about 2000 degrees of freedom. To obtain the ride index and the other response quantities the ANSYS program was externally coupled with TricaT post-processor developed at DAF Trucks NV.

TABLE 1. The results of the one- and two-level optimization of the truck

Design variable	2-D model					3-D model				Unit
	Definition			Result		Definition			Result	
	Lower bound	Upper bound	Initial value	Single level	Two levels	Lower bound	Upper bound	Initial value	Two levels	
y_1^a	400	2000	690	690	693	0.1	800	100	114	N/mm
y_2^a	1	50	35	24	23.5	0.5	25	16.9	7.7	Ns/mm
y_3^a	400	2000	600	1076	1079	100	500	150	100	N/mm
y_4^a	1	50	35	50	50	0.5	25	17	4.9	Ns/mm
y_1^c	50	160	90	100	104	25	80	45	45.5	N/mm
y_2^c	1	20	12.8	14.3	14.6	0.5	10	6.4	7.5	Ns/mm
y_3^c	50	185	150	163	157	25	92.5	75	74.4	N/mm
y_4^c	1	20	12.8	20	20	0.5	10	6.4	10	Ns/mm
y_5^c	8400	30000	8400	29567	27805	4200	20000	18000	20000	N/mm
y_1^e	7600	9900	7600	3590	3605	500	4950	2060	617	N/mm
y_2^e	8720	9900	8720	5650	5690	2500	4950	2600	2500	N/mm
y_3^e	24350	25000	24350	16754	16855	5000	12500	5400	6514	N/mm
z_1	1000	1250	1111	1000	1007	1000	1250	1111	1250	mm
z_2	100	600	600	583	584	100	600	600	537	mm
z_3	475	675	575	483	490	475	675	575	675	mm
z_4	700	1000	822	884	879	700	1000	822	700	mm
z_5	490	690	591	643	640	490	690	591	576	mm
z_6	890	915	906	893	895	890	915	903	910	mm
z_7	3250	3900	3250	3250	3250	3250	3900	3500	3900	mm
z_8	1100	1300	1201	1167	1170	1100	1300	1201	1206	mm
F_0	1.455			1.262	1.270			2.14	1.438	m/s²
N iterations				26						
N analyses				547	533					

The optimization started then from a feasible design (the realistic model from DAF Trucks) for which the ride index was 2.140 m/s^2 (Table 1). The optimum design was obtained after three cycles with a reduction of the ride index to 1.438 m/s^2 (Figure 5). It

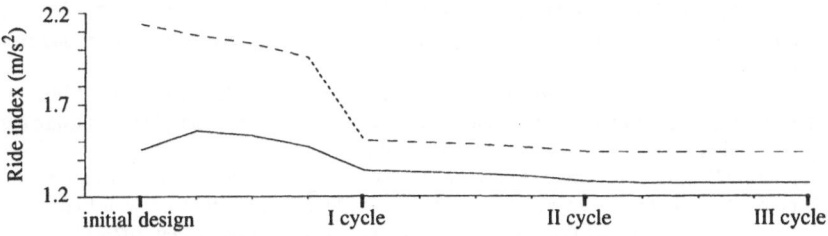

Figure 5. History of two-level optimization of the truck: ——— *2-D model,* - - - *3-D model*

should be noted that 2-D and 3-D model correspond to different trucks and the first one was used to test the optimization method only. From Figure 5 it is clear that changes in geometry strongly influence the objective function, and thus changes in the geometry represent a big reserve for further improvements of the existing vehicle.

6. Conclusions

The optimization of the dynamic characteristics of a linear mechanical system under stochastic load was done using a multipoint approximation technique. The results demonstrated that this technique could be efficiently used for dynamic problems. Moreover it can easily be coupled with a general-purpose finite-element software package.

Separating the sizing and geometrical design parameters the two-level optimization procedure for the above problem was proposed. Applied to the problem with the 2-D model of the truck it provided almost the same optimum design as the single-level optimization though it did not give considerable computational savings.

The results of optimization of the 3-D model of the truck and trailer combination showed robustness of the procedure for a complex systems with a large number of design variables and constraints.

Acknowledgments

The authors are grateful to the DAF Trucks NV (Technology/Technical Analysis Group) for giving the truck data and their help during this work.

7. References

1.   Barthelemy, J.-F.M., and Haftka R.T.: Approximation concepts for optimum structural design - a review, *Structural Optimization* 5 (1993), 129-144.
2.   Barthelemy, J.-F.M., and Riley, M.F.: Improved Multilevel Optimization Approach for the Design of Complex Engineering Systems, *AIAA Journal* 26 (1988), 353-359.
3.   Besselink, I., and van Asperen, F.: Numerical optimization of the linear dynamic behaviour of commercial vehicles, *Vehicle System Dynamics* 23 (1994), 53-70.
4.   Draper, N.R., and Smith, H.: *Applied regression analysis (2nd ed.)*, John Wiley & Sons, New York, 1981.
5.   ISO 2631/1-1985(E): *Evaluation of human exposure to whole body vibration, part 1: General requirements*, International Organization for Standardization, 1985.
6.   Haftka, R.T., Gürdal, Z.: *Elements of Structural Optimization, Third revised and expanded edition*, Kluwer Academic Publishers, Dordrecht, 1992.
7.   Kirsch, U.: Synthesis of structural geometry using approximation concepts, *Computers and Structures* 15 (1982), 305-314.
8.   Kirsch, U.: *Structural optimization; fundamentals and applications*, Springer, Berlin, 1993.
9.   Markine, V.L.: Optimization of the dynamic response of a linear mechanical system using a multipoint approximation method, Report LTM 1025, TU Delft, 1994.
10.   Newland D.E.: *An introduction to random vibrations and spectral analysis*, Longman, London, 1984.
11.   Toropov, V.V.: Multipoint Approximation Method in Optimization Problems with Expensive Function Values, in A. Sydow (ed.), *Computational System Analysis 1992*, Elsevier, pp. 207-212.
12.   *ANSYS, User's manual, Revision 5.0*, Swanson Analysis Systems Inc., 1992.

DIFFERENTIATION OF RELIABILITY FUNCTIONS

K. Marti
Federal Armed Forces University Munich,
Faculty of Aero-Space Engineering and
Technology,
D-85577 Neubiberg/München, Germany

1. Introduction

One of the main tools in reliability analysis of stochastic technical systems [4],[7],[10] are probability
functions of the type

$$P(x) := P(y_{\ell i} < y_i(a(\omega),x) < y_{ui}, \quad 1 \leq i \leq m). \qquad (1)$$

Here, $y_{\ell i} < y_i(a,x) < y_{ui}$, $i=1,\ldots,m$, are the basic
bahavioral constraints of the system, where $y = (y_1,\ldots,$
$y_i,\ldots,y_m)'$ are the response functions under consideration. The response variables y_i are functions

$$y_i = y_i(a,x), \quad i=1,\ldots,m, \qquad (2)$$

of a decision r-vector $x=(x_k)$ and a parameter ν-vector
$a = (a_j)$, where

$$a_j = a_j(\omega), \quad j=1,\ldots,\nu, \qquad (2.1)$$

are the **random** system parameters or coefficients. We assume that the random ν-vector

$$a(\omega) := (a_1(\omega),\ldots,a_\nu(\omega))' \qquad (2.3)$$

is defined on a probability space $(\Omega,\mathcal{O}\!\mathcal{L},P)$ with a given
distribution P. Finally, $y_\ell = (y_{\ell 1},\ldots,y_{\ell i},\ldots,y_{\ell m})'$,
$y_u = (y_{u1},\ldots,y_{ui},\ldots,y_{um})'$ are the m-vectors of lower,
upper bounds (margins) $y_{\ell i} < y_{ui}$, $i=1,\ldots,m$, for the
response vector y.

In most practical cases, the response variables
$y_i=y_i(a,x)$, $i=1,\ldots,m$, are complicated, highly nonlinear

197

D. Bestle and W. Schielen (eds.), IUTAM Symposium on Optimization of Mechanical Systems, 197–204.
© 1996 *Kluwer Academic Publishers.*

198   K. MARTI

functions of the vectors a and x. Hence, the **numerical evaluation** of P(x) and its first and higher order derivatives is in general a very difficult task, which was solved up to now only for special situations.

Especially, in reliability analysis and design of mechanical structures very efficient methods were developed for the computation of (**low**) probabilities of failure p_f=1-P(x), see [2],[3],[5],[6].

In the following, the computation of P(x) and its derivatives is achieved by using orthogonal function series expansions of the probability density f(.;x) of y=y(a(ω),x) and the derivatives of f(.;x) with respect to x.

2. Integral representations of P(x) and its derivatives

In the following we suppose that for all x under consideration the random m-vector
$$y_x(\omega) := y(a(\omega),x) = (y_1(a(\omega),x),\ldots,y_m(a(\omega),x))' \quad (3)$$
has a probability density
$$f = f(y;x). \quad (4)$$
The density f(y;x) exists under weak assumptions:

Lemma 2.1. Let m≤ν. For given x, suppose that the Jacobian $\frac{\partial y}{\partial a}(a,x)$ of the mapping a —> y(a,x) exists and has rank$\frac{\partial y}{\partial a}(a,x)$ = m for all a∈$\Re^\nu$ up to a set of $P_{a(\cdot)}$-measure zero, where $P_{a(\cdot)}$ denotes the probability distribution of a=a(ω). If the random parameter vector a=a(ω) has a probability density f_o=f_o(a), then the density f=f(y;x) exists.

Under the above assumptions the probability function P=P(x) can be represented now by the multiple integral
$$P(x) = \int_{y_\ell}^{y_u} f(y;x)dy \quad (5)$$
having the very simple **fixed** domain of integration
$$B = \{y\in\Re^m: y_\ell \le(<) y \le(<) y_u\}.$$
Under weak assumptions we may [1] interchange differ-

entiation and integration in (5), hence, the derivative $\frac{\partial P}{\partial x_k}(x)$ can be represented by the formula

$$\frac{\partial P}{\partial x_k}(x) = \int_{y_\ell}^{y_u} \frac{\partial f}{\partial x_k}(y;x)dy. \tag{6}$$

Density estimation and approximation of more general functions by **orthogonal function series expansions** is a well established technique, see e.g. [12-15].

Hence, in the following we consider estimates/approximations of P=P(x) and its derivatives based on **orthogonal function series expansions** of f(.;x), $\frac{\partial f}{\partial x_k}(.;x)$ resp., with respect to the orthonormal systems of Hermite functions (for the case that y(a(ω),x) has range $\mathfrak{R}^m$, unknown range, resp.).

3. Expansions in Hermite functions in case m=1

In the following, let x be an arbitrary, but fixed r-vector. For simplification we consider here only the case of a **scalar** response function $y(a,x) = y_1(a,x)$. Defining [12],[15] the Hermite polynomials H_j by

$$H_j(y) := (-1)^j e^{y^2} \frac{d^j}{dy^j} e^{-y^2}, \quad j=0,1,2,\ldots, \tag{7}$$

and the normalized Hermite functions φ_j [12],[15] by

$$\varphi_j(y) := (\pi^{1/2} 2^j j!)^{-\frac{1}{2}} e^{-\frac{y^2}{2}} H_j(y), \quad j=0,1,2,\ldots, \tag{7.1}$$

it is known [12],[15] that $\varphi_j \in L^1(\mathfrak{R}) \cap L^2(\mathfrak{R})$, $j=0,1,2,\ldots$, and the sequence (φ_j) forms a complete, orthonormal system in $L^2(\mathfrak{R})$. Furthermore, there exists a uniform constant $d_2 = \pi^{-\frac{1}{4}} 1.086435\ldots$, see [15], such that $|\varphi_j(y)| \le d_2$ for all $y \in \mathfrak{R}$ and each $j=0,1,2,\ldots$.

Assuming that

$$f(.;x) \in L^2(\Re), \quad \frac{\partial f}{\partial x_k}(.;x) \in L^2(\Re), \quad \text{resp.,} \tag{8}$$

then these functions can be expanded in the orthogonal series

$$f(y;x) = \sum_{j=0}^{\infty} c_j(x)\varphi_j(y) \tag{9}$$

$$\frac{\partial f}{\partial x_k}(y;x) = \sum_{j=0}^{\infty} c_{kj}(x)\varphi_j(y), \tag{9.1}$$

where the (Fourier) coefficients $c_j = c_j(x)$, $c_{kj} = c_{kj}(x)$, resp., are defined [12],[15] by

$$c_j(x) := \int f(y;x)\varphi_j(y)dy \tag{10}$$

$$c_{kj}(x) := \int \frac{\partial f}{\partial x_k}(y;x)\varphi_j(y)dy. \tag{10.1}$$

Since $f = f(y;x)$ is the probability density function of the random variable $y = y(a(\omega),x)$, see (3),(3.1), for $c_j(x)$ we have also the representation

$$c_j(x) = E\varphi_j(y(a(\omega),x)), \quad j=0,1,2,\dots . \tag{11}$$

Moreover, under weak assumptions we obtain the important relation

$$\frac{\partial}{\partial x_k}c_j(x) = \int \frac{\partial f}{\partial x_k}(y;x)\varphi_j(y)dy = c_{kj}(x). \tag{12}$$

Because of (5) and the series representation (9) of $f(.;x)$, for arbitrary finite bounds y_ℓ, y_u, $-\infty < y_\ell < y_u < +\infty$, we get

$$|P(x) - \sum_{j=0}^{n} c_j(x) \int_{y_\ell}^{y_u} \varphi_j(y)dy|$$

$$\leq (y_u - y_\ell)^{1/2} ||f(.;x) - \sum_{j=0}^{n} c_j(x)\varphi_j(.)||_2, \tag{13}$$

where $||.||_2$ denotes the norm of $L^2(\Re)$. In the same way from (6) and (9.1) we obtain

$$\left| \frac{\partial P}{\partial x_k}(x) - \sum_{j=o}^{n} c_{kj}(x) \int_{y_\ell}^{y_u} \varphi_j(y)\,dy \right|$$

$$\leq (y_u - y_\ell)^{1/2} \left\| \frac{\partial f}{\partial x_k}(.;x) - \sum_{j=o}^{n} c_{kj}(x)\varphi_j(.) \right\|_2 \qquad (13.1)$$

Since the series in (9),(9.1), resp., is convergent in the L^2-norm, from (13),(13.1) we obtain our first result:

<u>Theorem 3.1.</u> Based on the assumptions concerning the existence of the density $f=f(.;x)$ and its derivatives with respect to x mentioned in the above, suppose that $P(x)$, $\frac{\partial P}{\partial x_k}(x)$, resp., can be represented for a given vector x by (5),(6), and assume that the orthogonal series representation (9),(9.1), resp., holds. Then $P(x)$, $\frac{\partial P}{\partial x_k}(x)$, resp., can be represented for finite bounds $-\infty < y_\ell < y_u < +\infty$ by the following convergent series:

$$P(x) = \sum_{j=o}^{\infty} c_j(x) \int_{y_\ell}^{y_u} \varphi_j(y)\,dy, \qquad (14)$$

$$\frac{\partial P}{\partial x_k}(x) = \sum_{j=o}^{\infty} c_{kj}(x) \int_{y_\ell}^{y_u} \varphi_j(y)\,dy. \qquad (14.1)$$

3.1. THE INTEGRALS OVER THE BASIS FUNCTIONS AND THE COEFFICIENTS OF THE ORTHOGONAL SERIES

The integrals over the basis functions φ_j read

$$\int_{y_\ell}^{y_u} \varphi_j(y)\,dy = C_j \sum_{i=o}^{j} h_{ji} J_i \qquad (15)$$

where C_j, h_{ji}, $0 \leq i \leq j$, $j \geq 0$, are given, fixed coefficients, and for the definite integrals

$$J_i := \int\limits_{y_\ell}^{y_u} e^{-\frac{y^2}{2}} y^i dy, \quad i=0,1,\ldots, \tag{15.1}$$

by partial integration we find the recursion:

$$J_i = y_\ell^{i-1} e^{-\frac{y_\ell^2}{2}} - y_u^{i-1} e^{-\frac{y_u^2}{2}} + (i-1)J_{i-2}, \quad i\geq 2. \tag{15.2}$$

Mean value representations for the Fourier coefficients $c_{kj}(x)$ and can be obtained by interchanging differentiation and integration in the mean value representation (11) for $c_j(x)$:

$$c_{kj}(x) = \frac{\partial}{\partial x_k} E\varphi_j(y(a(\omega),x))$$

$$\tag{16}$$

$$= E\varphi_j'(y(a(\omega),x)\frac{\partial y}{\partial x_k}(a(\omega),x).$$

Based on the mean value representations (11),(16), the Fourier coefficients can be estimated by sampling methods, e.g. by the sample means

$$\hat{c}_j(x) := \frac{1}{n} \sum_{t=1}^{n} \varphi_j(y(a^t,x)), \tag{17}$$

$$\hat{c}_{kj}(x) := \frac{1}{n} \sum_{t=1}^{n} \varphi_j'(y(a^t,x))\frac{\partial y}{\partial x_k}(a^t,x), \tag{17.1}$$

where a^t, $t=1,\ldots,n$, denote independent realizations of $a=a(\omega)$.

3.2. ESTIMATION/APPROXIMATION OF P(X) AND ITS DERIVATIVES

Based on the expansions (9),(9.1), estimates of $P(x)$ and its derivatives can be defined by

$$\hat{P}(x) := \int\limits_{y_\ell}^{y_u} \hat{f}(y;x)dy = \sum_{j=o}^{q(n)} \hat{c}_j(x) \int\limits_{y_\ell}^{y_u} \varphi_j(y)dy, \tag{18}$$

$$\frac{\partial \hat{P}}{\partial x_k}(x) := \int_{y_\ell}^{y_u} \frac{\partial \hat{f}}{\partial x_k}(y;x)\,dy - \sum_{j=0}^{q(n)} \hat{c}_{kj}(x) \int_{y_\ell}^{y_u} \varphi_j(y)\,dy, \quad (18.1)$$

where $q=q(n)$ is a certain integer and $\hat{c}_j(x)$, $\hat{c}_{kj}(x)$ are unbiased estimators of the Fourier coefficients $c_j(x)$, $c_{kj}(x)$, resp., as defined by (17),(17.1).

We have the following **consistency result**:

Theorem 3.2. For a given, fixed vector x suppose that $f(.;x) \in L^2(\mathfrak{R})$, $\frac{\partial f}{\partial x_k}(.;x) \in L^2(\mathfrak{R})$, resp., and the second moments under consideration are finite. a) If $q=q(n)$ is chosen such that $q(n) \longrightarrow \infty$ and $\frac{q(n)}{n} \longrightarrow 0$ as $n \longrightarrow \infty$, then $E(P(x)-\hat{P}(x))^2 \longrightarrow 0$ as $n \longrightarrow \infty$.

   b) If $q=q(n)$ is selected such that $q(n) \longrightarrow \infty$ and $\frac{q(n)^2}{n} \longrightarrow 0$ as $n \longrightarrow \infty$, then $E(\frac{\partial P}{\partial x_k}(x)-\frac{\partial \hat{P}}{\partial x_k}(x))^2 \longrightarrow 0$ as $n \longrightarrow \infty$.

References

[1]   Barner, M., Flohr, F.: Analysis II. Walter de Gruyter, Berlin-New York 1983

[2]   Breitung, K., Hohenbichler, M.: Asymptotic approximations for multivariate integrals with applications to multinormal probabilities. J. of Multivariate Analysis 30 (1989), 80-97

[3]   Breitung, K.: Asymptotische Approximation für Wahrscheinlichkeitsintegrale. Habilitationsschrift, Fakultät für Philosophie, Wissenschaftstheorie und Statistik der Universität München 1990; to appear in the Springer Lecture Notes Series in Mathematics

[4]   Gajewski, A., Zyczkowski, M.: Optimal Structural Design under Stability Constraints. Kluwer, Dordrecht-Boston-London 1988

[5]   Gollwitzer, S., Rackwitz, R.: An efficient solution to the multinormal integral. Prob. Eng. Mech. 3(2) (1988), 98-101

[6] Hohenbichler, M., Rackwitz, R.: Non-normal depen-
     dent vectors in structural safety. J. Eng. Mech.
     Div. ASCE 107(6) (1981), 1227-1249
[7] Kall, P., Wallace, S.: Stochastic Programming. J.
     Wiley, Chichester (UK) 1994
[8] Marti, K.: Stochastic Optimization Methods in
     Structural Mechanics. ZAMM 70 (1990), T742-T745
[9] Marti, K.: Approximations and Derivatives of
     Probability Functions. In: G. Anastassiou, S.T.
     Rachev (eds.): Approximation, Probability and
     Related Fields. Plenum Press, New York 1994,
     367-377
[10] Melchers, R.E.: Structural Reliability Analysis and
     Prediction. J. Wiley, New York 1987
[11] Richter, H.: Wahrscheinlichkeitstheorie. Springer-
     Verlag, Berlin 1966
[12] Sansone, G.: Orthogonal Functions. Interscience
     Publishers Inc., New York - London 1959
[13] Schüler, L.: Schätzungen von Dichten und Vertei-
       lungsfunktionen mehrdimensionaler Zufallsvariabler
     auf der Basius orthogonaler Reihen. Dissertation
     Naturwissenschaftliche Fakultät der TU Braun-
     schweig, Braunschweig 1974
[14] Schüler, L., Wolff, H.: Schätzungen mehrdimensio-
     naler Dichten mit Hermiteschen Polynomen und li-
     neare Verfahren zur Trendverfolgung.
     Forschungsbericht BMVg-FBWT 76-23, TU Braunschweig,
     Institut für Angewandte Mathematik, Braunschweig
     1976
[15] Schwartz, S.C.: Estimation of Probability Density
     by an Orthogonal Series. Ann. Math. Statist. 38
     (1967), 1261-1265

A CONSTRAINT-BASED APPROACH TO THE DESIGN AND OPTIMISATION OF MECHANISM SYSTEMS

A.J. MEDLAND, G. MULLINEUX,
B.R. TWYMAN and A.H. RENTOUL

*School of Mechanical Engineering
University of Bath
Bath BA2 7AY
United Kingdom*

1. Introduction

The design synthesis, analysis and optimisation of mechanism systems tends to be a specialised discipline much researched by academic groups [1,2,3]. Within many industrial companies, designers meet such problems on an intermittent basis. This means that familiarisation with the techniques has to be refreshed each time a new design problem is encountered. Computer-based kinematic and dynamic programs are available on most advanced computer aided design systems. However, these are often means for analysing a given mechanism. Little help is available for synthesis of new mechanisms or for optimisation of existing ones.

A computer system called *SWORDS* is currently being generated by the authors. Its aim is to provide a more complete range of aids for the designer. The structure of a mechanism system can be defined in terms of the constraints which determine its required function and bound the range of possible design solutions. The constraints may, for example, prescribe the desired output path of the system or give velocity and acceleration constraints throughout the motion cycle. The software can resolve constraints automatically. It does so on the basis of internal optimisation techniques, which are currently based on direct search strategies. The use of an optimisation approach to constraint resolution permits "best compromise" configurations to be determined even when the constraints are mutually incompatible.

Help is provided with the synthesis aspects of design for function or path generation. The system is provided with libraries of standard mechanisms types. The system searches the libraries to find close approximations to specified outputs. Often these are good enough for further design development. If not, the built-in optimisation procedures can be used to vary the mechanism geometry to gain a better match to the desired output. Also at this stage, additional performance constraints can be added and the design optimised with respect to these.

The *SWORDS* software can perform kinematic analysis itself, together with simple kineto-static work. For a true dynamic analysis, the ability to communicate with an external analysis package (such as *ADAMS*) is provided. Here *SWORDS* prepares data for the external software, invokes it, and then retrieves the results.

D. Bestle and W. Schielen (eds.), IUTAM Symposium on Optimization of Mechanical Systems, 205–212.
© 1996 *Kluwer Academic Publishers.*

This can be carried out within an optimisation process, with *SWORDS* handling the optimisation search, and the external package handling the evaluation.

Once a design is complete, it can be archived either as a whole, or as a number of sub-designs. The aim here is to allow a designer to build up a collection of working designs which can be recalled and incorporated in future work. Since the constraints are archived along with the pure geometry of the design, a record is maintained of the function of the design and its limitations.

In the next section, we describe the underlying user interface language which is called *RASOR*. It is this that is used to define the geometry of mechanisms and the constraint rules by which it operates. From this we go on to discuss the selection and optimisation of mechanisms in section 3 and the interfacing with *ADAMS* in section 4. The ability within *SWORDS* to decompose a design task in described in section 5. Examples of the use of the software and underlying methodology are given throughout.

2.   RASOR Language

Underlying the *SWORDS* system is an interactive user language called *RASOR*. This is an interpretative language which has been used and enhanced over a number of previous versions of the system [4]. It is in this language that the variables and rules which describe a particular mechanism application are defined. We here review some of the features of the *RASOR* language as it helps to understand how the system works. In practice, for particular applications, many of the details can be hidden from the user via pre-defined menu options.

The language allows variables to be declared and manipulated. At the most basic level, these variables are of the same form as one would find in any programming language. Thus we have integer and real values, text strings and so on. The usual sort of mathematical operations can be applied to these to set and modify these values. The user can enter commands using the language directly into the system, or alternatively these can be read in from a text file.

One way in which *RASOR* differs from a conventional language is in its ability to define and resolve constraint rules. We consider a very simple example of solving a pair of simultaneous linear equations.

$$x + y = 3$$
$$x + 2y = 4$$

Within the language, these are set up within a user defined function as follows.

```
fnc solve
{     var x, y;
      rul   x + y - 3;
      rul   x + 2*y - 4;
}
```

Here it is assumed that the variables x and y have been declared previously as global variables. The *rul* command is used to define the constraint rule which is essentially an algebraic expression which is zero when it is true. Effectively, the (absolute) value of a rule expression is a measure of its falseness. The constraint resolution process is based upon optimisation. The sum of the squares of the rule values is taken and automatically treated as the function of the variables listed by the user in the *var* command. The optimisation code built into the language attempts to select values of these to minimise the combined rule value.

Currently the optimisation scheme used is based on Powell's direct search method [5]. Using an optimisation approach means that a result is obtained even when the constraint rules conflict with each other. It is in a sense the best possible solution. As the direct search process is iterative, there is a danger of finding false minima positions which do not represent a fully true solution. Also the system only attempts to find one solution; there could be several others (or even infinitely many). In practice, these problems have not been found to present serious drawbacks, as it is usually clear to the human user when an inappropriate solution has been found.

RASOR has simple geometric entities available to deal with mechanism work. These include points, straight line segments, and circular arcs. Free-form curves in the form of B-splines are also provided. All these entities are available in both two and three dimensions. (In a recent extension, solid objects are also available via the ACIS core modeller [6].) The entities are built up as structured types within the language and this means that the user has access to the individual pieces of data. For example, the end points of a line segment can be obtained as points, and then the individual coordinates of these accessed. Constraint rules can be applied to entities and their sub-parts.

In order to simplify the creation of mechanism within the *RASOR* language, the concept of model spaces has been introduced [7]. A model space is essentially a transformation matrix which is applied to one or more geometric entities. This means that the entities in question are defined with respect to some local coordinate system. Applying the transform gives the position of these either in world space or within another model space. A hierarchy of model spaces is allowed; this gives a tree structure in which the world space is the root node. If the transform at some point of the hierarchy is changed then this alters the position in world space of all the spaces further away from the root node. However the relationship of these spaces to each other is maintained.

Using model spaces allows certain parts of a mechanism's assembly and functionality to be set up. However, as loops within the hierarchy are not permitted, a full description of a mechanism usually requires the addition of at least one constraint rule.

As an example, we consider the construction of a four bar chain mechanism. Initially three lines are constructed, to represent the moving links of the chain, and two points, to represent the fixed pivot positions. Each link is embedded in its own model space, and additionally the space of the coupler link is embedded in that of the crank. This provides a first stage assembly of the mechanism. It also means that as the crank rotates, the coupler would rotate with it.

To complete the assembly, a single constraint rule needs to be invoked to bring the ends of the coupler and driven links together. The *RASOR* language is allowed to modify the rotation parts of the transforms for the coupler and driven links. Applying the rule completes the assembly of the mechanism. We can now cycle it by rotating the crank through a number of steps and resolving the rule at each stage.

3. Mechanism Selection and Optimisation

Thus far we have seen how the *RASOR* language can describe a mechanism whose geometry is already known. However, it is precisely this geometry that is initially unknown and a designer wishes to obtain. The mechanism is required to perform some function. For example, this may be: either, the provision of a one dimensional (linear or angular) motion in accordance with a given timing diagram; or the generation of a closed, two dimensional, motion curve from the coupler.

To facilitate these types of applications, a number of catalogues of standard mechanism types have been created containing the basic geometry and the output functions. The functions are stored in terms of Fourier coefficients [8] and this allows some compression of the stored information to take place.

Figure 1 shows an example of a required function that has been entered by the user by fitting a smooth curve through specified points. The system normalises the desired output, forms its Fourier coefficients and obtains the first few closest matches from the catalogues. Three possible mechanisms producing fairly close matches are also shown in the figure.

In practice, often one of the mechanisms directly from the catalogues serves the required purpose closely enough. If not, it is possible to use the built-in optimisation code of *RASOR* to improve matters. Here the system is allowed to vary the various geometric pieces of data for the mechanism. The objective function is essentially to obtain a good match between the (first few) Fourier coefficients of the desired curve and those of the mechanism's output.

An example of such an optimisation is also given at the foot of Figure 1. The starting mechanism is a geared five bar obtained from the catalogues. Various stages during the optimisation are shown together with the final version produced.

4. Link to ADAMS and Further Optimisation

As we have seen, the *SWORDS* system by itself can handle the kinematic description of a mechanism. To check that a proposed mechanism design will function correctly other types of analysis are required. There are often particular pieces of dedicated software to undertake such tasks. One examples is *ADAMS* which is an established commercial software package for carrying out dynamic analysis.

SWORDS carries the ability to communicate which such external software. It does this effectively by preparing data for the system, running that system, and then capturing some or all of the results generated. We consider in particular the link to *ADAMS* and its use in the design process.

Once a kinematically acceptable mechanism for a required function has been obtained by *SWORDS* from its catalogue, the very basic geometric details are known. These are arranged in a standard format. Within the user language of *ADAMS* exists the ability to create solid models of links in parametric form. From the data supplied by *SWORDS*, *ADAMS* can automatically produce solid shapes and hence determine the masses, centres of gravity and moments of inertia for the moving parts. These form the basis of the dynamic analysis which it can then undertake. Once this has taken place, other routines in the *ADAMS* user language are invoked to capture desired results and then export these back to *SWORDS*.

In order to handle the actual transfer of data between the two pieces of software, use is made of the fact that both run under the UNIX operating system. This means that data transfer via pipes is possible. These work in the same way as standard files as far as running programs are concerned.

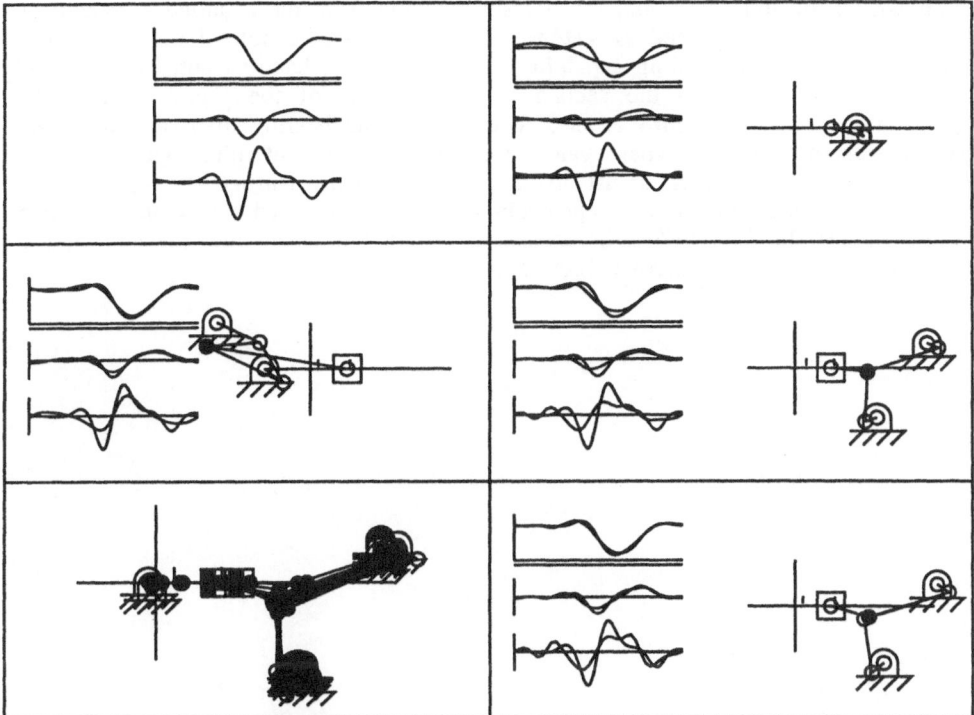

Figure 1: Selection and optimisation of a function generator

Functions are created in the *RASOR* language to invoke the data transfer and running of the external program. There is no reason why such a call cannot be made as part of a constraint rule. Thus we can specify any of the dynamic results as part of an expression to be minimised. Since the constraint resolution is based on a direct search strategy, *RASOR* simply treats the use of the external system as part of its way of evaluating the objective function. During an optimisation run, this means that the external program may be called very frequently. Naturally this can lead to run times taking several hours. However if the result is an improved design, the user is often prepared to wait.

In the case of *ADAMS*, it was found that some time could be saved by retaining the software in memory at all times and thus avoiding the overhead of reloading repeatedly. If this is done, then there is advantage in running *SWORDS* and *ADAMS* on two separate workstation. In this case, UNIX remote procedure calls need to be used in addition to pipes.

As an example of the link to *ADAMS*, we consider the case of obtaining a mechanism to generate a particular output path (to within a given tolerance) while minimising the bearing loads. Figure 2 gives some of the details. The path required in shown on the left. An initial mechanism is found from the catalogues. When the essential geometry is passed to *ADAMS*, this creates its version of the mechanism using its own solid modelling capability. This creation is handled automatically and the results are shown in the second and third parts of the figure. During the optimisation process, *ADAMS* is used to determine the bearing loads and these are passed back to *SWORDS*. These values are used in constraint rules together with a measure of the closeness of fit to the required path. During the optimisation processes new mechanisms are repeatedly proposed and tested (as in the first part of the figure). In fact, *SWORDS* checks to ensure that each new mechanism can cycle fully and does not invoke *ADAMS* for invalid ones.

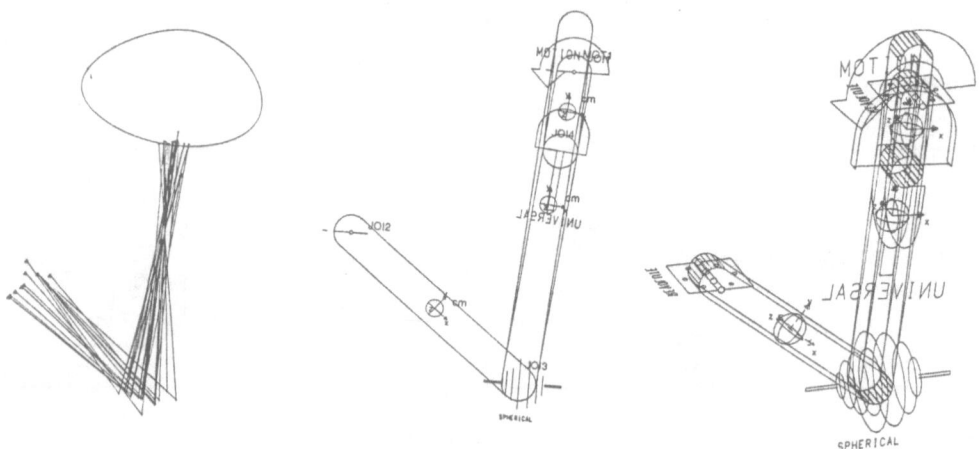

Figure 2:   Stages in link to external software

5. Design Decomposition

One of the features of *SWORDS* is the user's ability to work within what are termed "cells". Each cell contains the design variable and constraints associated with a particular sub-part of the overall design. Within a cell, individual constraints can be defined and these can be formed into groups in several ways. The aim here is to allow the designer to experiment with various different options and investigate their effects. It also means that extra constraints can be incorporated gradually thus reducing the need for any attempt at constraint resolution (with many free variables) which is initially far removed from a satisfactory solution.

The cells themselves are arranged on a hierarchical basis rather as the sub-directories of a UNIX file system. The intention here is not to impose a hierarchical structure of the design process itself but rather for convenience of implementation and to reflect the structure that exists within a design office. The cells at the lowest levels can relate to tasks undertaken by individual design personnel. Those further up, represent tasks undertaken by teams of designers and are made up of the sub-tasks of the team members. The top level represents the complete design.

Each cell and its contents can be "owned" by individual designers or teams. Constraints specified with any particular cell can reference design parameters held within others. However, by default, constraint resolution carried out within a cell is only permitted to change design variables within itself (and cells below it). Should conflict arise between cells, discussion between the owners is required. If this does not reach a satisfactory conclusion, the ultimate authority rests with the owner of that cell which is furthest down the hierarchy but still above the two conflicting cells.

As suggested above, one of the aims of using cells is to break a design task down into convenient sub-parts which can be handled easily. It is most desirable that constraint resolution be carried out only at the lower levels of the hierarchy. The danger of doing this higher up is that the number of free variables becomes unworkably large. An alternative is firstly to carry out a top-down design in which a few main design parameters are established. The details are then filled in by working within the lower cells. Regularly through the design process a check is made to see that there is no conflict within the overall scheme (but possibly without attempting automatic resolution).

Another advantage of the cell approach is that it enables individual parts of a design to be archived and re-used at a later stage in other design work. There is naturally a need to document what has been done and to try to establish conventions for such things as the names of variables. In this way a design group could create a library of standard mechanisms which have been designed and used successfully. By storing the constraint rules, information about the assembly and function can also be held.

6. Conclusions

We have seen how a computer system called *SWORDS* is being used to aid the design of mechanism and machine systems. It relies on an underlying user language called *RASOR* in which the design parameters are declared together with the constraint rules which govern their interaction and hence the operation of the system being designed.

The use of direct search optimisation for handling constraint resolution has been described. This has been used to help improve the performance of designs in two ways. The first relates to the synthesis of mechanisms to meet particular output requirements. Here a catalogue of standard mechanism types has been used. This provides an initial design from which the optimisation scheme can work to generate an improved version. The *SWORDS* system can work alongside existing design analysis packages by passing data forwards and backwards. The optimisation scheme can be used here to generate optimal solutions with respect to properties calculated by the external package.

The way in which *SWORDS* helps a design team to handle the individual sub-parts of a design task by breaking it into cells has been described. This also means that the optimisation scheme is used (in the first instance at least) to resolve constraints with relatively few degrees of freedom.

7. Acknowledgements

The enhancement of the constraint modelling methodology has and is being supported by the LINK Programme in the Design of High Speed Machinery; this funding is gratefully acknowledged. On the current project the authors are working with the University of Salford and a group of collaborating companies; the support of those involved is also acknowledged with thanks.

8. References

1. Kim, S.H. and Lee, K.: An assembly modelling system for dynamic and kinematic analysis, *Computer-Aided Design* **21** (1989), 2-12.
2. Ricci, R.J.: Spacebar: kinematic design by computer graphics, *Computer-Aided Design* **25** (1993), 727-735.
3. Srikanth, S. and Turner, J.U.: Toward a unified representation of mechanical assemblies, *Engineering with Computers* **6** (1990), 103-112.
4. Mullineux, G.: The introduction of constraints into a graphics system, *Engineering with Computers* **3** (1988), 201-205.
5. Walsh, G.R.: *Methods of Optimization*, Wiley, London, 1975.
6. Medland, A.J., Mullineux, G. and Rentoul, A.H.: Introducing solid objects into a constraint modelling system, *Engineering with Computers* **11** (1995) 27-35.
7. Leigh, R.D., Medland, A.J., Mullineux, G. and Potts, I.R.B.: Model spaces and their use in mechanism simulation, *Proc. I.Mech.E. Part B: Journal of Engineering Manufacture* **203** (1989), 167-174.
8. McGarva, J.R. and Mullineux, G.: A new methodology for rapid synthesis of function generators, *Proc. I.Mech.E. Part C: Journal of Mechanical Engineering Science* **206** (1992) 391-398.

MULTIBODY DYNAMICS OPTIMIZATION BY DIRECT DIFFERENTIATION METHODS USING OBJECT ORIENTED PROGRAMMING.

J.M. Pagalday and Iñaki Aranburu
Ikerlan
Pº J. M. de Arizmendiarrieta, 2. 20500 Mondragón (Guipúzcoa). Spain

Alejo Avello and Javier G. de Jalón
Centro de Estudios e Investigaciones Técnicas de Guipúzcoa
E.S.I.I. de San Sebastián (University of Navarra)
Pº Manuel Lardizabal, 15. 20009 San Sebastián. Spain.

1. Summary.

In this work, general purpose methodologies and formulations are developed for the optimization of the dynamic behaviour of 3D multibody systems. In order to achieve these goals, the latest advances in three fields are used: Dynamic formulations, object oriented programming languages and symbolic computation techniques.

On the first field, constraints are introduced as penalty forces, trying to reduce the complexity as well as the computational cost of the dynamic analysis. The object oriented programming is used to get an accurate representation of the physical system inside the computer, and a simpler and more compact code. The symbolic computation techniques are indispensable if a broad set of different cases is to be treated without rewriting the code, due to the great amount of different terms involved in the formulation. By this way, it is possible to treat examples of high complexity and practical interest, similar to those treated by commercial dynamic analysis codes existing nowadays.

2. Introduction.

The first research on optimization of the dynamic behaviour of mechanisms we have found are those of Haug and Arora (1978,1979) and Haug et al. (1980) on which the authors use the so called adjoint variable method, that keeps the computational cost of the problem at reasonable levels, but results on highly complex formulations when trying to be used as a general purpose method.

Other authors, as Krishnaswami and Bhatti (1984) proposed to use direct differentiation methods, that are the ones used in the present work. This approach is much simpler than the adjoint variable one, but in the previously mentioned work as well as in the one by Chang and Nikravesh (1985) the presented examples are relatively simple, due probably to the high complexity of the terms involved in the analysis.

D. Bestle and W. Schielen (eds.), IUTAM Symposium on Optimization of Mechanical Systems, 213–220.
© 1996 *Kluwer Academic Publishers.*

In order to solve the difficulties of the previous authors, Ashrafiuon and Mani (1990) used the symbolic computation letting the computer deal with the most complex formulae required in the sensitivity calculation, that is, as we will see, the heart of the direct differentiation methods.

3. Description of the problem.

Let the properties of a mechanism, both geometric and dynamic, be not fixed but function of a design parameter vector $\mathbf{b}=[b_1,...,b_k]^t$.

Although the objective function can be represented as a mere function of the design vector, it is often explicitly dependent on the state variables that are, on turn, dependent on the design vector. Chang (1985) proposed a different notation for ψ_0 that makes evident this fact.

$$\psi_0 = \psi_0\left(\mathbf{b}; t, \mathbf{q}, \dot{\mathbf{q}}, \ddot{\mathbf{q}}, \lambda\right) \tag{1}$$

where t represents the time and λ is the Lagrange multiplier vector, related to the reactions at the joints and the fixed points, and to the motor efforts.

The most usual objective functions are of the kind that might be called maximum and integral. The maximum type objective functions represent the maximum of a f function along time that represents a magnitude related to the mechanism, such as the reaction at a joint or an acceleration. The integral type objective functions often represent the mean values along time of magnitudes also related to the mechanism, and may be written as an integral of a function g along time, in a similar way to that expressed before.

If we consider the general form of objective function as an addition of both types before mentioned, we can treat a wide variety of cases.

$$\psi_0 = \psi_0^{max} + \psi_0^{integ} = \max_{t_{ini} \le t \le t_{end}} f\left(\mathbf{b}; t, \mathbf{q}, \dot{\mathbf{q}}, \ddot{\mathbf{q}}, \lambda\right) + \int_{t_{ini}}^{t_{end}} g\left(\mathbf{b}; t, \mathbf{q}, \dot{\mathbf{q}}, \ddot{\mathbf{q}}, \lambda\right) dt \tag{2}$$

Although the values of the design variables may be anyone, the most usual case in the engineering practice is that design constraints (different from kinematic constraints) make the values of such variables fulfil certain conditions. Following a notation similar to that used for the objective function, we may write them as

$$\psi_i = \psi_i\left(\mathbf{b}; t, \mathbf{q}, \dot{\mathbf{q}}, \ddot{\mathbf{q}}, \lambda\right) \le 0 \quad i=1,\cdots,m \tag{3}$$

Although some authors develop specific formulations for the design constraints, we have used the same formulation used for the objective function in a broad variety of design constraints.

The main difficulty when solving the optimization problems consists on the need for the calculation of the derivatives of the objective function and design constraints with relation to the design vector $\mathbf{b}$, required by most of the best existing optimization algorithms.

$$l_0 = \frac{d\psi_0}{d\mathbf{b}}, \quad l_i = \frac{d\psi_i}{d\mathbf{b}} \tag{4}$$

In the direct differentiation methods, these derivatives are calculated by simply using the chain rule, resulting (Krishnaswami and Bhatti, 1984)

$$\frac{d\psi_0}{db} = \frac{\partial\psi_0}{\partial q}q_b + \frac{\partial\psi_0}{\partial \dot{q}}\dot{q}_b + \frac{\partial\psi_0}{\partial \ddot{q}}\ddot{q}_b + \frac{\partial\psi_0}{\partial \lambda}\lambda_b + \frac{\partial\psi_0}{\partial b} \tag{5}$$

Taking for the objective function the above mentioned formulation, this derivative has the form

$$\frac{d\psi_0}{db} = \frac{d\psi_0^{max}}{db} + \frac{d\psi_0^{integ}}{db} = \left[f_q\,q_b + f_{\dot{q}}\,\dot{q}_b + f_{\ddot{q}}\,\ddot{q}_b + f_1\,\lambda_b\right]_{t=t_{max}} + \int_{t_{ini}}^{t_{end}} \left(g_q q_b + g_{\dot{q}}\dot{q}_b + g_{\ddot{q}}\ddot{q}_b + g_1\lambda_b\right)dt \tag{6}$$

In the previous equations, the subscripts represent partial derivatives, and t_{max} is the time at which the function f reaches its maximum value.

The main difficulty involved on the presented formulations is, rather than on the equations themselves, on the calculation of the partial derivatives of the state variables of the system with respect to the design variables. These derivatives are usually known as state variables sensitivities.

4. Sensitivity analysis

Bayo et al. (1988) presented a set of methods for the solution of the multibody dynamics that can be grouped under the common denomination of penalty methods. Basically, they consist on substituting the kinematic constraints by a set of external forces applied to the system, proportional to such kinematic constraints and their derivatives, resulting on an unconstrained system, whose analysis is much simpler than that of a constrained one. The process can also been seen (Pagalday, 1994) as the minimization of an objective function whose design constraints are the kinematic constraints, with or without adding the stabilization terms proposed by Baumgarte (1972). Anyway, the motion equations of the system result to be

$$\left(M + \Phi_q^T \alpha \Phi_q\right)\ddot{q} = f - \Phi_q^T \alpha\left(\dot{\Phi} + 2\xi\omega\dot{\Phi} + \omega^2\Phi\right) \tag{7}$$

where M is the mass matrix of the mechanism, Φ is the kinematic constraint vector, α is a large penalty factor and ξ and ω are "stiffness" and "damping coefficients that are usually set to get critical damping on the penalty forces. Equation (7) is a set of second order ordinary differential equations, that can be integrated along time to obtain the vectors q and $\dot{q}$ of dependent positions and velocities of the system.

If we differentiate the previous equation with respect to the generic design variable b, we obtain the equation

$$\left(M' + \Phi_q^{'T}\alpha\Phi_q + \Phi_q^T\alpha\Phi_q'\right)\ddot{q} + \left(M + \Phi_q^T\alpha\Phi_q\right)\ddot{q}_b =$$
$$= f' - \Phi_q^{'T}\alpha\left(\ddot{\Phi} + 2\xi\omega\dot{\Phi} + \omega^2\Phi\right) - \Phi_q^T\alpha\left(\ddot{\Phi}' + 2\xi\omega\dot{\Phi}' + \omega^2\Phi'\right) \tag{8}$$

This equation is a new set of differential equations, that can be used to integrate vector $\ddot{\mathbf{q}}_b$ of dependent acceleration sensitivities, obtaining thus vectors $\mathbf{q}_b$ and $\dot{\mathbf{q}}_b$.

In equation (8), $\mathbf{x}'$ represents the total derivative of magnitude $\mathbf{x}$ with respect to the generic design variable b in an specific time step.

5. The need for symbolic computation.

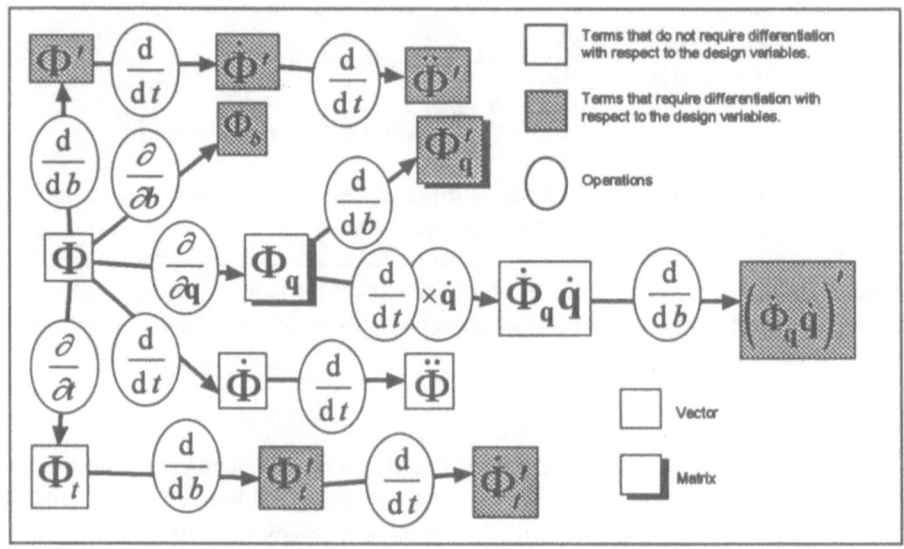

Figure 1: Terms derived from the constraint vector.

In the same way that happened when obtaining derivatives for the objective function, the complexity of the formulations for the sensitivity analysis is not mainly due to the complexity of the used equations, but to the high amount of different terms that have to be considered. Figure 1 shows the terms that are deviated from the constraint vector Φ, and the operations that are required to obtain them.

Among the 14 considered terms, the 8 highlighted in grey have a different form depending on the kind of design variable to be considered, since they require differentiations with respect to the design variables. So, if we take into account these 8 terms, and consider the number of different design variables that can be used in a general purpose multibody optimization program can be more than ten, and that the number of different kinds of kinematic constraints can be more than 30, it results that the number of different terms that will have to be programmed by hand will be more than two thousand. The enormous size of this job, and the systematic way on which the required operations can be performed make it specially suited to be performed by the computer.

The fact of manipulating the equations automatically by computer is usually known as symbolic computation. When using symbolic computation, the numbers are substituted by data structures that keep the original information contained in the equations.

- Although several commercial codes for symbolic computation are nowadays available (Mapple, Theorist, Mathematica,...), a set of classes able to represent the equations used in the sensitivity computation has been developed in this work.

The equations of the multibody dynamics may be written, in a general form, as:

$$M\ddot{x} = f \tag{9}$$

where M is a generalized mass matrix and f is a generalized force vector whose formulae depend on the method to be used, and $\ddot{x}$ is the dependent or independent acceleration vector, also depending on the used method. In a similar way, we may write the sensitivity equations by simply derivating equation 11 with respect to the generic design variable b, to obtain

$$M\ddot{x}_b = f' - M'\ddot{x} \tag{10}$$

Taking this into account, we can use three approaches in order to use the symbolic computation:

- **Totally numerical**: The basic components of matrix M and vector f and their derivatives are obtained simbolically, by using the expressions of equation (7) and (9) or those corresponding to other methods, and the generated equation system is also solved numerically, and the accelerations and their sensitivities are calculated.
- **Totally symbolic**: This approach is opposite to the previous one. Matrix M and vector f are calculated symbolically, and the equation system (9) is also solved symbolically, obtaining an explicit expression for vector $\ddot{x}$, that can be symbolically differentiated to obtain the acceleration sensitivity vector $\ddot{x}_b$.
- **Mixed**: This is an intermediate approach, on which matrix M and vector f and their derivatives are obtained simbolically, and are also simbolically differentiated to obtain M' and f'. Once this is done, equation systems (9) and (10) can be numerically solved to obtain the accelerations and their sensitivities.

6. The object oriented programming on the multibody system optimization.

The OOP has reached nowadays a level of maturity that makes it possible to take it as a valid alternative for the development of engineering codes. Nevertheless, its use in the area of multibody analysis is still reduced, although in recent years some interesting works have been published, among which we could mention those from Anantharaman (1993), Keckskemety and Hiller (1993) or Koh and Park (1994).

Although describing the whole of the developed class library would be very long, Figure 2 shows a representative example of the organization of the library.

First of all, we can mention the large size classes, such as the **MECHANISM** class, that represents the mechanism as a whole, acting as data containers, forming usually lists of smaller size classes (such as those representing joints or bodies of the multibody system containing themselves lists of even smaller classes), or in the form of vectors, such as vector q that we can see in Figure 2. On the other side, these classes perform the most complex tasks, such as the solution of the dynamic equations, but these complex tasks are often passed, totally or partially, to objects or lists of lower level.

On the other hand, the small size classes, such as the **COORDINATE** class, that represents any of the dependent coordinates of the system, contain pointers that point to data that are often contained in bigger classes, such as the coordinate position, that points to a position of the q vector contained in the **MECHANISM** class. In addition, these classes perform more basic tasks, which are not passed to other objects, but performed by themselves.

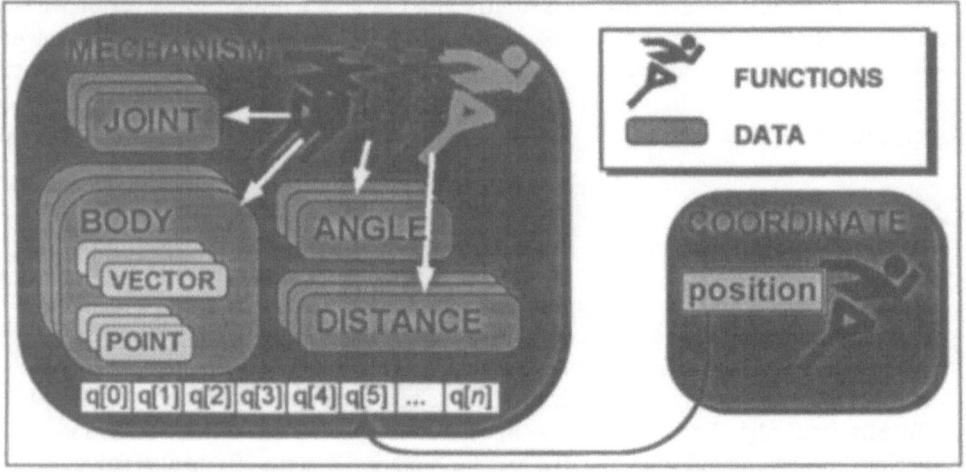

Figure 2: Class example.

7. Example

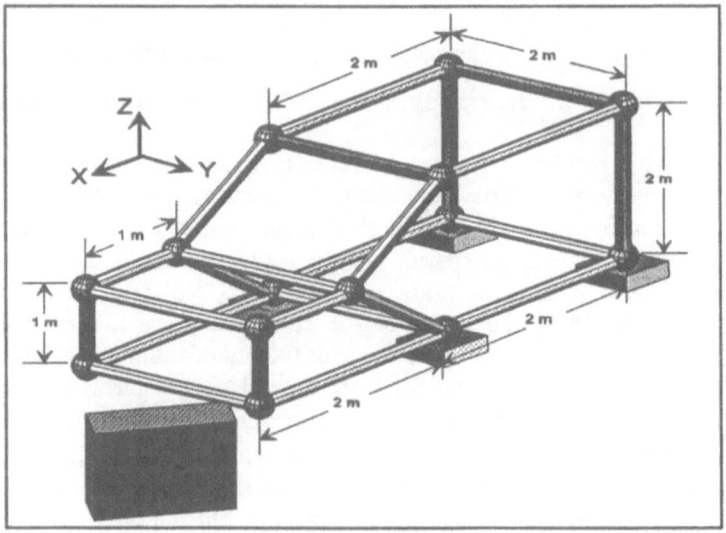

Figure 3: Vehicle crash.

The example, that we can see in Figure 3, represents the crashing of a simplified model of a 3D vehicle against a wall that forms an angle of 45° with the trajectory of the vehicle. The flexibility of the vehicle is lumped on the spherical joints between the elements, that have been considered to be linear. The mechanism has 23 bodies, 52 relative angles and 90 dependent coordinates.

The relative values of the stiffnesses at the joints have been considered as design parameters. Taking into account symmetry considerations and grouping the joints that could be considered similar depending on their location, seven different stiffness parameters have been defined. As a result of that, we may write the **b** design vector as

$$\mathbf{b} = \left\{ k_1 / k_{10} \quad k_2 / k_{20} \quad \cdots \quad k_7 / k_{70} \right\} \tag{11}$$

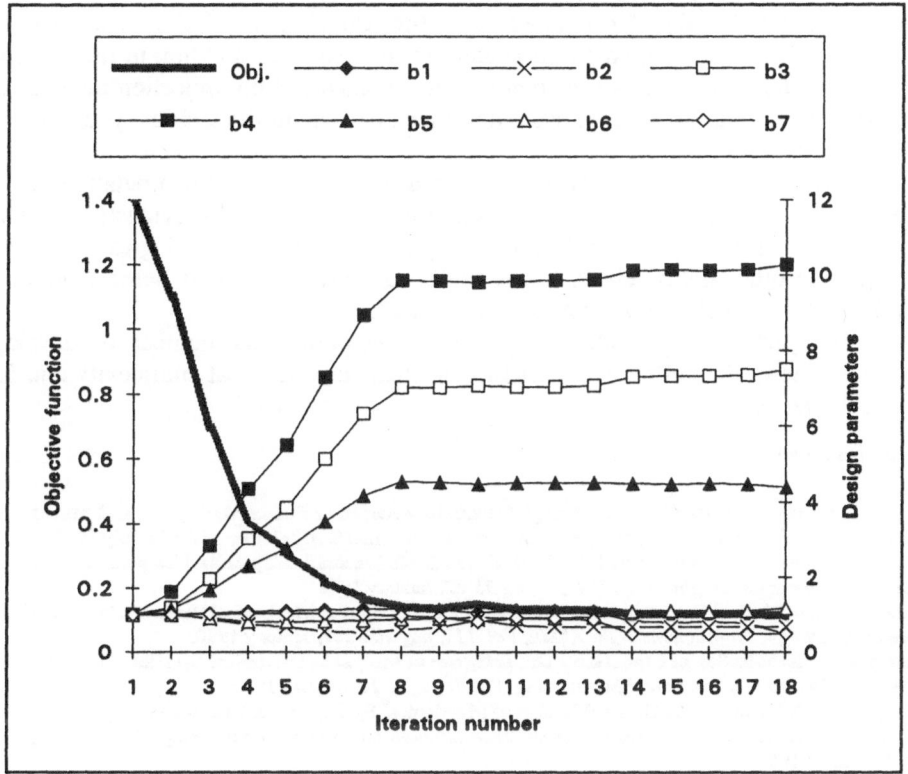

Figure 4: Example results.

On the other hand, although the objective function could have been written in terms of accelerations, it has been written to achieve that the deformation of the cockpit is minimum. This deformation has been formulated as a function of the difference of the value of the angles that form each couple of bars with respect to the original value. By doing so, the objective function results to be.

$$\psi_0 = \max_{t=0}^{t=t_{end}} \left(\sum_{i=1}^{numAngles} \left(\theta_i - \theta_{i0} \right)^2 \right) \tag{12}$$

The simulation results, that can be seen in Figure 4, show how the optimization program behaves in a similar way to a human designer, making the cockpit stiffer and maintaining or reducing the rest of the stiffnesses.

The model can be easily improved, for example by introducing elastic-plastic joints, or increasing the number of points used to model the vehicle, but it is enough to show the abilities of the developed program to treat cases of high complexity and size.

8. Conclusions.

In this paper, formulations as well as methodologies for the optimization of 3D mechanisms in a general way have been developed.

The formulations are divided in two steps. First of all, the derivatives of the objective function and the design constraints are calculated by direct differentiation methods. Secondly, the formulations for the calculation of the state variables sensitivities are developed, based on penalty formulations. In addition to the presented formulations for dynamic analysis, other formulations based on projection methods and on purely lagrangian approaches were developed but due to the lack of space were not included here.

The methodologies are focused on the use of the symbolic computation and the object oriented programming as tools that simplify greatly the previously mentioned task that has, in itself, a relatively high complexity. These methodologies can be used separately in other fields related to the multibody systems, as it is being done at the present such as vehicle simulation and biomechanics.

The presented example, that is only one of a great number of developed examples, shows the ability of the method to treat cases of great complexity and high practical interest.

9. References.

Haug, E.J., Wehage, R. & Barman, N.C., **Design Sensitivity Analysis of Planar Mechanism and Machine Dynamics,** *Journal of Mechanical Design, Transactions of the ASME, ASME pág 560-570,* July 1980.

Haug,E.J. & Arora, J.S., **Design Sensitivity Analysis of Elastic Mechanical Systems,** *Computer Methods in Applied Mechanics and Engineering, Vol. 15, pág 35-62,* January 1978.

Ashrafiuon, H. & Mani, N.K., **Analysis and Optimal Design of Spatial Mechanical Systems,** *Journal of Mechanical Design, Transactions of the ASME, Vol. 112, pág 200-207,* January 1990.

Baumgarte, J., **Stabilization of Constraints and Integrals of Motion in Dynamical Systems,** *Computer Methods in Applied Mechanics and Engineering, Vol. 1, pág. 1-16,* January 1972.

Chang, C.O. & Nikravesh, P.E., **Optimal Design of Mechanical Systems With Constraint Violation Stabilization Method.,** *Journal of Mechanisms, Transmissions and Automation in Design, Vol. 107, pág 493-498,* December 1985.

Anantharaman, M., **Flexible Multibody Dynamics. An Object Oriented Approach.,** *Proceedings of the NATO-ADvanced Study Institute on Computer Aided Analysis of Rigid and Flexible Mechanical Systems. Volume II: Contributed Papers, pág. 383-402,* June 1993.

Pagalday, J. M., **Optimización del comportamiento dinámico de mecanismos.,** *Ph.D. Thesis, E.S.I.I. de San Sebastián.,* July 1994.

Bayo, E., García de Jalón, J., and Serna, M. A., **A Modified Lagrangian Formulation for the Dynamic Analysis of Constrained Mechanical Systems.,** *Computer Methods in Applied Mechanics and Engineering, Vol. 71, pág. 183-195,* January 1988.

Kecskemethy, A., and Hiller, M., **Object Oriented Approach for an Effective Formulation of Multibody Dynamics,** *Second U.S. National Congress on Computational Mechanics, Washington, D.C.,* January 1993.

Haug, E. J., and Arora, S.J., **Applied Optimal Design,** *John Wiley & Sons.,* January 1979.

Krishnaswami, P., and Bhatti, M. A., **A General Approach for Design Sensitivity Analysis of Constrained Dynamic Systems.,** *ASME Paper 84-DET-132,* January 1984.

OPTIMIZATION OF PLANAR MECHANISM KINEMATICS WITH SYMBOLIC COMPUTATION

V.J. PESCH, C.L. HINKLE AND D.A. TORTORELLI
Department of Mechanical and Industrial Engineering
University of Illinois at Urbana-Champaign
Urbana, IL 61801, USA

1. Introduction

In mechanism design, the driving speeds and geometry of each member are varied and the mechanism is re-analyzed until it meets the necessary design specifications. Common design specifications monitor link positions, velocities, accelerations, joint reaction forces, and actuator power requirements. Each trial design requires an analysis whereupon the results are inspected to determine if the re-design is adequate. In the present work, this iterative design process is automated. Symbolic computation is used to generate the analysis and sensitivity analysis *FORTRAN* source code. The final software package is able to optimize a general multi-body system, hence it differs from the methods that generate source code on-the-fly. [1–4] The software is used to design a wheel loader bucket to maximize load capacity for a given material load cycle.

Some research has attempted to automate the mechanism design process. One methodology seeks analytical solutions to position synthesis problems. [5] A finite set of points are provided through which a given point on the mechanism must pass. Other variants of this problem are concerned with orienting bodies at given instances of time. Solutions for these problems are restricted to a few simple mechanisms. In another position synthesis problem, the desired trajectory is provided through which a given point on the mechanism must pass. A least squares error function is then defined to quantify the difference between the desired trajectory and the trajectory of the current design. [6–8] The design is subsequently modified to minimize this error function. Dynamic and force constraints are considered in the optimal design of mechanisms by Haug and his colleagues [9,10]. These

D. Bestle and W. Schielen (eds.), IUTAM Symposium on Optimization of Mechanical Systems, 221–230.
© 1996 *Kluwer Academic Publishers.*

last two approaches use computer simulation programs, design sensitivity analysis, and numerical optimization in much the same way as it is used here.

2. Design Sensitivity Analysis

Efficient optimization algorithms use sensitivities to systematically search the design space to determine the optimal design parameter values, s, that will minimize a cost function while satisfying the constraint functions. Efficient algorithms are desired because every optimization iteration requires one or more computationally expensive simulations. Zero-order optimization algorithms which do not require sensitivity analysis, e.g. Neural Networks, Genetic Algorithms, and Simulated Annealing, are computationally intensive as they may require thousands of simulations. [11–14]

There are several different means of extracting sensitivity information, the finite difference, the adjoint, and the direct differentiation methods. The finite difference method is computationally unattractive due to the additional analysis requirements, furthermore, finite difference sensitivities are susceptible to round-off and truncation errors. The adjoint method has been used with great success [9, 15], however, for simplicity we adopt the direct method.

The Cartesian coordinate formulation, developed by, e.g., Nikravesh, Haug and Chace [16–18], is used to determine the position, velocity, acceleration, and reaction forces of kinematically driven planar mechanisms. Throughout, we use the notation of [17].

In design sensitivity analysis we consider the change in a performance measure G, with respect to a design parameter, s. The performance measure is expressed in a general form as

$$G(\mathbf{q}_i(t,s), \dot{\mathbf{q}}_i(t,s), \ddot{\mathbf{q}}_i(t,s), \mathbf{Q}_i^k(t,s), s), \quad i = 1, n_b, \quad k = 1, n_j \qquad (1)$$

where n_b and n_j denote the number of bodies and joint/driver constraints, respectively. In the above, the response quantities $\mathbf{q}_i$, $\dot{\mathbf{q}}_i$, $\ddot{\mathbf{q}}_i$, and $\mathbf{Q}_i^k$ are expressed as functions of the design parameter since their values change as the design parameter is varied. Assuming sufficient smoothness, the gradient, or design sensitivity, is given by

$$\frac{DG}{Ds} = G_{,\mathbf{q}_i} \, \mathbf{q}_{i,s} + G_{,\dot{\mathbf{q}}_i} \, \dot{\mathbf{q}}_{i,s} + G_{,\ddot{\mathbf{q}}_i} \, \ddot{\mathbf{q}}_{i,s} + G_{,\mathbf{Q}_i^k} \, \mathbf{Q}_{i,s}^k + G_{,s} \qquad (2)$$

where the arguments have been suppressed for conciseness.

In the direct differentiation method, the response sensitivities $\mathbf{q}_{i,s}$, $\dot{\mathbf{q}}_{i,s}$, $\ddot{\mathbf{q}}_{i,s}$, and $\mathbf{Q}_{i,s}^k$ are evaluated so that Equation 2 may be computed. [16, 17]

To evaluate the response sensitivities, the constraint vector is re-expressed to denote its dependence on the design i.e.,

$$\mathbf{\Phi}(\mathbf{q}(t, s), t, s) = \mathbf{0} \tag{3}$$

The sensitivity, $\mathbf{q}_{,s} = [\mathbf{q}_{1,s}^T, \mathbf{q}_{2,s}^T, \mathbf{q}_{3,s}^T, ..., \mathbf{q}_{nb,s}^T]^T$ is obtained by differentiating the global constraint equation with respect to the design parameter.

$$\mathbf{\Phi}_{,\mathbf{q}}\, \mathbf{q}_{,s} = -\mathbf{\Phi}_{,s} \tag{4}$$

Similarly, differentiation of the generalized velocity and acceleration equations yields equations for $\dot{\mathbf{q}}_{,s}$ and $\ddot{\mathbf{q}}_{,s}$

$$\mathbf{\Phi}_{,\mathbf{q}}\, \dot{\mathbf{q}}_{,s} = -\left(\mathbf{\Phi}_{,\mathbf{q}}\,\dot{\mathbf{q}}\right)_{,\mathbf{q}} \mathbf{q}_{,s} - \mathbf{\Phi}_{,\mathbf{q}s}\,\dot{\mathbf{q}} + \boldsymbol{\nu}_{,\mathbf{q}}\,\mathbf{q}_{,s} + \boldsymbol{\nu}_{,s} \tag{5}$$

$$\mathbf{\Phi}_{,\mathbf{q}}\, \ddot{\mathbf{q}}_{,s} = -\left(\mathbf{\Phi}_{,\mathbf{q}}\,\ddot{\mathbf{q}}\right)_{,\mathbf{q}} \mathbf{q}_{,s} - \mathbf{\Phi}_{,\mathbf{q}s}\,\ddot{\mathbf{q}} + \boldsymbol{\gamma}_{,\mathbf{q}}\,\mathbf{q}_{,s} + \boldsymbol{\gamma}_{,\dot{\mathbf{q}}}\,\dot{\mathbf{q}}_{,s} + \boldsymbol{\gamma}_{,s} \tag{6}$$

where $\dot{\mathbf{q}}_{,s} = [\dot{\mathbf{q}}_{1,s}^T, \dot{\mathbf{q}}_{2,s}^T, \dot{\mathbf{q}}_{3,s}^T, ..., \dot{\mathbf{q}}_{nb,s}^T]^T$ and $\ddot{\mathbf{q}}_{,s} = [\ddot{\mathbf{q}}_{1,s}^T, \ddot{\mathbf{q}}_{2,s}^T, \ddot{\mathbf{q}}_{3,s}^T, ..., \ddot{\mathbf{q}}_{nb,s}^T]^T$.

To obtain the reaction forces sensitivities, we must first evaluate the sensitivities of the Lagrange multiplier vector from

$$\mathbf{\Phi}_{,\mathbf{q}}^T\, \boldsymbol{\lambda}_{,s} = -\left[\left(\mathbf{\Phi}_{,\mathbf{q}s}^T\,\boldsymbol{\lambda}\right) + \left(\mathbf{\Phi}_{,\mathbf{q}}^T\,\boldsymbol{\lambda}\right)_{,\mathbf{q}}\mathbf{q}_{,s} + \right.$$
$$\left. \mathbf{M}_{,s}\,\ddot{\mathbf{q}} + \mathbf{M}\ddot{\mathbf{q}}_{,s} - \mathbf{Q}^A{}_{,s} - \mathbf{Q}^A{}_{,\mathbf{q}}\,\mathbf{q}_{,s}\right] \tag{7}$$

Once $\boldsymbol{\lambda}_{,s}$ is know, the reaction force sensitivities are computed from

$$\mathbf{Q}_{i,s}^k = -\left[\mathbf{\Phi}_{,\mathbf{q}_i s}^{k\,T}\,\boldsymbol{\lambda}^k + \left(\mathbf{\Phi}_{,\mathbf{q}_i}^{k\,T}\,\boldsymbol{\lambda}^k\right)_{,\mathbf{q}_i} \mathbf{q}_{i,s} + \left(\mathbf{\Phi}_{,\mathbf{q}_i}^{k\,T}\,\boldsymbol{\lambda}^k\right)_{,\mathbf{q}_j} \mathbf{q}_{j,s} + \mathbf{\Phi}_{,\mathbf{q}_i}^{k\,T}\,\boldsymbol{\lambda}_{,s}^k\right] \tag{8}$$

Note that the sensitivity analyses require only the formulation of the right hand side vectors in Equations 4-7 as the coefficient matrix of $\mathbf{q}_{,s}$, $\dot{\mathbf{q}}_{,s}$, $\ddot{\mathbf{q}}_{,s}$, and $\boldsymbol{\lambda}_{,s}$ is the Jacobian, which has previously been decomposed in the position analysis. Thus, the direct differentiation sensitivities require far fewer computations to evaluate than the finite difference sensitivities which require a complete re-analysis for each design parameter. Further, the direct differentiation sensitivities are not subject to round-off and truncation errors.

3. Computer Implementation

To compute the response and its sensitivities, a modular computer environment is used. For a given time, the response, i.e. position, velocity, acceleration, Lagrange multipliers and reaction forces, is first computed and then the response sensitivity with respect to each design parameter is calculated. Finally, each performance measure and its sensitivity is evaluated from Equations 1 and 2

To compute the sensitivities at each time, t, we first note that two types of terms are present in the sensitivity analyses of Equations 4-7: explicit derivatives (e.g. $\mathbf{\Phi}_{,s}$, $\mathbf{\Phi}_{,qs}$) and implicit response sensitivities (e.g. $\mathbf{q}_s$, $\dot{\mathbf{q}}_s$). The explicit derivatives are readily evaluated quantities, because the joint/driver equations are known. The response sensitivities, however, are implicitly defined and are evaluated from Equations 4-7. To modularized the program, we group the terms which only contain explicit derivatives as these are only non-zero for the individual joint/driver constraint which the design parameter affects. On the other hand, terms which contain response sensitivities must be included for all of the joints and drivers. Consequently, when evaluating the right-hand side of Equations 4-7, we compute the explicit derivatives, (e.g. $-\mathbf{\Phi}_{,qs}\dot{\mathbf{q}}+\boldsymbol{\nu}_{,s}$ of Equation 5) for only the joint/driver constraint which is affected by the design parameter, s. Meanwhile, we compute the terms which contain the implicit response derivatives for all joints (e.g. $-\left(\mathbf{\Phi}_{,\mathbf{q}}\dot{\mathbf{q}}\right)_{,\mathbf{q}}\mathbf{q}_{,s}+\boldsymbol{\nu}_{,\mathbf{q}}\mathbf{q}_{,s}$ of Equation 5).

Once the appropriate right-hand sides are assembled, the response sensitivities, (e.g. $\dot{\mathbf{q}}_{,s}$ of Equation 5) are computed by back substitution into the previously decomposed Jacobian.

The global constraint vector $\mathbf{\Phi}$ is the assemblage of all the individual joint/driver constraint equations, $\mathbf{\Phi}^k$ and likewise for the Jacobian, $\boldsymbol{\nu}$, $\boldsymbol{\gamma}$, and the right-hand sides of Equations 4-7. The assembly process is conducive to modular program design. A subroutine is written for each joint/driver which outputs the joint/driver constraint equation $\mathbf{\Phi}^k$, its Jacobian, $\left[\mathbf{\Phi}^k_{,\mathbf{q}_i}\ \mathbf{\Phi}^k_{,\mathbf{q}_j}\right]$, etc.

In our analysis, the joint constraints are expressed as,

$$\mathbf{\Phi}^k(\mathbf{q}_i(t,s),\mathbf{q}_j(t,s),s) = \mathbf{0} \tag{9}$$

and the driver constraints are expressed as

$$\mathbf{\Phi}^k(\mathbf{q}_i(t,s),\mathbf{q}_j(t,s),t,s) = \alpha^k(\mathbf{q}_i(t,s),\mathbf{q}_j(t,s),s) - C^k(t,s) \tag{10}$$

where α^k is a joint-like constraint and C^k is the associated driving function. To automate the software implementation, the joint/driver constraint, i.e. $\mathbf{\Phi}^k/\alpha^k$, is symbolically expressed using the symbolic mathematics program *Mathematica*. *Mathematica* commands, in turn, are used to evaluate all of the remaining joint/driver derivatives, e.g. $\left[\mathbf{\Phi}^k_{,\mathbf{q}_i}\ \mathbf{\Phi}^k_{,\mathbf{q}_j}\right]$ and then to generate the *FORTRAN* source code which is used to evaluate these expressions. [1]

---

[1]It is also possible to generate C source code for these expressions.

In the symbolic mathematical implementation, variable names are assigned for the generalized body coordinates and the design parameters, so that all of the derivatives in the primal and sensitivity analyses can be systematically expressed. For each joint/driver, the generalized coordinates and design parameters are expressed as

```
q = {roi[1], roi[2], phii, roj[1], roj[2], phij}
```

```
s = {spi[1], spi[2], sqi[1], sqi[2],
     spj[1], spj[2], sqj[1], sqj[2]}
```

where $\texttt{roi[1]} = x_i$, $\texttt{roi[2]} = y_i$, and $\texttt{phii} = \phi_i$ are the generalized coordinates, i.e. $\mathbf{q}_i$, and $\texttt{spi[1]} = s_{ix}^{\prime P}$, $\texttt{spi[2]} = s_{iy}^{\prime P}$, $\texttt{sqi[1]} = s_{ix}^{\prime Q}$, $\texttt{sqi[2]} = s_{iy}^{\prime Q}$ are the joint coordinates (expressed with respect to the body i) which define joint k on body i, i.e. $\mathbf{s}_i^{\prime P}$ and $\mathbf{s}_i^{\prime Q}$. The $\texttt{roj}$, $\texttt{phij}$, $\texttt{spj}$, and $\texttt{sqj}$ quantities are analogously defined for body j. Each joint/driver constraint equation is defined in terms of these variables. For example, the revolute joint of Equation 3.3.10 from Haug [17] is defined from the following statements.

```
ai = {{Cos[phii], -Sin[phii]}, {Sin[phii], Cos[phii]}}
aj = {{Cos[phij], -Sin[phij]}, {Sin[phij], Cos[phij]}}
```

```
spii  = {spi[1],spi[2]}
spjj  = {spj[1],spj[2]}
```

```
ri = {roi[1], roi[2]}
rj = {roj[1], roj[2]}
```

$$ (11) $$

```
phi = rj + aj.spjj -(ri + ai.spii)
```

To generate the source code which evaluates the constraint vector, $\mathbf{\Phi}^k$ for each joint/driver we use

```
FormPhi[PHI_List,array_,fname_String]:=
        Module[{arr,fnamestring,lp = Length[PHI]},
        arr = Flatten[Table[array[i,j],{i,lp},{j,1}]];
        fnamestring = Flatten[Table[fname,{i,lp},{j,1}]];
        MapThread[generateFortran,{arr,PHI,fnamestring}]]
```

where PHI is the *Mathematia* variable name given to the symbolic joint/driver constraint equation, **array** is the *FORTRAN* source code array name given to the constraint equation, and **fname** is the file to which the source code is written. **generateFortran** is itself a routine that uses the intrinsic *Mathematica* command **FortranForm** to express the symbolic expressions as *FORTRAN* source code. For example, for the revolute joint constraint represented by Equation 11, the command
FormPhi[phi,data,"filename.f"] produces the *FORTRAN* source code

```
      data(1,1)=-roi(1) + roj(1) - Cos(phii)*spi(1) + Sin(phi
    & i)*spi(2) + Cos(phij)*spj(1) - Sin(phij)*spj(2)
      data(2,1)=-roi(2) + roj(2) - Sin(phii)*spi(1) - Cos(phi
    & i)*spi(2) + Sin(phij)*spj(1) + Cos(phij)*spj(2)
```

in the file "filename.f".

Likewise, to generate the source code which evaluates the Jacobian matrix for each joint/driver we use

```
FormJacobian[PHI_List,q_List,array_,fname_String]:=
        Module[{jac,arr,fnamestring,lp = Length[PHI],
        lq = Length[q]},
        jac = Flatten[Transpose[Map[D[PHI,#]&,q]]];
        arr = Flatten[Table[array[i,j],{i,lp},{j,lq}]];
        fnamestring = Flatten[Table[fname,{i,lp},{j,lq}]];
        MapThread[generateFortran,{arr,jac,fnamestring}]]
```

where the arguments are the same as those in FormPhi with the addition of the symbolic generalized coordinate vector, q. For the constraint of Equation 11, the command FormJacobian[phi,q,data,"filename.f"] generates

```
      data(1,1)=-1.d0
      data(1,2)=0.d0
      data(1,3)= dsin(phii)*spi(1) + dcos(phii)*spi(2)
      data(1,4)=1.d0
      data(1,5)=0.
      data(1,6)=-(dsin(phij)*spj(1)) - dcos(phij)*spj(2)
      data(2,1)=0.
      data(2,2)=-1.
      data(2,3)=-(dcos(phii)*spi(1)) + dsin(phii)*spi(2)
      data(2,4)=0.
      data(2,5)=1.
      data(2,6)=dcos(phij)*spj(1) - dsin(phij)*spj(2)
```

The source code which is used to evaluate the remaining terms required for the primal and sensitivity analyses is generated in a similar manner.

4. Example

To demonstrate the present methodology, a wheel loader mechanism (See Figure 1) is designed to maximize lift capacity and satisfy design constraints on the bucket trajectory and the hydraulic cylinder pressures.

The mechanism is modeled with 8 rigid bodies, 9 revolute joints, and 2 translational distance drivers (See Figure 1). Three pins of interest are the G-Pin that attaches the tilt cylinder to the tractor frame, the K-Pin that

Figure 1.   Wheel Loader Mechanism

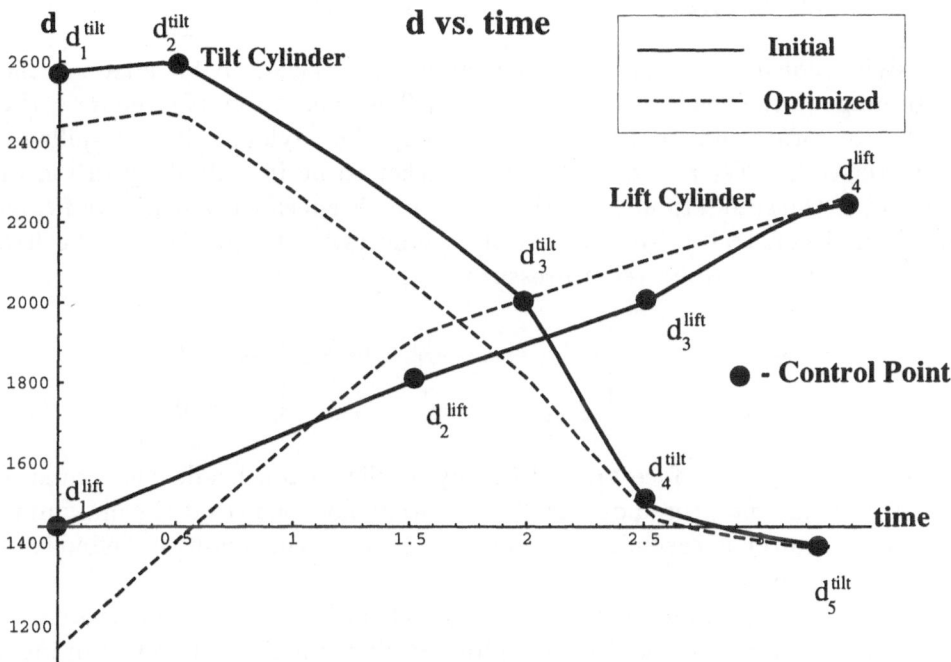

Figure 2.   Initial and Optimal Cubic Spline Driver Parameters

connects the lift cylinder to the main arm, and the C-Pin that attaches the bucket to the tilt cylinder. The drivers consist of 2 hydraulic cylinders that control both the bucket altitude and orientation. The driving functions, i.e. C^k, are cubic splines with 4 control points for the lift cylinder and 5 control points for the tilt cylinder as seen in Figure 2. An applied load in the vertical direction, i.e. $\mathbf{Q}^A = Q_y^A \hat{\jmath}$, is applied to the bucket to simulate the weight of the payload.

The cost and constraint functions are expressed via Equation 1. To quantify the payload to be maximized we use

$$G_{cost} = \left\| \mathbf{Q}^A \right\| = -Q_y^A \tag{12}$$

The bucket trajectory constraints are expressed as

$$
\begin{aligned}
G_1(t_{start}, \mathbf{s}) &= y_{bucket}(t_{start}, \mathbf{s}) - y_{fill}(t_{start}) \leq 0 \\
G_2(t_{start}, \mathbf{s}) &= \phi_{bucket}(t_{start}, \mathbf{s}) - \phi_{fill}(t_{start}) = 0 \\
G_3(t_{middle}, \mathbf{s}) &= \phi_{rack}(t_{middle}) - \phi_{bucket}(t_{middle}, \mathbf{s}) \leq 0 \\
G_4(t_{end}, \mathbf{s}) &= x_{dump}(t_{end}) - x_{bucket}(t_{end}, \mathbf{s}) \leq 0 \\
G_5(t_{end}, \mathbf{s}) &= y_{dump}(t_{end}) - y_{bucket}(t_{end}, \mathbf{s}) \leq 0
\end{aligned}
\tag{13}
$$

Satisfying constraints G_1 and G_2 enables the bucket to be filled at the beginning of the cycle when $t_{start} = 0$. The constraint G_3 requires the bucket to rotate sufficiently enough to carry the payload without spillage. The constraints G_4 and G_5 allow the bucket to be in a dump position at the end of the simulation. Note the cycle time is specified by t_{start} and t_{end}. Additional constraints restrict the hydraulic cylinders pressure to be less than a given prescribed relief pressure, P_{relief}.

$$
\begin{aligned}
G_6(t, \mathbf{s}) &= \left\| Q_{lift}^k(t_{cycle}, \mathbf{s}) \right\| - P_{relief} A_{head} \leq 0 \\
G_7(t, \mathbf{s}) &= \left\| Q_{tilt}^k(t_{cycle}, \mathbf{s}) \right\| - P_{relief} A_{head} \leq 0
\end{aligned}
\tag{14}
$$

Here A_{head} is the surface area of the hydraulic piston head. The pressure constraints are monitored during the entire simulation so that the maximum pressure is never exceeded. Thus, one G_6 and G_7 constraint is defined for each time step.

The design parameter set, $\mathbf{s}$, has 16 elements: the applied load Q_y^A, nine control points in the 2 cubic spline driving functions, two G-Pin local coordinates relative to the frame, two K-Pin coordinates relative to the lift arm, and two C-Pin coordinates relative to the bucket (see Figures 1 and 2). The splines are included in the optimization to obtain the most effective combination of the mechanism's hydraulic pump and mechanical advantage.

$$
\begin{aligned}
\mathbf{s} = \Big[& Q_y^A, d_1^{lift}, d_2^{lift}, d_3^{lift}, d_4^{lift}, d_1^{tilt}, d_2^{tilt}, d_3^{tilt}, d_4^{tilt}, d_5^{tilt}, \\
& s_x^{K-Pin}, s_y^{K-Pin}, s_x^{C-Pin}, s_y^{C-Pin}, s_x^{G-Pin}, s_y^{G-Pin} \Big]^T
\end{aligned}
\tag{15}
$$

The sensitivity of the cost function is trivially obtained, because it is an explicit function of the design parameters, i.e. $\frac{\partial G_{cost}}{\partial s} = [-1, 0, 0, \cdots, 0]$. The sensitivities of the constraints, i.e. Equations 13 and 14, are obtained using the direct differentiation method.

The optimization is performed using the sequential quadratic algorithm of DOT [19]. The optimization requires 213 function calls and 106 gradient calls to converge. Initially, two of the position constraints are violated thus giving an infeasible initial design. The optimization required many iterations to converge in part because the sensitivities are discontinuous due to the cubic splines. Furthermore, the pressure constraints of functions G_6 and G_7 are active at different times, thus these constraints are also discontinuous (see Hsieh and Arora [20] for discussion of these time dependent constraints). A larger move limit may also be used to hasten the convergence.

The results of the optimization are encouraging. The applied load increases significantly from the arbitrary initial value of 6000 to over 55,000,000. To achieve this performance increase the control points on the lift cylinder all increase with the exception of the first point and all of the tilt cylinder driver function control points decrease (see figure 2). The G-Pin moves up and out with respect to the frame; the K-Pin moves intuitively closer to the bucket, giving the lift cylinder a better mechanical advantage; and the C-Pin moves higher and farther out on the bucket, hence, the bucket may have to be modified to accommodate the new pin position (see the arrows in figure 1).

5. Conclusions

A computer-aided optimal design methodology is presented to automate the mechanism design process. An integral element of this methodology is the ability to efficiently and accurately calculate design sensitivities. The sensitivities are computed accurately and efficiently because they use the decomposed Jacobian from the analysis. The $FORTRAN$ source code used to perform the analysis and sensitivity analysis is generated symbolically via the program $Mathematica$. The effectiveness of the method is illustrated by optimizing a wheel loader bucket mechanism to maximize the payload capacity subject to position and force constraints.

References

1. Dieter Bestle. *Analyse und Optimierung von Mehrkorpersystemen.* Springer, Berlin, 1994.
2. H. Ashrafiuon and N.K. Mani. Analysis and optimial design of spatial mechanical systems. *ASME Journal of Mechanical Design*, 112:200–207, 1990.

230   V. J. PESCH, C. L. HINKLE AND D. A. TORTORELLI

3. A.K. Dhingra and N.K. Mani. Computer-aided mechanism design: A symbolic-computing approach. *Computer-Aided Design*, 25(5):300–309, 1993.
4. Dieter Bestle and Peter Eberhard. Analyzing and optimizing multibody systems. *Mechanical Structures and Machines*, 20(1):67–92, 1992.
5. George Sandor and Arthur Erdman. *Advance Mechanism Design - Analysis and Synthesis*. Prentice-Hall, Englewood Cliffs, New Jersey, 1984.
6. Michael Hansen. A general procedure for dimensional synthesis of mechanisms. *Mechanism Design and Synthesis*, 46:67–71, 1992.
7. John Hansen. Synthesis of spatial mechanisms using optimization and continuation methods. In M.S. Pereira and J.A.C. Ambrósio, editors, *Computer Aided Analysis of Rigid and Flexible Mechanical System*, volume 2 of *NATO ASI*, pages 423–439, Troia, Portugal, June 1993. IDMEC.
8. Tyler Bruns. Design of planar, kinematic, rigid body mechanisms. Master's thesis, University of Illinois at Urbana-Champaign, Urbana, IL, 1992.
9. Edward Haug. Elements and methods of computational dynamics. In E.J. Haug, editor, *Computer Aided Analysis and Optimzation of Mechanical System Dynamics*, volume F9 of *NATO ASI*, pages 3–40, Iowa City, USA, 1984. Springer-Verlag.
10. E.J. Haug and V.N. Sohoni. Design sensitivity analysis and optimization of kinematically driven systems. In E.J. Haug, editor, *Computer Aided Analysis and Optimization of Mechanical System Dynamics*, volume F9 of *NATO ASI*, pages 499–554, Iowa City, USA, 1984. Springer-Verlag Berlin Heidelberg.
11. E. Pisino, P. Giacomin, and P Campanile. Numerical investigation of the influence of the shock absorber on the vertical force tranmissiblity of a mcpherson suspension. In M.S. Pereira and J.A.C. Ambrósio, editors, *Computer Aided Analysis of Rigid and Flexible Mechanical System*, volume 2 of *NATO ASI*, pages 485–504, Troia, Portugal, June 1993. IDMEC.
12. Lawrence Davis. *Handbook Of Genetic Algorithms*. Van Nostrand Reinhold, New York, 1991.
13. David Goldberg. *Genetic Algorithms in Search, Optimization and Machine Learning*. Addison-Wesley, Reading, MA, 1989.
14. P.J.M. Laarhoven. *Simulated Annealing: Theory and Applications*. Dordrecht, Boston, 1988.
15. E.J. Haug, R.A. Wehage, and N.C. Barman. Design sensitivity analysis of planar mechanisms and machine dynamics. *ASME Journal of Mechanical Design*, 103:560–570, 1981.
16. Parviz Nikravesh. *Computer Aided Analysis of Mechanical Systems*. Allyn and Bacon, Englewood Cliffs, New Jersey, 1988.
17. Edward Haug. *Computer Aided Kinematics and Dynamics of Mechanical Systems*. Allyn and Bacon, Boston, 1989.
18. M. A. Chace. Methods and experience in computer aided design of large- displacement mechanical systems. In E.J. Haug, editor, *Computer Aided Analysis and Optimization of Mechanical System Dynamics*, volume F9 of *NATO ASI*, pages 347–361, Iowa City, USA, 1984. Springer-Verlag Berlin Heidelberg.
19. VMA Engineering. *DOT User's Manual, Version 3.00*. Vanderplaats, Miura and Associates, Inc., Goleta, CA, 1992.
20. C.C. Hsiseh and J.S. Arora. Design sensitivity analysis and optimization of dynamic response. *Computer Methods in Applied Mechanics and Engineering*, 43(1):195–219, 1983.

OPTIMIZATION OF STRUCTURAL PARAMETERS AND GAITS OF A PIPE-CRAWLING ROBOT

F. PFEIFFER AND T. ROSSMANN
Institute B of Mechanics of Munich Technical University,
21 Arcisstrasse, Munich 80333, Germany

AND

F.L. CHERNOUSKO AND N.N. BOLOTNIK
Institute for Problems in Mechanics of Russian Academy of Sciences,
101 prosp. Vernadskogo, Moscow 117526, Russia

Abstract. The pipe-crawling robot is an eight-legged walking machine that moves inside pipe-lines and can be used for inspection, maintenance, and repair. Optimization of structural parameters and possible gaits of the robot is discussed. The results obtained by computer simulation show a considerable sensitivity of operation characteristics of the robot with respect to its geometrical and kinematical parameters.

1. Introduction

In this paper, we present some investigations concerning an eight-legged pipe-crawling robot intended for motion inside pipes. Such a machine can be used for inspection, maintenance, and repair of pipe-lines. The robot was designed and developed by Prof. F. Pfeiffer and his colleagues in the Institute B of Mechanics at Munich Technical University. The results on optimization described in the paper were obtained in the course of the joint work of research teams from the Institute B of Mechanics, Munich TU, and the Institute for Problems in Mechanics of the Russian Academy of Sciences in Moscow.

To choose structural parameters of the robot (e.g., dimensions of the legs, types of actuators, etc.) and such characteristics of the gait as the

D. Bestle and W. Schielen (eds.), IUTAM Symposium on Optimization of Mechanical Systems, 231–238.
© 1996 *Kluwer Academic Publishers.*

length of the step and positions of the legs relative to the robot body during the step, we use the optimization techniques combined with a computer simulation of the robot dynamics. For the simulation, a software has been developed which makes it possible to investigate the behaviour of the robot, depending on the design variables, a control algorithm, and operation conditions (e.g., the angle of crawling). We consider two performance criteria to be maximized: the velocity of motion of the robot along the pipe and the thrusting force produced by the legs. Both the criteria are found to be sensitive to the variation of the lengths of the legs links, the length of the step, and positions of the feet on the pipe surface.

Having simulated the behaviour of the robot crawling at different angles to the horizontal and using the optimization techniques, we offer recommendations on the rational choice of the robot structural parameters and gait characteristics. Some qualitative conclusions are drawn. We established, for example, that in a number of cases, the second links (containing feet) of the legs should be made longer than the first links. It is shown also that a considerable gain in thrusting forces created by the legs can be achieved by the optimal placing of support legs with respect to the robot body. Optimal distribution of driving torques between the actuators is discussed.

2. Structure and Kinematics of the Robot

The robot (Fig. 1(a)) consists of a body and eight identical two-link legs attached to it by revolute joints (hip joints). The axes of four of the joints (numbered from 1 to 4) are normal to a plane π_1 and intersect this plane at points A_1, A_2, A_3, and A_4. The axes of the other joints (numbered from 5 to 8) are normal to a plane π_2 and intersect the latter at points A_5, A_6, A_7, and A_8. In what follows, we identify the points A_i $(i = 1, ..., 8)$ with the corresponding joints. The planes π_1 and π_2 are orthogonal to each other. We will call the line of their intersection the axis of the robot and assume the mass centre C of the robot body to belong to this axis. We associate with the robot body a reference frame $Cxyz$, with the x-axis directed along the robot axis, the y-axis belonging to the π_1 plane, and the z-axis belonging to the π_2 plane. The points A_i form rectangles $A_1A_2A_3A_4$ and $A_5A_6A_7A_8$ shown in Fig. 1(a). The rectangles have a common symmetry axis x and match each other if we rotate the plane π_2 with respect to π_1 about the x-axis by the angle of $\pi/2$. The positions of the points A_i are specified by three numbers: a and d (the x- and y-coordinates of the point A_2) and b, the x-coordinate of the point A_3. We assume that the robot body is contained inside the cylinder $y^2 + z^2 \leq d^2$.

The links of the legs are rigid rods connected by means of revolute joints (knee joints) B_i $(i = 1, ..., 8)$ whose axes are parallel to the axes of the corresponding hip joints A_i. On the end of the second link, there is a foot P_i. Thus, the leg can be considered as a planar two-member linkage $A_iB_iP_i$ (Fig. 1(b)).

To describe configurations of the legs we introduce reference frames $A_ix_iy_iz_i$ fixed relative to the robot body. The x_i-axis is collinear to the x-axis. The y_i-axis belongs to the plane π_1 or π_2 in which the point A_i is located and is directed towards the robot axis. The z_i-axis completes the right-hand triad of coordinate axes. The configuration of the ith leg is described by two angles q_{1i} and q_{2i} measured as shown in Fig. 1(b). The coordinates of the foot P_i are given by

$$x_i = l_1 \sin q_{1i} + l_2 \sin(q_{1i} - q_{2i}) \tag{2.1}$$

$$y_i = l_1 \cos q_{1i} + l_2 \cos(q_{1i} - q_{2i})$$

Here, $l_1 =\mid A_iB_i \mid$ ($l_2 =\mid B_iP_i \mid$) is the length of the first (second) links of the legs.

The robot is controlled by driving torques M_{1i} and M_{2i} applied to the axes of joints A_i and B_i, respectively. The positive directions of the torques are shown by arrows in Fig. 1(b).

We consider a regular motion of the robot inside a circular cylindrical pipe, in which the robot's body travels translationally at a constant speed v, with the x-axis coinciding with the pipe axis. The pipe axis is inclined at a certain angle δ to the horizontal plane while the robot body is oriented in such a way that the y-axis forms an angle α with the vertical plane containing the pipe axis.

3. Equations of Motion

The first group of equations describes the motion of the robot as a whole and is given by

$$M\mathbf{g} + \sum_{i=1}^{8} \mathbf{F}_i = M\ddot{\mathbf{R}} \tag{3.1}$$

$$M\mathbf{R}_C \times \mathbf{g} + \sum_{i=1}^{8} \mathbf{R}_i^P \times \mathbf{F}_i = \dot{\mathbf{L}}_C \tag{3.2}$$

Here, $M = m + 8(m_1 + m_2)$ is the total mass of the robot (m is the mass of the robot body, while m_1 and m_2 are masses of the leg links), $\mathbf{R}_C$ is the position vector of its mass centre with respect to the point C, $\mathbf{R}_i^P$ is the position vector of the point P_i with respect to the point C, $\mathbf{g}$ is the gravity acceleration vector, $\mathbf{L}_C$ is the angular momentum of the robot with respect

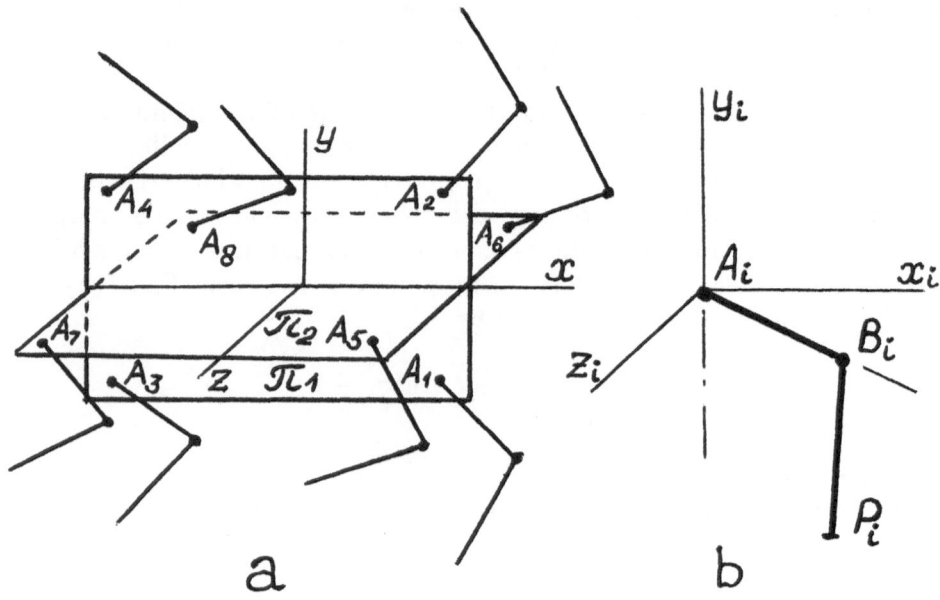

Figure 1.

to the point C, and $\mathbf{F}_i$ is the contact force exerted on the foot P_i by the pipe surface.

Two vector equations (3.1) and (3.2) can be represented as six scalar differential equations of the second order with respect to q_{1i} and q_{2i}.

Denote by F_i, N_i, and Φ_i the x_i-, y_i-, and z_i-components of the force $\mathbf{F}_i$ in the $A_i x_i y_i z_i$ reference frame. The component N_i is the normal reaction applied to the foot P_i by the pipe surface, while F_i and Φ_i are the components of the friction force. We assume that there is no slippage between the feet of support legs and the pipe and that only dry friction is present here. Therefore, the inequalities

$$N_i \geq 0, \; F_i^2 + \Phi_i^2 \leq \mu N_i^2, \qquad (3.3)$$

where μ is the friction coefficient, must hold.

The other group of equations includes 16 Lagrangian equations describing the motion of the legs of the robot relative to the body. These equations are

$$\frac{d}{dt}\frac{\partial K_i}{\partial \dot{q}_{1i}} - \frac{\partial K_i}{\partial q_{1i}} + \frac{\partial \Pi_i}{\partial q_{1i}} = -M_{1i} + F_i \left[l_1 \cos q_{1i} + l_2 \cos(q_{1i} - q_{2i}) \right] \quad (3.4)$$

$$+ N_i \left[l_1 \sin q_{1i} + l_2 \sin(q_{1i} - q_{2i}) \right]$$

$$\frac{d}{dt}\frac{\partial K_i}{\partial \dot{q}_{2i}} - \frac{\partial K_i}{\partial q_{2i}} + \frac{\partial \Pi_i}{\partial q_{2i}} = M_{2i} - F_i l_2 \cos(q_{1i} - q_{2i})$$

$$- N_i l_2 \sin(q_{1i} - q_{2i})$$

Here, $K_i = K_i(q_{2i}, \dot{q}_{1i}, \dot{q}_{2i})$ is the kinetic energy of the ith leg and $\Pi_i = \Pi_i(q_{1i}, q_{2i})$ is the potential energy due to gravity. Besides, the kinetic and potential energies depend on geometrical and inertial parameters of the robot.

4. Gaits

We confine ourselves to the gaits satisfying the following conditions.

1. At each time instant, all legs belonging to one of the planes π_1 or π_2 are support ones, while the legs belonging to the other plane are in the transfer phase.

2. All support legs and all transferred legs move synchronously and have identical configurations at each time instant, i.e.,

$$q_{s1}(t) = q_{s2}(t) = q_{s3}(t) = q_{s4}(t);$$

$$q_{s5}(t) = q_{s6}(t) = q_{s7}(t) = q_{s8}(t), \quad s = 1, 2.$$

3. The gait is periodic, i.e.,

$$q_{si}(t + T) = q_{si}(t), \quad s = 1, 2, \quad i = 1, ..., 8. \quad (4.1)$$

with some positive T for any t. The minimal T satisfying (4.1) is called the period of the gait.

4. The legs belonging to the planes π_1 and π_2 repeat the motions of each other with a half-period time shift.

5. The transfer phase of each leg includes two time intervals τ during which the foot stays on the pipe surface but exerts no pressure on it. One of the intervals immediately follows the support phase of a current step, while the other precedes the support phase of the next step. These *safety intervals* are introduced to avoid the loss of contact between the robot and the pipe when changing support legs.

The gaits can be described as time histories either of the angles q_{1i} or q_{2i} or of the feet coordinates x_i and y_i, see (2.1). In the latter case, we assume the knees to be bent in the direction of motion as shown in Fig. 1.

To describe a gait satisfying conditions *1 - 4* it is sufficient to specify the motion of one leg during one step ($0 \leq t \leq T$). Denote by x the position of the foot (related to the corresponding $A_i x_i y_i z_i$ frame) at the beginning of the support phase ($t = 0$) and by $s = vT$, the length of the step. For time periods $[0, T/2 + \tau]$ and $[T - \tau, T]$ during which the foot stays on the pipe surface, we have

$$x_i(t) = x - vt, \ t \in [0, T/2 + \tau]; \ x_i(t) = x - v(t - T), \ t \in [T - \tau, T];$$

(4.2)

$$y_i(t) \equiv -h = -(\rho - d),$$

where ρ is the radius of the pipe. Let the motion of the foot in the transfer phase between the instants of contact be governed by the equation

$$\ddot{x}_i = \frac{s}{(T/4 - \tau)^2} \text{sign}\,(3T/4 - t),$$

(4.3)

$$y_i(t) = -h + \frac{\Delta}{2}\left[1 - \cos\frac{\pi(t - T/2 - \tau)}{T/4 - \tau}\right], \ t \in [T/2 + \tau, T - \tau],$$

where Δ is the minimum lift of the foot from the pipe surface.

5. Determination of control torques

Using (2.1) where $x_i = x_i(t)$ and $y_i = y_i(t)$ we find the functions $q_{1i}(t)$ and $q_{2i}(t)$ and substitute them into (3.1) and (3.2) to obtain the contact forces $\mathbf{F}_i$. Note that 12 components F_i, N_i, and Φ_i of these forces for four support feet are not uniquely determined from six equations (3.1) and (3.2). To avoid the nonuniqueness we impose the additional conditions

$$\Phi_{2j-1} = -\Phi_{2j} = \xi_j P \cos \delta \sin \alpha_j$$

(5.1)

$$F_{2j-1} = F_{2j} = (P \sin \delta + D(t))/4$$

$$N_{2j-1} - N_{2j} = \nu_j P \cos \delta \cos \alpha_j$$

$$\alpha_j = \alpha \text{ for } j = 1, 2; \ \alpha_j = \pi/2 - \alpha \text{ for } j = 3, 4; \ j = 1, 2, 3, 4.$$

Here, P is the weight of the robot, $D(t)$ is the inertial force acting along the x-axis, and ξ_j and ν_j are unknown coefficients; $j = 1, 2$ (3, 4) when the legs belonging to the plane π_1 (π_2) are support legs. The function $D(t)$ is expressed through $q_{1i}(t)$ and $q_{2i}(t)$ for a chosen gait. Relations (5.1) are valid only for support legs; for transferred legs, $F_i = N_i = \Phi_i = 0$.

Substituting (5.1) into (3.1) and (3.2) we uniquely determine ξ_i and ν_i. Thus, the problem is reduced to choosing N_1 and N_3 for any instant t. We determine N_1 and N_3 by minimizing the maximum normal reaction $N = \max_i N_i$ for support feet under constraints (3.3) and (5.1).

6. Optimization of Parameters

We consider optimization with respect to two performance criteria: the speed v of motion of the robot body and the thrusting force $\mathbf{F}_i$ in the direction of motion (in the latter case, the constancy of speed is not supposed). The parameters to be varied are l_2, s, and x. Other parameters of the robot and the pipe are fixed. For examples described below, we take

$$a = 0.4\text{m}, \quad b = -0.4\text{m}, \quad d = 0.113\text{m}, \quad l_1 = 0.15\text{m}, \quad \rho = 0.375\text{m},$$

$$m = 5\text{kg}, \quad m_1 = m_2 = 1\text{kg}, \quad \mu = 1, \quad M_1^* = 68\text{N} \cdot \text{m}, \quad M_2^* = 35\text{N} \cdot \text{m},$$

$$\dot{q}_{01} = 1.363\text{rad/s}, \quad \dot{q}_{02} = 2.726\text{rad/s}.$$

Here, M_1^* and M_2^* are maximum allowable torques at the joints A_i and B_i, respectively; and $\dot{q}_{01}$ and $\dot{q}_{02}$ are maximum allowable angular velocities of the corresponding links.

The variable parameters l_2, s, and x are subject to a number of constraints implying the implementability of the gait. These *parametric constraints* ensure that the links of the legs do not touch the robot body and the pipe. Besides, certain constraints are imposed on control torques and angular velocities of the legs links. These *drive constraints* are determined by the chosen electric motors and gears. Both the parametric and drive constraints are not presented here.

Problem 1. Find the length l_2 of the second links of the legs, and the gait parameters s and x maximizing the speed v of motion of the robot under the parametric and drive constraints.

Problem 1 was solved by direct computer simulation and variation of parameters within the constraints imposed. Here, we present only some typical results and conclusions.

Investigations reveal a considerable sensitivity of the maximum velocity v to variation of s and l_2. For example, let us fix $\alpha = 0$, $\delta = 0$, and $l_2 = 0.15$ m (the links of the legs are identical). When varying s from 0.01m up to 0.46m, for $x = 0.11$m, the maximum allowable velocity v changes from 0.004m/s to 0.085m/s. When changing l_2 from 0.15m to 0.25m for $x = 0.11$m and $s = 0.24$m, the maximum allowable velocity v changes from 0.087 m/s to 0.196m/s.

Let us present some examples showing efficiency of the optimization. In these examples, we fix $\alpha = 0$ and $\delta = 0$. Investigation show that the dependence of v on these angles is weak, and hence the data given below are representative.

First, we fix $l_2 = 0.15$m, $s = 0.30$m, and $x = 0.08$m. In this case, the maximal reachable velocity is $v = v_1 = 0.079$m/s.

If we fix $l_2 = 0.15$m and vary s and x, we find that the optimal velocity $v = v_2 = 0.085$m/s is reached at $s = 0.26$m and $x = 0.11$m.

By varying all the three parameters l_2, s, and x, we achieve the maximum velocity $v = v_3 = 0.245\text{m/s}$ at $l_2 = 0.25\text{m}$, $s = 0.12\text{m}$, and $x = 0.15\text{m}$.

The comparison of ratios $v_2/v_1 = 1.08$, $v_3/v_1 = 3.1$, and $v_3/v_2 = 2.88$ shows a considerable sensitivity of the maximum velocity to the variation of l_2, s, and x (especially, of l_2). Therefore the adjustment of these parameters when designing and operating the robot is advisable.

Consider now the problem of maximizing the thrusting force of the robot. For simplicity, we confine ourselves to the case where the mass of the legs is much less than that of the body, and hence the methods of statics are applicable.

Problem 2. Given s, find x and l_2 maximizing the guaranteed maximum of the thrusting force during the support phase

$$J(x, l_2, s) = \min_{x_i \in [x-s/2, x]} \max_{M_{1i}, M_{2i}} \sum_i F_i$$

under the constraints $|M_{1i}| \leq M_1^*$, $|M_{2i}| \leq M_2^*$, (3.3), and the parametric constraints.

Omitting details of the solution, we give some numerical examples showing advisability of the optimization. In these examples we take $M_1^* = 68\text{N·m}$, $M_2^* = 27\text{N·m}$, $s = 0.28\text{m}$, $l_1 = 0.15\text{m}$, $h = 0.24\text{m}$, and $\mu = 1$.

Let us fix $x = 0.14\text{m}$ and $l_2 = 0.15\text{m}$. Then we have $J = J_1 = 642\text{N}$.

The variation of only one parameter x gives $J = J_2 = 786\text{N}$ reached at $x = 0.105\text{m}$.

The variation of both x and l_2 gives the maximum thrusting force $J = J_3 = 1047\text{N}$ at $x = 0.02\text{m}$ and $l_2 = 0.21\text{m}$.

Comparison of the ratios $J_2/J_1 = 1.22$, $J_3/J_1 = 1.63$, and $J_3/J_2 = 1.33$ indicates high sensitivity of the thrusting force to the choice of positions of the support feet and the length of the second links of the legs.

7. Conclusion

The results of calculations show that the characteristics of the pipe-crawling robot depend significantly on the length of the second link of its legs: the longer the link, the greater are the speed and the thrusting force. These characteristics depend also on the position of the feet at the beginning and at the end of the support phase. Thus, optimization of parameters is recommended when designing the structure of the robot and planning its gaits.

Acknowledgements

We thank Dr. G.V. Kostin who took part in this research, especially in developing the software for computer simulation.

This work is supported by the Koerber Foundation (Hamburg, Germany).

OPTIMIZATION METHODS FOR PARAMETER ADAPTION IN MECHATRONIC SYSTEMS

K. POPP AND K.-D. TIESTE
Institute of Mechanics, University of Hannover
Appelstrasse 11, D-30167 Hannover, Germany
email: tieste@ifm.uni-hannover.de

1. Introduction

Machine tools require linear guides with high slide velocity and very high position accurancy. The three tasks of a linear guide — supporting, guiding and driving — shall be realized by means of active magnetic bearings (AMB). The resulting linear magnetically levitated (maglev) guide has to accomplish the following characteristics: high stiffness, good damping and low noise as well as low heat production.

First research on a one degree-of-freedom (DOF) support magnet unit aimed at the development of components and efficient control strategies for the linear maglev-guide. The actual research is directed to realise a five DOF maglev linear-guide for machine tools without drive. The quality of the state feedback controller of the maglev guide depends on the accuracy of the model parameters. Thus, automatic parameter identification is required.

The parameter identification and -adaption using nonlinear optimization techniques is the subject of this paper.

2. Control of the Maglev Guide

2.1. ONE-DOF AMB CONTROL

The one-DOF magnetic bearing shown in fig. 1 consists of a pair of electromagnets in differential arrangement and a mass of 11kg guided by roller bearings. The forces of the electromagnets counteract due to the differential arrangement. Premagnetisation of the electromagnets is used to liearize the force-current-characteristic of the electromagnets.

The linear model shown in fig. 2 is used for the voltage controlled magnetic bearing. The model consists of a linear electromagnet with the co-

239

D. Bestle and W. Schielen (eds.), IUTAM Symposium on Optimization of Mechanical Systems, 239–246.
© 1996 *Kluwer Academic Publishers.*

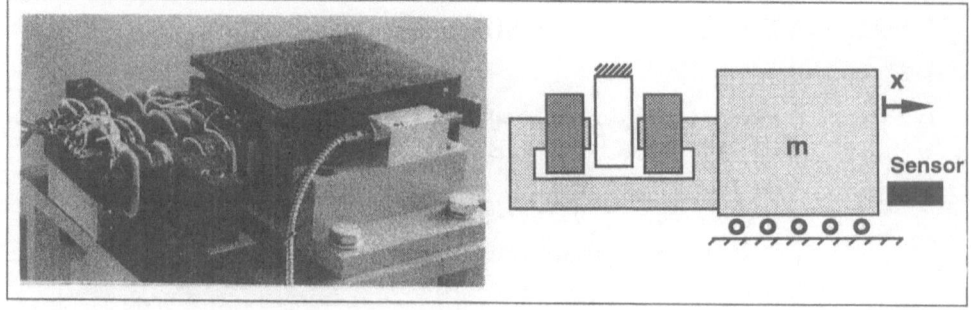

Figure 1. Photo and schema of the one-DOF magnetic bearing

efficients L, R, the force-current coefficient k_i, the mass m, the induction
coefficient k_v and the stiffness k_c which is responsible for the instability of
an uncontrolled magnetic bearing. All coefficients were gained by measurements.

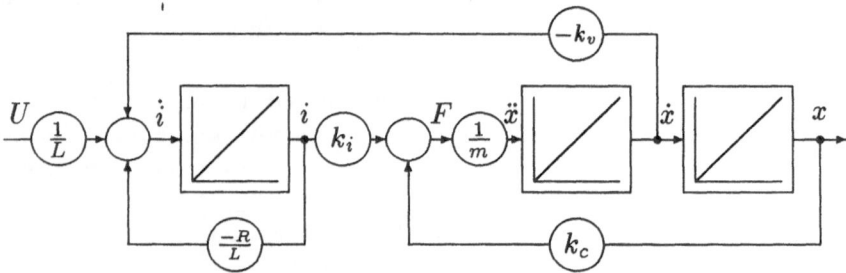

Figure 2. Model of the magnetic bearing

The corrosponding set of differential equations reads

$$\dot{i} = \frac{-R}{L}\, i - k_v\, \dot{x} + \frac{1}{L}\, U, \tag{1}$$

$$\ddot{x} = \frac{1}{m}\, (k_c\, x + k_i\, i) \tag{2}$$

which can be written in state space presentation as

$$\frac{d}{dt} \begin{pmatrix} x \\ \dot{x} \\ i \end{pmatrix} = \begin{pmatrix} 0 & 1 & 0 \\ \frac{k_c}{m} & 0 & \frac{k_i}{m} \\ 0 & -k_v & \frac{-R}{L} \end{pmatrix} \begin{pmatrix} x \\ \dot{x} \\ i \end{pmatrix} + \begin{pmatrix} 0 \\ 0 \\ \frac{1}{L} \end{pmatrix} U. \tag{3}$$

The structure of the controller for the one-DOF magnetic bearing is
shown in fig. 3. The three state variables of the electromagnet are gained

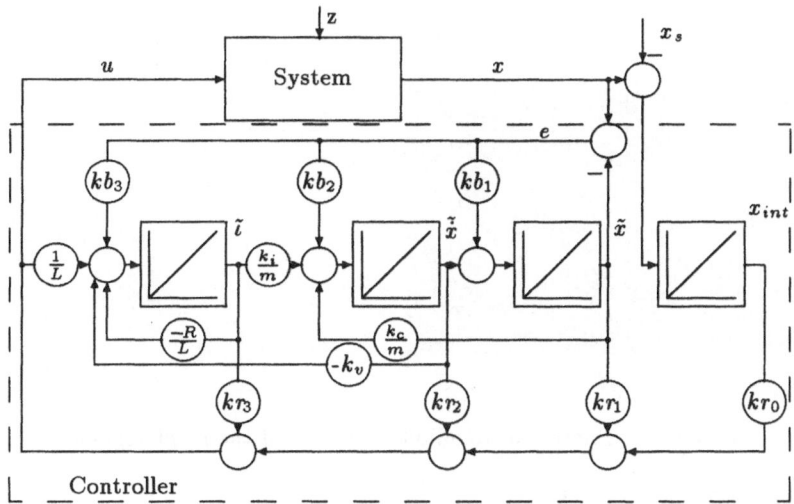

Figure 3. One-DOF magnetic bearing with observer and state control

by means of an observer. The control variable is generated by state feedback of the observed states supplemented by an additional integral stage.
The static accuracy of the magnetic bearing is very high because of the
additional integral state in the control system. The dynamical behaviour of
the magnetic bearing is defined by the location of the poles of the observer
and of the controller. The feedback gains have been calculated by pole assignment. It turned out that a configuration of four equal real poles give
best results.

2.2. MULTI-DOF MAGLEV GUIDE CONTROL

The maglev linear guide consists of a guide block with six pairs of electromagnets (two horizontal magnets and four vertical magnets) in differential
arrangement and a guideway with three rails cf. fig. 4. The nomimal air
gap of the guide is only 0.35 mm wide to achieve high stiffness. The characteristics of the electromagnets are: pole area: $A = 10$ cm^2, coil: 500 turns,
wire 0.75mm in diameter. The mass is $m = 30.5$ kg and the moments of
inertia are $J_x^{(c)} = 0.24$ kg m^2, $J_y^{(c)} = 0.45$ kg m^2 and $J_z^{(c)} = 0.64$ kg m^2. The
lever of the magnet forces are: $b = 0.123$m, $c = 0.125$m.

The maglev guide is controlled by a centralised degree-of-freedom (DOF)
control, where each DOF or mode has its own single-DOF controller using a
centralised measuring system. The advantages are: simple pole assignment
and independent adjustment of the stiffness and the damping of each DOF.

The coordinate system and the free body diagram shown in fig. 4 are
the basis for the equations of motion. The following simplifications have

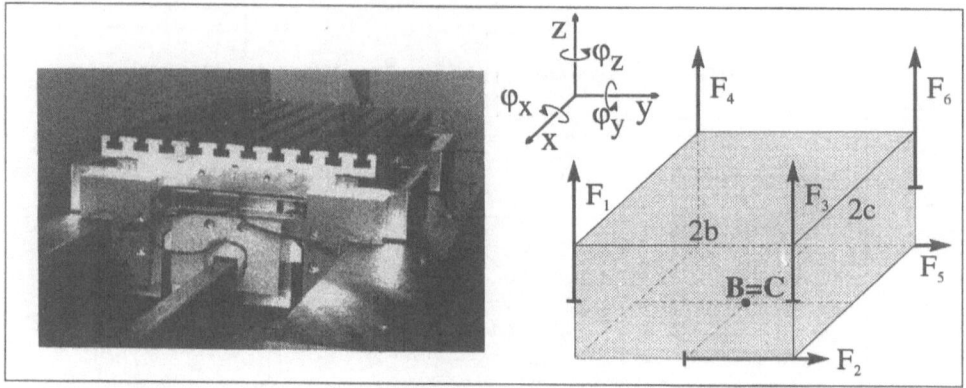

Figure 4.   Photo and free-body-diagram of the maglev guide

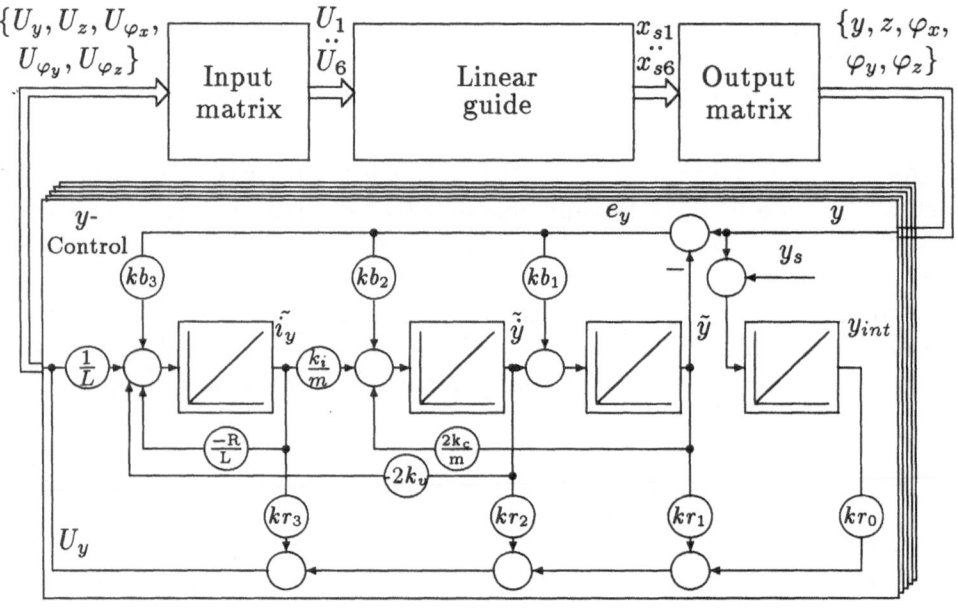

Figure 5.   Control structure of the maglev guide

been made: All magnets are equal, the guide has no payload (at this time) so that the centre of mass C is located at the point B in the middle of the slide and the principal axes coincide with the coordinate axes.

The guide coordinate vector $\mathbf{x}^{(B)} = \{y, z, \varphi_x, \varphi_y, \varphi_z\}$ for the reference point B is ganied from the six sensor values by the output matrix. Each of the five coordinate values of the guide coordinate vector is fed into a single-DOF controller generating a virtual voltage which is transformed into the

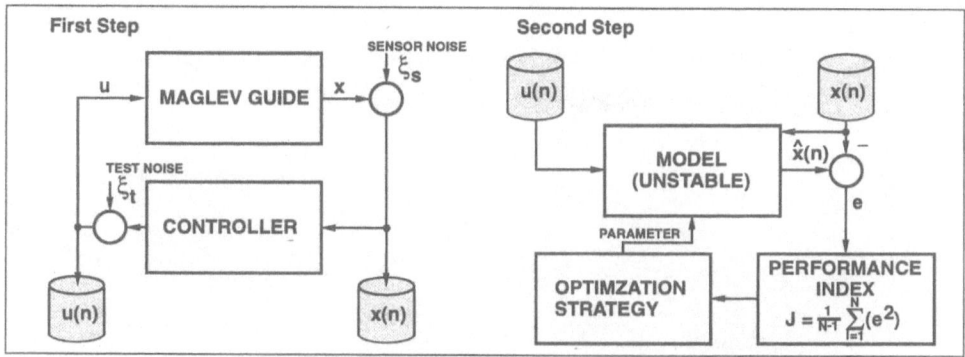

Figure 6. First and second step of the model fitting method

real voltages for the six electromagnets by the input matrix. The block diagram of the maglev guide controller is shown in fig. 5. For details see (tie94).

The control of the maglev guide is programmed in the high level language PEARL and is implemented on a process computer with Motorola 68040 CPU. The sample frequency is 3.3kHz. The real time multi tasking operating system RTOS-UH supervises the control activities and interprocess communications. This enables on the one side a very fast motion control of the linear guide and on the other side identification, adaption and other supervisory tasks operating at lower sample frequencies.

3. Parameter Identification

The quality of the control clearly depends in the parameters of the model the observer relies on. For stabilizing the maglev guide these parameters were first gained from their theoretical values. First optimization of the control was done by trial and error. With a stable maglev guide control optimization techniques were used to improve the control quality.

3.1. THE PERFORMANCE INDEX

Identification of the system parameters is difficult because of the high noise level, the instability of the system and the high number of parameters.

A quality index is gained by measuring the input and the output signals of each DOF of the maglev guide using a least-squares model-fitting method, cf. fig. 6. Each DOF has its own model which must be fitted separately. In a first step sampled input values $u(n)$ and output values $x(n)$, $n = 1(1)N, N = 20000$, of the maglev guide are gained by the controller task. An additional gaussian test noise $\xi(n)$ is added to the volt-

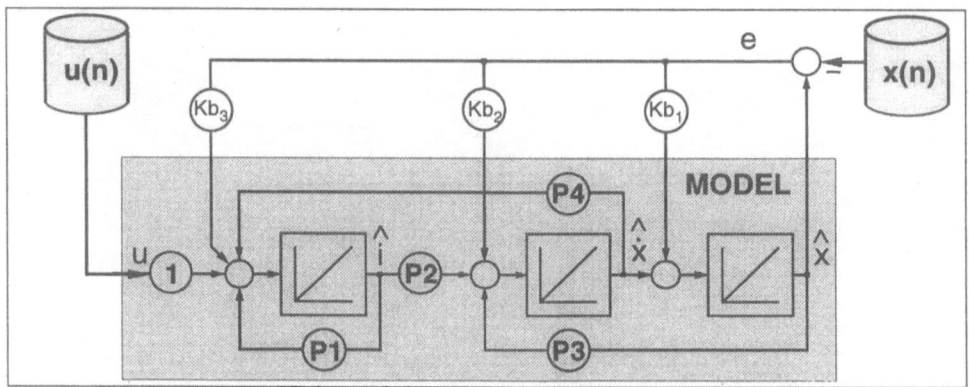

Figure 7. Stabilization of the model

age to get a better performance index. The input values $u(n)$ and output values $x(n)$ are fed into the model, cf. fig. 6. The performance index $J(\mathbf{p}) = \frac{1}{N-1} \sum_{n=1}^{N} (\hat{x}(n) - x(n))^2$, the variance, is calculated from the difference between the output values $\hat{x}(n)$ of the model and the sampled output values $x(n)$. The instability of an uncontrolled magnetic bearing requires a stabilization of the model which is carried out by an observer-like feedback, cf. fig. 7.

3.2. THE OPTIMIZATION PROCEDURE

The minimization of the performance index $J(\mathbf{p})$ is done with the four-dimensional parameter vector $\mathbf{p}$ of the model, where each DOF has its own model. Thus, 20 parameters have to be optimized all together. The performance index can only be calculated by a performance index call which takes about 0.4 seconds. There are no gradients available; gradients can only be approximated by unreliable differential quotients.

With respect to these conditions only optimization methods can be used which need no gradients and no Hesse matrix. The performance index depending on two parameters each is shown in fig. 8.

Two optimization methods have been tested for the parameter identification of the maglev guide: i) A line search algorithm and ii) the simplex method of NELDER and MEAD.

i) Line search algorithm: The minimum is approached by sequentially searching the local minimum along one dimension in the four dimensional parameter space. The one dimensional line search uses an algorithm which is based on the division of the search space in proportions due to the golden mean. The convergence speed is poor because the different eigenvalues of

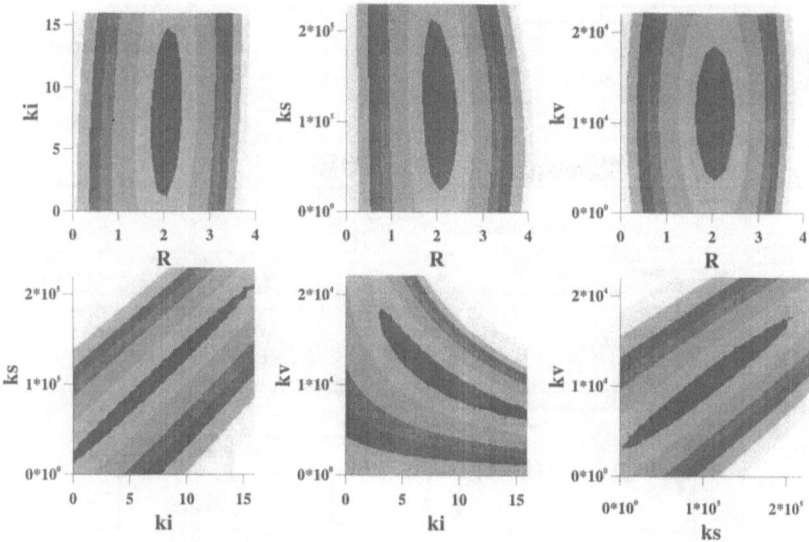

Figure 8.   Performance index depending on two parameters each

the Hesse matrix are leading to a zigzag run. However, the line search algorithm gives a good set of parameters which improves the quality in the control of the maglev guide but the calculation time takes about 15 minutes for each DOF.

ii) The simplex method of NELDER and MEAD, cf. (bes94): The basic idea of the simplex method of NELDER and MEAD is to include the minimum inside a simplex which is a polyeder of $n + 1$ cornerpoints in the n dimensional search space. Only a few rules which are controlled by the order of the cost function values of the corner points describe the simplex strategy. In a first phase the simplex expands using reflection and expansion steps until the minimum finds itself inside the simplex. Then the contraction phase follows using contraction and replacement steps to reduce the volume of the simplex. The quality and the convergence speed of the simplex method are very good; the calculation time takes about 40 seconds (cf. fig. 9).

4. Extension

The end of the iteration procedure is reached when the volume of the simplex decreases beyond a limit. Instead of this the simplex can be kept open by adding a random vector to the corner points of the simplex with the aim of a parameter adaption:

After updating the observer of the maglev control with the identified parameters new input and output values are sampled (cf. step 1) and fed

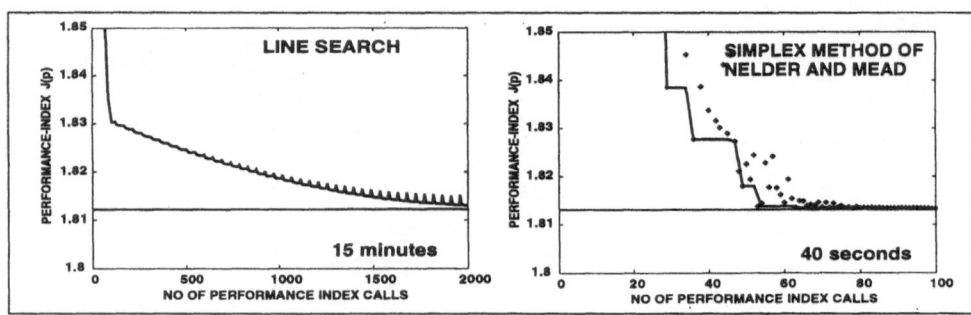

Figure 9. Convergence properties

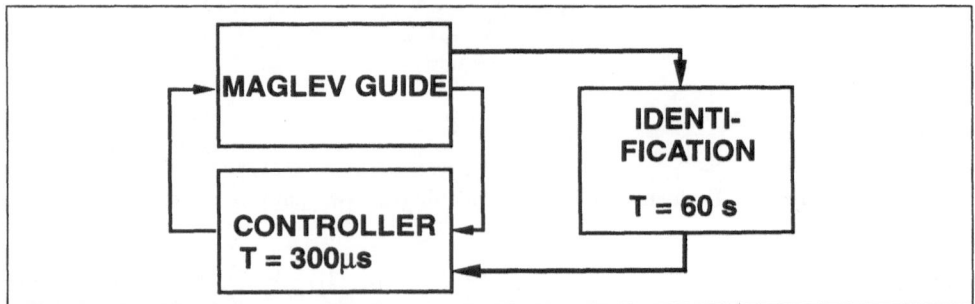

Figure 10. Parameter adaption

into the model to generate the new performance index (cf. step 2). The simplex which is held open by the random vector finds then the minimum in the new dataset. Fig. 10 illustrates the adaption technique. The controller works at a sample time of $300\mu s$, the identification gives an update for the parameters all 60s.

5. Further research

Further research will be directed to implement an adaption of the model parameters and to extend the control system to adapt itself to different payloads (workpieces) and to the corresponding center of mass.

References

Tieste,K.-D., Popp,K.: Magnetlager für den Einsatz bei schnellen Translationsbewegungen, *Schwingungen in rotierenden Maschinen*, Viehweg 1993

Tieste,K.-D.,Popp,K.: Dynamical Behaviour of a Linear Maglev Support Unit for Fast Tooling Machines, 4th *International Symposium on Magnetic Bearings*, ETH Zürich 1994

Bestle, D.: *Analyse und Optimierung von Mehrkörpersystem*, Springer 1994

PSI METHOD AND MULTI-CRITERIA OPTIMIZATION OF OBJECTS WITH THE USE OF FINITE ELEMENT MODELS.

A.V.PTCHELINTSEV*, V.S.SHENFEL'D**

*   *Building Research Institute,*
    1 Tatehara, Tsukuba-City,
    Ibaraki Prefecture, 305, Japan

** *Blagonravov Mechanical Engineering Research Institute*
    ul. Griboedova 4, Moscow, 101830, Russia

1. Introduction

Presently, general-purpose finite-element programs -- ANSYS, NASTRAN, I-DEAS, NISA, etc, are widely used in the most diverse spheres of industry for designing and analyzing of automobiles, planes, ships, machine tools and so on. At the same time, when one considers such mass-produced objects, as for example, automobiles it is not enough to be able to conduct static, dynamic or other type of analysis. It is necessary to receive the optimum recommendations simultaneously on all main characteristics of studied objects.

Usually the finite element optimization in these programs uses single criterion. Such approach does not reflect the specific character of the tasks considered. The multicriteria optimization [1--3] satisfies to the above requirements in the most extent. The experience gained in solving engineering optimization and optimal design problems shows, that a designer cannot state the problem correctly: to define design variables, functional and criteria constraints [1]. Unfortunately, the known optimization techniques are only of slender assistance. Just therefore the PARAMETER SPACE INVESTIGATION method (PSI method) has been created for correct statement and solution of engineering optimization problems. This method has received the wide application. PSI method takes into account the features of problems from the class under consideration and in the end finds all alternate variants of optimum solutions. So, for example, in problems of dynamics and strength the criteria of optimization are: mass, maximal stresses in the most dangerous points of structure, dimensions, dynamic forces, acting in structure; the frequencies, amplitude of displacements (speeds, accelerations) and many other important characteristics of object are optimized also.

D. Bestle and W. Schielen (eds.), IUTAM Symposium on Optimization of Mechanical Systems, 247–252.
© 1996 *Kluwer Academic Publishers.*

2. PSI Method

The PSI method is based on the exploration of the search domain, defined by inequalities

$$a_j^{**} \le \alpha_j \le \alpha_j^{**}, j = 1,2,....,r$$

using uniformly distributed points [3]. For variable vectors α^i corresponding to these points, the values of functional constraints are calculated. If these constraints are satisfied, the values of perfomance criteria $\Phi_v(\alpha^i)$, $v = 1,2,....k$ are calculated. The correct definition of design-variable, functional, and criterion constraints is performed by a special algorithm developed in [2,3]. Here, heavily borrowing from [1], we briefly report its version that better takes into account the capabilities of a designer and the correctness of the definition of feasible solution set.

The starting point of the algorithm is the same as in [2,3,5]. At this point, the search is performed in three steps. In the first step, the test tables are obtained for each criterion. They contain the values of $\Phi_v(\alpha^1)$ ..., $\Phi_v(\alpha^N)$ in ascending order (assuming that all criteria are to be minimized). In the second step, the designer performs a preselection of the criterial constraints Φ_v^{**}, $v = 1,2,...k$.

Φ_v^{**} are the maximal values of the criteria $\Phi_v(\alpha)$ that guarantee an acceptable functioning of the object. If Φ_v^{**} is not taken as maximal, many interesting solutions may be lost, because of contradictory criteria. As a rule, a designer may define some $\Phi_v(\alpha)$, that is definitely feasible, as the Φ_v^{**}. If the maximal values are taken for Φ_v^{**}, one should switch to the third step.

In the third step, the problem solvability is verified. The vectors α^i that simultaneously satisfy all constraints $\Phi_v(\alpha^i) \le \Phi_v^{**}$, $v = 1,2,...k$ are determined. If the set of these vectors is nonempty, it is possible to construct the feasible solution set. If not, either the values Φ_v^{**} should be corrected or the first step should be run once more, with more points, so that the test table of the second step are larger.

The process repeats until the feasible solution set D becomes nonvoid. Then, the Pareto solution set P is constructed according to [4].

Let us consider the case when the designer meets with difficulties in defining the maximal Φ_v^{**}. Usually, the values of $\Phi_v(\alpha)$ within the interval $\Phi_v(\alpha) \le \Phi_v(\alpha) \le \Phi_v^{**}$ are indefinite in terms of feasibility. Here, Φ_v^{**} is the value of the v-th criterion, for which the values $\Phi_v(\alpha) > \Phi_v^{**}$ are a fortiori inadmissible. In other words, the designer is not sure whether $\Phi_v(\alpha) > \Phi_v(\alpha)$ are feasible or not. In this situatoin, as before, one should pass to the third step and construct the admissible set D and the Pareto optimal set P under the constraints $\Phi_v^{**} = \Phi_v(\alpha)$. Then, one should obtain the set D under the constrain Φ_v^{**}, $v = 1,2,...,k$ and the corresponding optimal Pareto set P. Then, $\Phi(P)$ and $\Phi(P)$ are compared. If vectors in $\Phi(P)$ have no considerable advantage over the vectors in $\Phi(P)$, the values $\Phi_v^{**} = \Phi_v(\alpha)$ may be taken as criterion constraints. Otherwise, if the advantage is considerable, the criterion values may be taken equal to Φ_v^{**}. One should verify the obtained optimal solution for its admissibility. If the designer cannot verify the feasibility of the solution, then the former $\Phi_v^{**} = \Phi_v(\alpha)$ are taken as the criterion constraints. (All possible values of $\Phi_v (\alpha)$ and Φ_v^{**} fit into this scheme).

The feasible set approach, in most cases, gives results that are inaccessible to the above-mentioned program packages and also, interesting for designers (see, e.g., the design of an optimal structure in [6], of an automobile casing in [7], and of automobile impact strength in [8]).

3. An example to illustrate two methods of optimization

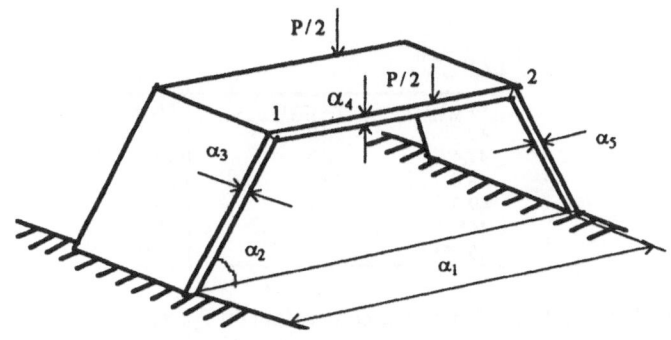

Figure 1.

The purpose of example proposed below is to compare two different approaches - multi-criteria, based on the PSI method , and single-criterion one for the search of best solutions in engineering optimization problems. Not pretending on the completeness of comparison, we will put only some reasons on this occasion.

The example: the finite element model of a structure shown on Fig. 1 consists of linear four-node shell elements. The load P is applied on distance L/3 from point 2, where L - length of plate, P = 1200 N; L = 1.2 m.

The design-variable constraints define parallelepiped Π_1: the distance between support $1.8 \le \alpha_1 \le 2.2$ (m), angle of inclination $45° \le \alpha_2 \le 65°$, thickness of three plates $2.5 \le \alpha_i \le 4.5$ (mm), i= 3, 5.

The functional constraints:

- maximal stress should not be greater than allowable,

$$f_1 (\alpha) \le C_1 = 250 \text{ MPa}$$

- vertical displacement of points 1 and 2, accordingly

$$f_i (\alpha) \le C_i \text{ (m)}, i = 2,3.$$

$$f_4 (\alpha) = | f_2 (\alpha) - f_3 (\alpha) | \le C_4 \text{ (m)}$$

- value of the first eigenfrequency

$$f_5(\alpha) \le C_5 \quad \text{(Hz)}.$$

The performance criteria set included the mass of structure Φ_1 (kg), the maximum deflection of the horizontal plate Φ_2 (mm) and the second eigenfrequency Φ_3 (Hz).

Note, that in the general case the significance of the criterion Φ_2 is not less, than that of Φ_1 and both of them are in contradiction. The criteria Φ_1 and Φ_2 were minimized, and Φ_3 - maximized.

TABLE 1.

F1 (kg)	F2 (mm)	F3 (Hz)
45.7	30.2	8.39
47.4	28.3	8.73
48.8	27.4	8.87
49.5	26.7	8.86
46*	31*	8.37*
45.6**	31**	8.39**

Using the PSI method, the functional constraints on $f_2(\alpha)$ - $f_5(\alpha)$ were found: $C_2 = C_3 = 0.033$ m, $C_4 = 0.01$ m, $C_5 = 3.4$ Hz. For the determination of these constraints functions $f_2(\alpha)$ - $f_5(\alpha)$ were submitted in kind of pseudo-criteria $\Phi_4(\alpha)$ - $\Phi_7(\alpha)$. Thus for generation of feasible set of decisions the tables of tests have been constructed [1], where Φ_1 - Φ_3 are quality criteria , Φ_4 - Φ_7 are pseudo-criteria.

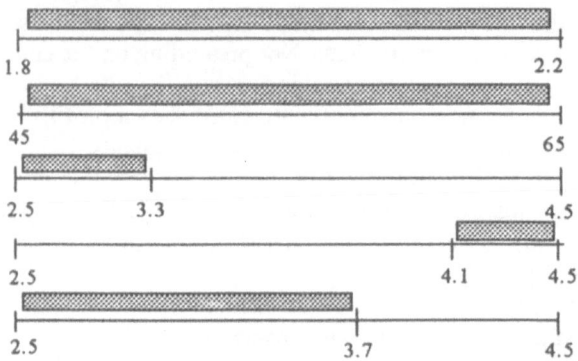

Figure 2. Distribution of the parameters α_1 - α_5 of feasible solution, found in parallelepiped Π_1 .

Four best results, found in parallelepiped Π_1 , are shown in the Table 1. The histograms that show distribution of the parameters of the feasible solutions [3], were constructed (Fig. 2). Each histogram represents the section $[\alpha_j , \alpha_j]$, $j = 1, 5$ on which zones of the feasible solution are filled. So for example, feasible solution on parameters α_3 and α_5 are located on left ends of sections $[\alpha_3 , \alpha_3]$ and $[\alpha_5 , \alpha_5]$, on α_4 - on

right end of section [α_4 , α_4], but on α_1 and α_1 they are located along the whole lengths of appropriate sections. On the basis of histograms analysis at
the expense of correction of initial design-variable constraints a new parallelepiped Π_2 was formed:

$$1.6 \leq \alpha_1 \leq 2.4 \quad (m);$$
$$40° \leq \alpha_2 \leq 70° \quad ;$$
$$2.2 \leq \alpha_3 \leq 3.7 \quad (mm);$$
$$3.7 \leq \alpha_4 \leq 5.0 \quad (mm);$$
$$2.2 \leq \alpha_5 \leq 3.9 \quad (mm).$$

Taking into account former functional and criteria constraints, previously obtained solutions have been considerably improved. Five such solutions are shown in the Table 2. The general purpose finite element code ANSYS 5.0 and MOVI program were used

TABLE 2.

F1 (kg)	F2 (mm)	F3 (Hz)
44.7	29.6	8.5
45.6	25.9	8.8
47	26	8.9
48.8	24.6	9.1
49.9	22.6	9.4

for the calculation and analysis of the performance criteria. The finite element modeling aimed to receive the output values for the vector of parameters α^i was implemented with the ANSYS program. These results are described above and are shown in Table 1, except for the last two lines.

The single-criterion optimization has been carried out based on the functional and criteria constraints determined by the PSI method. Two general-purpose finite-element codes -- ANSYS and I-DEAS were used for this purpose. These well known programs have good optimizing modules, based on various methods of nonlinear programming and sensitivity.

The best results are shown in the last two lines of the Table 1:
* - corresponds to the I-DEAS, ** - corresponds to the ANSYS.

4. Conclusions

1. Series of solutions, found with the PSI method, as well as with the ANSYS and the I-DEAS programs on set of criteria are practically identical: see four last lines in the Table 1. However, as one could expect, another interesting pareto-optimal results have been obtained with the PSI method. A few of them are shown in the Table 2.
2. With the PSI method the correction of the initial parallelepiped Π_1 was carried out and the new one Π_2 was constructed. The results of optimization in Π_1 have been considerably improved in Π_2 .

3. Although the number of loops at applications of single-criterion methods has appeared less, than at use of PSI method, hardly this factor is possible to admit as decisive one. The main factor which should be considered when choosing the optimization method is the result to be achieved. And it is more often possible in engineering optimization problems to obtain the best result only if a designer has the total information about all the most interesting and important solutions-pareto-optimal variants. We see the main purpose of the PSI method from this point of view.

Computer program MOVI (Multicriteria Optimization and Vector Identification) which is capable to implement the PSI Method, was created, directed on statement and solving of engineering optimization problems. The MOVI can be easily incorporated into any general-purpose finite-element program.

5. Acknowledgements

The authors thank Prof. R.B. Statnikov for his invaluable advices and comments and Dr. M.N. Nikolaenko for the assistance in the preparing of this paper.

6. References

1. Statnikov, R.B., Matusov I.B. *Doklady Akad. Nauk* (Russia*), 1994*, vol. 336, no. 4, 481 - 484.
2. Sobol', and Statnikov, R.B., Formulation of Some Problems of Optimal Design with a Computer, *Preprint od Inst. of Applied Mathematics, USSR Acad. Sci., Moscow, 1977, no.24*
3. Sobol' I.M., Statnikov R.B.: *Optimal Parameters Choice in Multicriteria Problems*, Nauka, Moscow, 1981.
4. Statnikov R.B., Matusov I.B.: *Multicriteria Design of Machines*, Znanie, Moscow, 1989.
5. Statnikov R.B., Uzvolok Yu.Yu.: Determination of parameters variations boundaries in the problems of optimal design and vector identification., *Doklady AN USSR 5*, v. 315 .(1990).
6. Statnikov R.B., Matusov I.B., and Mioduushevskii P.V., *Doklady AN* (Russia) 1, v. 329, (1993).
7. Velikhov E.P., Betelin V.B., and Stavitskii A.I., *Mashinovedenie 5*, (1986), 3-8
8. Bondarenko M.I., et. al., *Rossiiskaya Akademiya Nauk 6* (Russia), vol. 335, (1994).

OPTIMAL DESIGN OF STRUCTURES
SUBJECT TO NONCONSERVATIVE FORCES

U. T. RINGERTZ
Department of Lightweight Structures
Royal Institute of Technology
S-100 44 Stockholm, Sweden

Abstract. The essential difficulties with optimization of nonconservative systems are considered. Model problems are used to show that the functions of the optimization problem may be nonsmooth and possibly also discontinuous functions of the design variables. It is further demonstrated that considering a slightly more complicated structural model with damping may simplify design optimization. Using a model with damping, it is possible to pose the optimization problem in the form of matrix inequalities. The resulting problem may then be solved using a barrier method with smooth objective and constraint functions.

1. Introduction

Optimal design of structures subject to nonconservative forces is significantly more difficult compared to the design of structures subject to conservative forces. The main difference is that stability analysis in the former case involves analysis of unsymmetric eigenvalue problems, whereas the latter case involves eigenvalues of symmetric matrices, usually in the form of the matrix of second derivatives of a potential energy function.

Structural optimization subject to stability constraints often leads to an optimal design with several simultaneously critical stability modes. Stability constraints are usually posed by considering the eigenvalues of the appropriate equations of motion such that simultaneous stability modes implies coalescing eigenvalues. This causes significant difficulties when the external force is nonconservative since the eigenvalues are nondifferentiable functions of the design parameters when they coalesce [1].

D. Bestle and W. Schielen (eds.), IUTAM Symposium on Optimization of Mechanical Systems, 253–260.
© 1996 Kluwer Academic Publishers.

Although the problem of structural optimization subject to stability constraints is frequently studied, see for example Olhoff and Taylor [2], and Zyczkowski [3], many issues remain unresolved. In particular the non-smoothness of the optimization problem is often ignored even though such an approach may be very unreliable. The main result of the present paper is to demonstrate that it is possible to pose the problem of optimal design of a structure subject to a nonconservative force in smooth form.

2. Design optimization difficulties

A simple model problem, known as Beck's column [4], will be used to illustrate some of the difficulties involved in optimal design of structures subject to nonconservative forces. The column is clamped at one end and is subject to a follower force at the other end. The force remains tangential as the column deforms causing the mechanical system to be nonconservative.

The discretized linear equations of motion for a simple nonconservative system, such as Beck's column, are given by

$$M\ddot{u} + Ku - pAu = 0, \tag{1}$$

where M denotes the consistent mass matrix, K the stiffness matrix, and A the matrix defining the influence of the load on system stability. The matrices K and M are symmetric positive definite and the matrix A is real unsymmetric. The scalar parameter p defines the magnitude of the load and the vector $u \in R^m$ denotes the nodal rotations and displacements.

The solution to the differential equation (1) is assumed to be of the form

$$u = \tilde{u}\, e^{i\omega t}, \tag{2}$$

where ω is a vibration frequency and $\tilde{u} \in R^m$ an eigenvector.

Introducing this assumption into the equations of motion gives

$$(K - pA - \omega^2 M)\tilde{u} = 0. \tag{3}$$

This is obviously an eigenvalue problem depending on the scalar parameter p. The eigenvalue ω^2 is the square of the vibration frequency and may be real or complex since the matrix A is unsymmetric. The column is stable if all the eigenvalues ω^2 are real and positive. A real but negative eigenvalue corresponds to a divergence instability and a complex eigenvalue corresponds to a flutter type instability meaning oscillations of increasing amplitude.

A stability analysis of the column involves finding the critical load p_c for which the column becomes unstable. This means that all the eigenvalues ω^2 must be real and positive for all values of p in the interval $[0, p_c]$. In practice it is only possible to compute the eigenvalues for a finite number of values of p in the interval $[0, p_c]$. Let $\mathcal{K}$ denote the index set of values p_k for which the eigenvalues are computed. It is now possible to define the critical load as

$$p_c = \{p \mid \omega_i^2(p_k) \text{ real and positive}, \ i = 1, ..., m, \ \forall k \in \mathcal{K}\}. \qquad (4)$$

Turning now to design optimization, it is obvious that the eigenvalues ω^2 must depend on the shape of the column since the matrices M and K in the eigenvalue problem depend on the shape of the column. Assume that the shape of the column is circular symmetric and that the cross-sectional area varies as a piecewise linear function along the length of the column.

The optimal design problem of finding the shape of the column maximizing the critical load for a given amount of material may be posed as

$$\max_{\underline{x} \leq x \leq \overline{x}} p_c(x) \qquad (5)$$

$$c^T x = v_0, \qquad (6)$$

where x_j denotes the cross-sectional area at nodal point j, v_0 the prescribed volume. $\underline{x}_j$ and $\overline{x}_j$ the upper and lower limit of the cross-sectional area. The volume constraint is linear with coefficients c_j.

It is now of interest to investigate the properties of the objective function in the optimization problem. This is accomplished by performing a linear interpolation between two different columns with the same volume, as done in [5]. The first column x_0 is uniform with constant cross-sectional area and the other column $\hat{x}$ is the one obtained by Gutkowski et al. [4] using an optimization method. Intermediate columns are given by

$$x(\xi) = (1 - \xi)x_0 + \xi\hat{x}. \qquad (7)$$

The column $x(\xi)$ satisfies the linear constraint for any ξ and satisfies the constraint on positive cross-sectional area for values of ξ in the interval $[-1.408, 1.318]$. The objective function p_c is plotted as a function of the design parameter ξ in Figure 1. The problem is in the same dimensionless form as described in [4]. It is obvious that the objective function is discontinuous for $\xi = 1$ making optimization difficult if not impossible.

The reason for the discontinuity can be illustrated by investigating how the eigenvalues ω^2 change with p for two designs in the neighbourhood of $\xi = 1$. The eigenvalues are shown in Figure 2 for $\xi = 0.995$ and $\xi = 1.005$. For $p = 0$ all the eigenvalues are real and positive since K and M are real symmetric and positive definite. As p is increased, the eigenvalues start changing. For $\xi = 0.995$ the column becomes unstable at $p \simeq 70$ when two real eigenvalues coalesce and form a complex conjugate pair with the

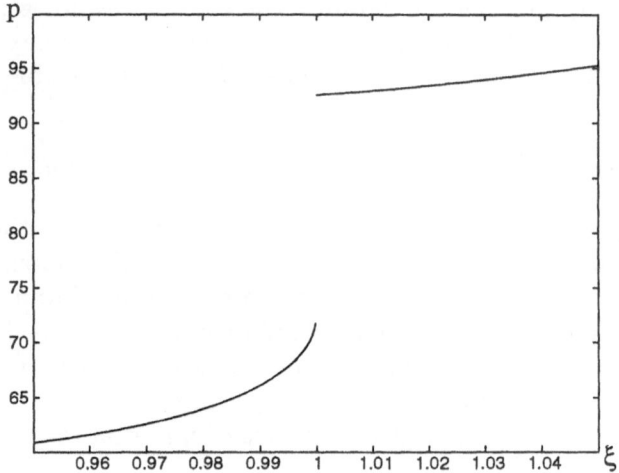

Figure 1.   Critical load versus design parameter.

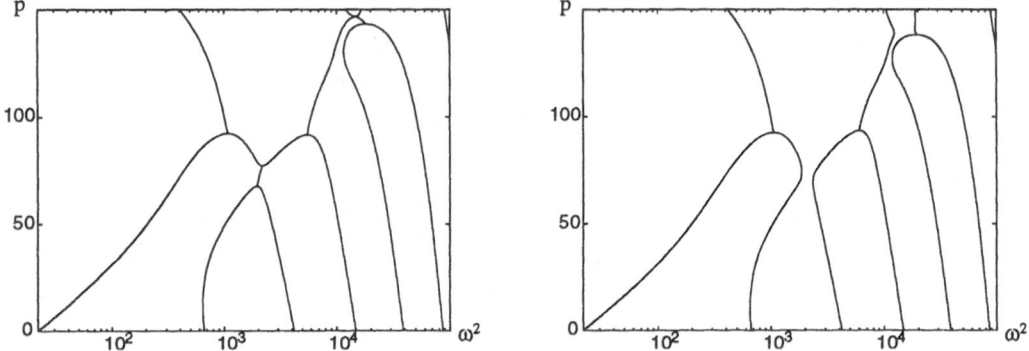

Figure 2.   The load p vs ω^2 for $xi = 0.995$ and $\xi = 1.005$.

same real part. For $p > 70$ only the common real part are shown for these two eigenvalues. However, for $\xi = 1.005$ the column remains stable until p reaches 92.6 when two pairs of eigenvalues simultaneously coalesce and form two pairs of complex conjugate eigenvalues. Obviously, the discontinuity in Figure 1 is caused by the change in shape of the curves in Figure 2 as ξ changes. For $\xi = 1.005$ then eigenvalues are well separated until the load reaches 92.6, but changing the design to $\xi = 0.995$ causes two eigenvalues to coalesce for a significantly lower load.

It is important to remember that it is not the eigenvalues themselves that are discontinuous functions. The eigenvalues are continuous but possibly nonsmooth functions of x and p. It is when a function such as (4) is defined using eigenvalues that depend on a parameter that the discontinuity appears. It is obviously not very good to formulate the optimization problem as (1)-(3).

2.1. AEROELASTICITY

The equations of motion for a simple aeroelastic system, given by

$$M\ddot{u} + Ku - qAu = 0, \tag{8}$$

are similar to the equations of motion for Beck's column (1). The load is in this case the dynamic pressure q which is proportional to the square of the speed of flight. The resulting eigenvalue problem is more difficult to analyze since the matrix of aerodynamic forces A is usually complex and frequency dependent. This causes the eigenvalue problem to be nonlinear in the eigenvalue making analysis difficult.

An interesting aspect is that the nonlinearity of the eigenvalue problem may simplify optimization. As is frequently observed in structural optimization with stability constraints, simultaneous stability modes are common for an optimal design. Fortunately, in aeroelastic systems the simultaneous stability modes for a given flight speed often appear for different frequencies which does not cause the eigenvalues to be nondifferentiable [6].

However, the special case of a static aeroelasticity may cause difficulties in design optimization. If the frequency is assumed zero, the static stability is analyzed by solving the linear eigenvalue problem

$$K\tilde{u} - qA\tilde{u} = 0. \tag{9}$$

The smallest real eigenvalue q of (9) gives the critical speed of flight for which a divergence instability may appear. But since the matrix of aerodynamic forces A is real and unsymmetric, the eigenvalues can be real but may also come in complex conjugate pairs.

The dynamic pressure is defined as $q = \rho v^2/2$, where ρ is the air density and v the speed of flight. Obviously, negative and complex eigenvalues have no physical meaning. The eigenvalues of (9) may change from being real and positive to being complex. This means that a constraint on divergence speed may not be differentiable and in some sense not even continuous, as discussed in [6].

3. Numerical optimization methods

Despite the numerous difficulties involved in structural optimization of non-conservative systems, substantial improvements are possible provided that the problem is properly posed and solved. Consequently, efficient numerical methods for these apparently nonsmooth problems are highly desirable.

Most studies concerning optimal design of structures subject to nonconservative forces simply ignore the possible nonsmoothness when solving the optimization problem. Although this strategy may work in some cases it can not be considered as a reliable and efficient approach.

A possible approach for minimum weight design of Beck's column is to ensure that the eigenfrequencies ω^2 remain distinct and positive. This keeps the column stable since at least two eigenvalues must coalesce before they can form a complex conjugate pair. The minimum weight design problem may be posed as

$$\min_{\underline{x} \leq x \leq \overline{x}} c^T x , \quad x \in R^n \tag{10}$$

$$\omega_1^2(p_k, x) \geq 0 , \quad k \in \mathcal{K}, \tag{11}$$

$$\omega_{i+1}^2(p_k, x) - \omega_i^2(p_k, x) \geq 0 , \quad i = 1, ..., m - 1, \ k \in \mathcal{K}. \tag{12}$$

In this case the maximum load p_c is fixed. Using an interior point method for solving (10)-(12) ensures a sequence of strictly feasible columns x with decreasing structural weight as demonstrated in [5].

This may appear to be a reasonable approach but unfortunately there are difficulties involved. The eigenvalue problem becomes increasingly ill-conditioned as two eigenvalues approach each other, which is typical for an optimal design. Consequently, it becomes more and more difficult to compute ω^2 and its derivatives as the solution is approached.

3.1. DAMPING IN THE STRUCTURAL MODEL

Possibly, it may be useful to consider a slightly more complicated structural model where damping is considered. In the case of Beck's column, the equations of motion become

$$M\ddot{u} + D\dot{u} + Ku - pAu = 0, \tag{13}$$

where D denotes the damping matrix.

The solution to (13) is assumed to be of slightly different form

$$u = \tilde{u}\, e^{\lambda t}, \tag{14}$$

where λ is the eigenvalue and $\tilde{u} \in R^m$ an eigenvector.

This assumption gives the eigenvalue problem

$$(\lambda^2 M + \lambda D + K - pA)\tilde{u} = 0, \tag{15}$$

which can be transformed to standard form

$$(B(p) - I)\hat{u} = 0 \tag{16}$$

with

$$B(p) = \begin{pmatrix} 0 & I \\ -M^{-1}(K - pA) & -M^{-1}D \end{pmatrix} \tag{17}$$

and $\hat{u} \in R^{2m}$, by a simple transformation.

The structure is now stable when the eigenvalues λ have negative real part meaning that the solution decays in magnitude with time.

The minimum weight design problem may be posed as

$$\min_{\underline{x} \le x \le \overline{x}} c^T x , \quad x \in R^n \tag{18}$$

$$\Re\{\lambda_i\} \le 0 , \quad i = 1, ..., m, \ k \in \mathcal{K}, \tag{19}$$

This problem is unfortunately also nonsmooth since the eigenvalues λ are nonsmooth functions of x.

Fortunately, this somewhat different structural model (13) and stability condition makes it possible to use a different but equivalent stability condition known as Lyapunov's stability condition [7]. Lyapunov's theorem states that there exists a symmetric positive definite matrix P such that the matrix inequality

$$B^T P + PB \prec 0 \tag{20}$$

holds if and only if the eigenvalues of $B(p)$ have negative real part. This corresponds to the constraint (19) in the optimization problem. The matrix inequality $E \succ F$ simply means that $E - F$ is positive definite.

Using Lyapunov's condition, (18)-(19) may be rephrased as

$$\min_{\underline{x} \le x \le \overline{x}, P} c^T x \tag{21}$$

$$B^T(x)P + PB(x) \prec 0 \tag{22}$$

$$P \succeq 0. \tag{23}$$

The essential difference is the dramatic increase in the number of independent variables, since the elements of the symmetric matrix P are now variables. There may be multiple sets of constraints (22)-(23), each corresponding to a different load p_k.

The significant advantage is that the stability constraints are now posed in terms of symmetric matrix inequalities for which barrier functions can be used. Applying a logarithmic barrier transformation to (21)-(23) gives the optimization problem

$$\min_{\underline{x} \le x \le \overline{x}, P} c^T x - \mu \log \det(-B^T(x)P - PB(x)) - \mu \log \det P, \tag{24}$$

where μ is a barrier parameter. The barrier subproblem (24) is solved for a
sequence of μ going to zero and it can be shown [8] that a solution to (24)
approaches a solution to (21)-(23) as $\mu \to 0$.

The main advantage of this formulation is that it is now a smooth optimization problem which can be solved using standard methods for unconstrained optimization, such as Newton's method. This approach has been
successfully used for design optimization of structures subject to conservative forces, see [9], and can using Lyapunov's stability criterion also be used
for nonconservative systems even though B is unsymmetric.

4. Final remarks

Some of the essential difficulties in optimal design of structures subject
to nonconservative forces have been discussed in this paper. It has also
been shown that difficulties may be serious involving both nonsmooth and
possibly discontinuous functions. However, is was also demonstrated that
it is indeed possible to formulate this type of optimal design problem in
smooth form using matrix inequalities and barrier functions. The most
significant drawback is the increased number of independent variables

Acknowledgement

This research was financially supported by the Swedish Research Council
for Engineering Sciences (TFR).

References

1. A. P. Seyranian and P. Pedersen. On interaction of eigenvalue branches in non-
    conservative multi-parameter problems. DCAMM report 478, Department of Solid
    Mechanics, The Technical University of Denmark, 1993.
2. N. Olhoff and J. E. Taylor. On structural optimization. *J. Appl. Mech.*, 50:1139–1151,
    1983.
3. M. Zyczkowski and A. Gajewski. Optimal structural design under stability con-
    straints. In *Collapse: The buckling of structures in theory and practice*. Cambridge
    University Press, 1982.
4. W. Gutkowski, O. Mahrenholtz, and M. Pyrz. Minimum weight design of structures
    under nonconservative forces. In *NATO ASI E 231*, pages 1087–1099, 1991.
5. U. T. Ringertz. On the design of Beck's column. *Structural Optimization*, 8:120–124,
    1994.
6. U. T. Ringertz. On structural optimization with aeroelasticity constraints. *Structural
    Optimization*, 8:16–23, 1994.
7. S. Boyd, L. El Ghaoui, E. Feron, and V. Balakrishnan. *Linear matrix inequalities in
    system and control theory*. SIAM, 1994.
8. Y. Nesterov and A. Nemirovsky. *Interior point polynomial methods in convex pro-
    graming*, volume 13 of *Studies in applied mathematics*. SIAM, 1993.
9. U. T. Ringertz. An algorithm for optimization of nonlinear shell structures. *Int. J.
    Num. Meth. Eng.*, 38:299–314, 1995.

SIMULATION, SENSITIVITY ANALYSIS AND OPTIMIZATION OF CONSTRAINED MULTIBODY SYSTEMS WITH IMPACTS BASED ON MASS-ORTHOGONAL PROJECTIONS

M. SCHULZ AND H. BRAUCHLI

Institut für Mechanik, ETH-Zentrum

CH-8092 Zürich, Switzerland

Abstract

In this paper the optimization of the transient dynamical behaviour of multibody systems with impacts and unilateral constraints is treated. As a basis the projection method for the simulation of multibody systems is used. The optimization of the dynamical behaviour of multibody systems is reduced to a nonlinear programming problem, we also use extended semi-analytical methods of sensitivity analysis presented in [7]. These methods serve as a link between simulation and general purpose optimization codes. The theory is applied to the optimization of the geometry of a circuit breaker.

1. Simulation of constrained multibody systems with impacts

The projection method [1,3,5] for the simulation of multibody systems is used as a basis for treating dynamical optimization problems. This method allows the description of multibody systems by redundant coordinates resulting in a system of purely differential equations for the motion and an algebraic equation for the reaction forces. But this system of differential equations is by no means uniquely determined. In [8] a choice of a particular formulation guided by numerical considerations was presented:

$$\dot{q} = A\alpha^T p \, , \tag{1}$$

$$\dot{p} = \alpha^T \hat{Q} + \alpha^T \Omega^T A p - \Omega A \alpha^T p \quad . \tag{2}$$

Here q and p denote the generalized coordinates and the canonical momenta, respectively. A is the inverse mass matrix and $\hat{Q}$ stands for the sum of generalized forces and the derivative of the kinetic energy with respect to the coordinates. Furthermore, α is the mass-orthogonal projector and Ω is a matrix occuring in the expression for the derivative of α with respect to time. In [7] Newton's impact hypothesis was applied yielding a very simple expression for the jumps in the values of the canonical momenta produced by an impact at time t_i :

D. Bestle and W. Schielen (eds.), IUTAM Symposium on Optimization of Mechanical Systems, 261–268.
© 1996 *Kluwer Academic Publishers.*

$$\left.\underline{p}\right|_{t_i+0} = \left.\underline{p}\right|_{t_i-0} - \frac{(1+\varepsilon_i)\,\underline{c}_i^T\Delta\underline{p}\big|_{t_i-0}}{\underline{c}_i^T\Delta\underline{\alpha}^T\underline{c}_i}\underline{\alpha}^T\underline{c}_i \ . \tag{3}$$

Here ε_i is the restitution coefficient and $\underline{c}_i$ a vector containing the geometry concerning the impact.

Unilateral constraints active at the beginning of the motion are modelled as bilateral constraints. They are removed when the corresponding reaction force changes its sign. Unilateral constraints becoming active during the motion of the system can be treated by defining an impact. For impacts obeying Newton's impact hypothesis and occuring repeatedly with decreasing relative velocity of the colliding bodies it seems useful to switch to another treatment of the impact when the relative velocity is lower than a certain small value. In this case we add a linear spring and a damper at the contact point when

$$d\left(\underline{q}\big|_{t_i}\right) = 0 \ , \tag{4}$$

where d is the distance of the surfaces which can collide. The damper is removed when the relative velocity of the contact points becomes zero, and the spring should be removed when the contact force (i.e. spring force) becomes zero. This is equivalent to equation (4). But, for numerical reasons we remove the spring, if

$$d = \varepsilon \ , \tag{5}$$

where ε is a very small constant greater than zero. In this way we can model unilateral constraints which are inactive at the beginning and become active during the motion of the system.

2. Optimization of constrained multibody systems with impacts

The optimization of the transient dynamical behaviour of multibody systems is considered. This task can be reduced to a nonlinear programming problem based on the simulation model described above and supplemented by one or multiple optimization criteria. The design model is defined by the choice of the design variables and the relationships between the design- and analysis variables.

Our task is to find an optimal design by minimizing a performance measure (objective function of optimization) which evaluates the dynamical behaviour of the system. A rather general form of the performance measure is presented e.g. in [2,4]. In order to account of impacts, we propose

$$\psi(b_k) = \sum_{i=0}^{n-1} \int_{(t_i+0)}^{(t_{i+1}-0)} f(t, \underline{q}, \underline{p}, \underline{\dot{p}}, b_k)\, dt \ , \tag{6}$$

where b_k $(k = 1, ..., m)$ are the design variables and $t_0 = t_0(b_k)$ and $t_n = t_n(b_k)$ are the initial and the final time. $t_i = t_i(b_k)$ $(i = 1, ..., n-1)$

denotes the time of discontinuities of the integrand f, e.g the time of impacts or the time when a spring-damper element is added to model an unilateral constraint.

2.1. PERFORMANCE MEASURE FOR THE MOTION OF A CIRCUIT BREAKER

In paragraph four we present the optimization of the opening process of a circuit breaker. Typical time histories of the distance between the stationary and the free movable contact are shown in figs. 4 c) and d). To optimize the opening process is equivalent to the minimization of the deviation from an ideal motion. Characteristic for such an ideal opening is that no rebouncing takes place and that the switching time is equal to zero. A corresponding performance measure is given by

$$\psi = \text{konst} \cdot \sum_{i=0}^{n-1} \int_{(t_i+0)}^{(t_{i+1}-0)} ((d(\underline{q}, b_k) - d_0)^2 \cdot w(t - t_0)) \, dt \quad , \tag{7}$$

where the initial time $t_0 = t_0(b_k)$ is the time when the contact opens (unilateral constraint is removed), the final time is $t_n = \infty$ and $w(t - t_0)$ is a weighting function. If a rebouncing lower than a critical distance d_{crit} is irrelevant, one may use

$$\psi = \text{konst}$$

$$\cdot \sum_{i=0}^{n-1} \int_{(t_i+0)}^{(t_{i+1}-0)} (h(d_{crit} - d(\underline{q}, b_k))) \cdot (d(\underline{q}, b_k) - d_{crit})^2 \cdot w(t - t_0)) \, dt \ , \tag{8}$$

wherein h is the Heavyside function.

2.2. CONTINUITY OF THE PERFORMANCE MEASURE

In general the performance measure defined by equation (6) as a function of the design variables is discontinuous. Small changes in the values of the design variables may result in qualitatively different motions. As an example we consider a system with an elbow lever. Depending on the length of the levers the elongated position (position where the levers are parallel) may or may not be reached during the motion leading to totally different toggle motions. This simple example also shows that the performance measure will be continuous and has continuous derivatives up to a given order, if one imposes appropriate constraints on the design variables. In the case of our example we only have to make sure that the toggle is never elongated.

3. Sensitivity analysis of constrained multibody systems with impacts

Efficient optimization algorithms require the gradient of the performance measure. Therefore, the problem of sensitivity analysis of multibody systems with impacts is considered. Differentiation of (6) with respect to the design variables yields the gradient

$$\frac{d\psi}{db_k} = -f|_{(t_0 + 0)} t_{0, b_k} + f|_{(t_n - 0)} t_{n, b_k} + \sum_{i=1}^{n-1} (f|_{t_i - 0} - f|_{t_i + 0}) t_{i, b_k}$$

$$+ \sum_{i=0}^{n-1} \left(\int_{(t_i + 0)}^{(t_{i+1} - 0)} \left(\frac{\partial f}{\partial q} \right)^T q_{, b_k} dt + \int_{(t_i + 0)}^{(t_{i+1} - 0)} \left(\frac{\partial f}{\partial p} \right)^T p_{, b_k} dt \right.$$

$$\left. + \int_{(t_i + 0)}^{(t_{i+1} - 0)} \left(\frac{\partial f}{\partial \dot{p}} \right)^T \dot{p}_{, b_k} dt + \int_{(t_i + 0)}^{(t_{i+1} - 0)} \frac{\partial f}{\partial b_k} dt \right) . \tag{9}$$

The quantities t_{i, b_k}, $q_{, b_k}$, $p_{, b_k}$ and $\dot{p}_{, b_k}$ in equation (9) are unknown and depend implicitly on the design variables. Mainly two semi-analytical methods are known for the computation of the gradient (9): the direct differentiation method and the adjoint variable method [2,4]. The extension of these methods to systems with discontinuous variables of state, e.g. multibody systems with impacts, has been presented in [7].

4. An example optimization

The opening process of the circuit breaker shown in figure 1 shall be optimized. The mechanism consists of four levers (reference numeral 1 to 4). They are interconnected by three revolute joints 7, 8, 9. Two revolute joints 6, 10 connect parts of the mechanism to the inertial system. Furthermore, the system contains a spring 5 providing the driving force of the motion and a movable contact 13 facing a stationary contact 12 to form a pair of switch electrodes. 14 and 15 are stop pieces, i.e. unilateral constraints (regions of impact) which are not active in the closed configuration shown in fig. 1.

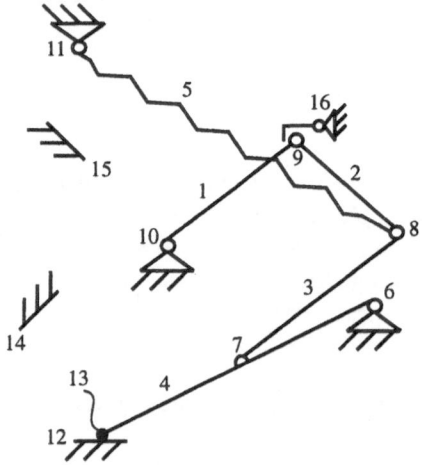

Figure 1.   Circuit breaker

The mechanism stays in the closed position as long as body 1 is held by the lock 16. Turning the lock to the right by a tripping mechanism (not shown) starts the transition from "closed" to "open".

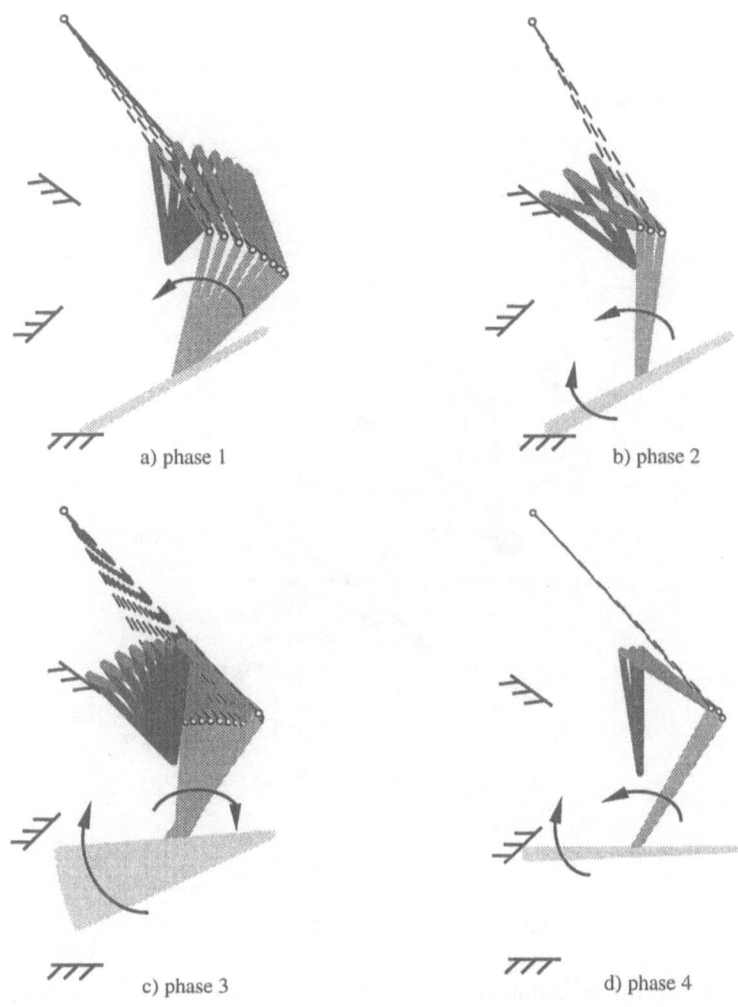

a) phase 1   b) phase 2

c) phase 3   d) phase 4

Figure 2.   Opening of the circuit breaker

What happens during this transition can be described roughly as follows: The force of the elongated tension spring acting on joint 8 leads via body 2 and joint 9 to a torque acting counterclockwise on body 1. Body 2 turns clockwise until the line of application of the tension spring changes from the left hand side of joint 9 to the right hand side. It then turns counterclockwise and the toggle formed by the levers 2 and 3 is lifted, so that the movable contact 13 moves and the circuit breaker adopts its open state. The open state is defined by the activity of unilateral joints 14 and 15. Then, lever 1 and stop piece 15 as

well as lever 4 and stop piece 14 are in contact. A typical opening and its different phases
are depicted in figure 2. During phase 1 the contact lever 4 does not move. The rebouncing of lever 1 and of lever 4 characterize the motion after phase 4.

4.1. DESIGN MODEL

We have chosen ten design variables $b_1,...,b_{10}$ (figure 3) describing the geometry of
the circuit breaker with lower and upper bounds:

$$\alpha < b_1 < \pi + \alpha \, , \quad 0 < b_2 < \pi \, , \quad 0 < b_3 < \pi \, , \quad 0 < b_4 < \pi \, , \quad 0 < b_5 < 2\pi \, , \quad 0 < b_6 < \infty \, ,$$

$$0 < b_7 < \infty \, , \quad 0 < b_8 < \infty \, , \quad 0 < b_9 \leq 1 \, , \quad 0 < b_{10} < \infty \, .$$

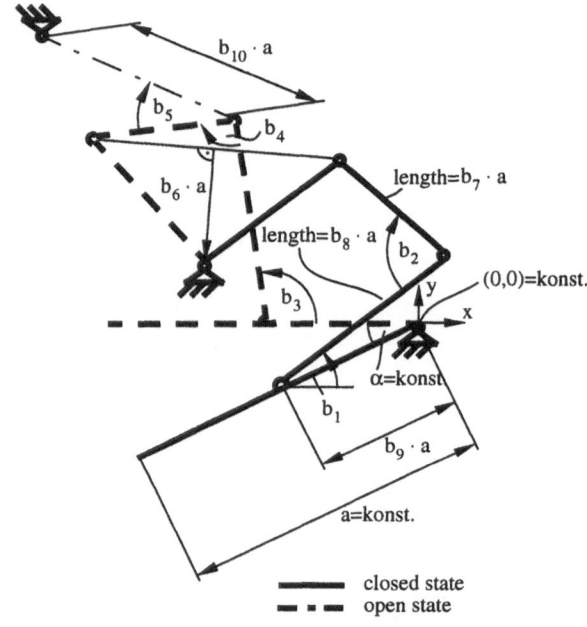

Figure 3.   Design variables

The mass of a lever is assumed to increase linearly with increasing length. The
unloaded spring length amounts p_1 times the length of the spring in the open state of the
circuit breaker ($p_1 = 0.95$ chosen). Furthermore, we choose $\alpha = \pi/6$. Eight constraints have been taken into account to ensure that the mechanism works.

4.2. RESULTS

We have optimized the circuit breaker shown in fig. 1 by applying DYNAMITE and
using the performance measure (7) with w=1. DYNAMITE is a computer program for
the simulation and optimization of multibody systems based on the theory and equations
presented above. All equations (equations of motion, equations of sensitivity analysis)

are generated in numerical form. The user can choose among various optimization algorithms (SQP, Feasible Directions, Conjugate Directions). A special choice of design variables is provided, but it is also possible to define one's own design model by a user-defined subroutine. Concerning the sensitivity analysis the direct differentiation method extended to multibody systems with impacts and unilateral constraints has been implemented. Of course, we can also approximate the gradient of the performance measure by finite differences.

It turns out, that several local optima exist and the corresponding function values differ substantially. Therefore, we have determined 14 initial designs and have chosen the best local optimum as an approximation for the global optimum. One initial design, the best locally optimal design and the time histories of the distance d are shown in fig. 4. The reduction of the value of the performance measure is about 96 per cent.

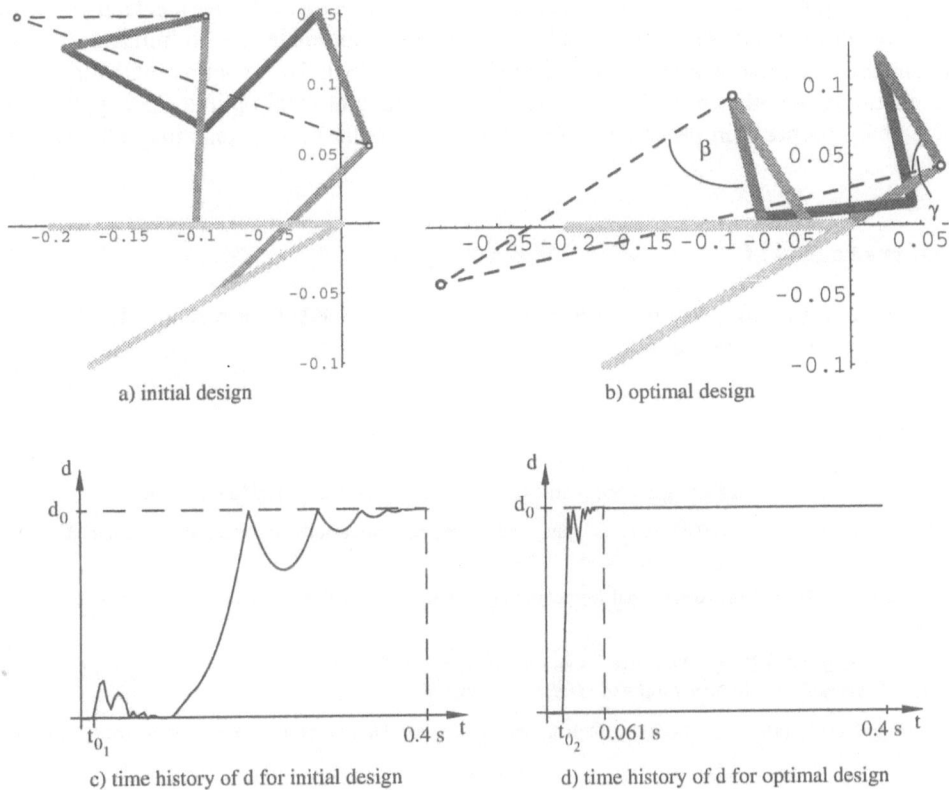

a) initial design

b) optimal design

c) time history of d for initial design

d) time history of d for optimal design

Figure 4. Optimization results

The levers 1 to 3 of the optimal design are as short as possible so that due to the low mass high accelerations are achieved. Furthermore, a characteristic of the optimal design is the wide area covered by the motion of the 4-hinged arch formed by lever 1 to 3. Also, the angles β and γ are big. If γ is as big as possible the contact 12/13 (fig. 1) will open very fast at the beginning of phase 2. A big angle β leads to a high torque acting on lever 2, so that the toggle is lifted very fast during phase 2 and phase 3 of the motion. On the other hand, if β is too small, the spring will move to the left hand side of joint 9 in cause of the rebouncing of lever 1 during phase 3 of the motion. The contact 12/13 may then close again as is the case in the opening process of the initial design (fig. 4 c)).

Concluding remarks

A circuit breaker consisting of four bodies has been chosen as an example for optimization of multibody systems with impacts and unilateral constraints. It turns out that the choice of suitable design parameters for which the constraints of the programming problem can be easily formulated is of consequence. The constraints have to ensure that the mechanism works. Therefore, the choice of the design parameters depends on the underlying principle of the mechanism. Due to the great variety of existing mechanisms with different underlying principles it will be difficult to automate the optimization of the transient dynamical behaviour of multibody systems. This is in contrast with structural optimization. Another problem is the existence of several local optima. The application of global optimization procedures [6] embodying stochastic elements may be advantageous.

Acknowledgement

The authors gratefully acknowledge the support of the KWF (a research fund of the Swiss Federal Department of Economic Affairs).

References

1. Bach, D. N. (1993) Ein Beitrag zur Starrkörperdynamik, thesis, Diss. ETH Nr. 10367, Zurich.

2. Bestle, D. and Eberhard, P. (1992) Analyzing and Optimizing Multibody Systems, *Mech. Struct. & Mach.* **20(1)**, 67-92.

3. Brauchli, H. (1991) Mass-orthogonal formulation of equations of motion for multibody systems, *ZAMP* **42**, 169-182.

4. Haug, E. J. (1987) Design Sensitivity Analysis of Dynamic Systems, *NATO ASI Series* **F 27**, *Computer Aided Optimal Design: Structural and Mechanical Systems.*

5. Melliger, O. F. (1994) Numerisch effiziente Methoden der Mehrkörperdynamik, thesis, Diss. ETH Nr. 10755, Zurich.

6. Rinnooy Kan, A. H. G. and Timmer, G. T. (1987) Stochastic global optimization methods, part I: Clustering methods and part II: Multi level methods, *Mathematical Programming* **39**, 57-78.

7. Schulz, M. and Brauchli, H. (1994) Simulation and optimization of constrained multibody systems based on mass-orthogonal projections, *CISS-First Joint Conference of International Simulation Societies Proceedings*, Halin, J. (ed.), Zurich, 317-321.

8. Sofer, M., Melliger, O. and Brauchli, H. (1992) Numerical Behaviour of Different Formulations for Multibody Dynamics, *Numerical Methods in Engineering*, 277-284.

PARAMETER ESTIMATION IN DISCONTINUOUS DESCRIPTOR MODELS

R. VON SCHWERIN, M. WINCKLER, V. SCHULZ
Interdisciplinary Center for Scientific Computing (IWR)
University of Heidelberg, Im Neuenheimer Feld 368
D-69120 Heidelberg, Germany

1 Introduction

Parameter estimation is an important tool for deriving models for multibody systems, which not only describe the dynamical system in a qualitatively correct way, but are also quantitatively correct. A number of efficient methods for treating such problems for general ordinary differential equations and differential–algebraic equations (DAE) have been developed [3, 6, 7]. The problem class of multibody systems (MBS) in descriptor form is especially challenging due to the high index of the DAE and the typical presence of discontinuities. The focus of the present paper is on the reliable computation of *derivatives* of integration schemes for MBS descriptor models with *discontinuous effects* which is a key task within the parameter estimation process. [1] It is based on the theoretical background about the exploitation of invariants presented in [7] and has been implemented extending the Runge–Kutta integrator MBSD85 from the integrator library MBSSIM [8]. The paper is organized in three sections. First some basic facts about MBS and parameter identification as well as the efficient solution of initial value problems for MBS descriptor models are recalled. The second section provides the basic methods for the efficient and accurate computation of derivatives of IVP solutions of MBS. Finally numerical results illustrate the practical applicability of the methods developed.

Multibody descriptor models. It is well–known that the choice of redundant coordinates for multibody models leads to linearly implicit second order DAE of index three:

$$\begin{aligned} M(p,q)\ddot{p} &= f(t,p,\dot{p},q) - G^T(p,q)\lambda \\ g(p,q) &= 0. \end{aligned} \tag{1}$$

Here p and v denote the (generalized) positions and velocities and q denotes a vector of system parameters such as masses, moments of inertia, damping and stiffness coefficients, etc. Furthermore $g(p,q)$ denotes the constraints on position level and $G(p,q) := \frac{\partial}{\partial p} g(p,q)$ is the constraint matrix, $M(p,q)$ is called the mass matrix and $f(t,p,\dot{p},q)$ are the explicit force terms. The Lagrange-multipliers λ are the algebraic variables of the DAE system.

---

[1]This work is funded by the German Department of Science and Technology within the research grant GESAMTFAHRZEUGSIMULATION and conducted under the advice of Prof. Dr. H.G. Bock and Dr. J. Schlöder.

D. Bestle and W. Schielen (eds.), IUTAM Symposium on Optimization of Mechanical Systems, 269–276.
© *1996 Kluwer Academic Publishers.*

In MBSSIM (1) is treated by index reduction to index one:

$$\dot{p} = v$$

$$\mathcal{A}(p,q)\begin{pmatrix} \dot{v} \\ \lambda \end{pmatrix} := \begin{pmatrix} M(p,q) & G(p,q)^T \\ G(p,q) & 0 \end{pmatrix}\begin{pmatrix} \dot{v} \\ \lambda \end{pmatrix} = \begin{pmatrix} f(t,p,v,q) \\ \gamma(p,v,q) \end{pmatrix} \qquad (2)$$

$$\text{where} \quad \gamma(p,v,q) := -v^T \tfrac{\partial}{\partial p} G(p)v.$$

If necessary, its numerical solution is subsequently projected onto the constraints:

$$\begin{aligned} g_p(p,q) &:= g(p,q) = 0 & \text{(position constraint)} \\ g_v(p,v,q) &:= \dot{g}_p(p,q) = G(p,q)v = 0. & \text{(velocity constraint)} \end{aligned} \qquad (3)$$

Mathematically, these are interpreted as *invariants* [2, 7].

MBSSIM also offers the opportunity for a mathematically correct treatment of discontinuities in the state variables or their time derivatives, which appear frequently in MBS applications. For instance, they occur naturally in the modeling of *impact, back-lash, Coulomb friction, hysteresis, discrete time controllers*, and through the use of *tabulated data* to approximate, e.g., non-linear force laws by piecewise smooth functions. The time instances t_s at which the model changes are given implicitly by the zeros of switching functions

$$s(t_s, p(t_s), v(t_s), q) = 0.$$

All the integrators in MBSSIM are equipped with a safe and efficient state–event handler to localize and deal with such switching points. This allows them to be easily extended for the safe generation of derivative information needed for the solution of optimization problems.

Parameter estimation.   The parameter estimation task is to identify the unknown system parameters q from given measured data of the state history. The measured data are assumed to be perturbed values of functions ϕ_i of an exact solution of the system equations,

$$\psi_{i,j} = \phi_i(p(t_j), v(t_j), q) + \varepsilon_{i,j}$$

with measurement errors $\varepsilon_{i,j}$ distributed according to a normal distribution with mean value 0 and known variances $\sigma_{i,j}^2$. A Maximum Likelihood estimator for the parameters is the minimizer of the weighted output functional

$$\ell_2(p,v,q) := \sum_{i,j} \sigma_{i,j}^{-2}\left(\psi_{i,j} - \phi_i(p(t_j), v(t_j), q)\right)^2.$$

This functional is minimized subject to the model DAE and additional boundary conditions. The resulting problem is an overdetermined multipoint boundary value problem (MPBVP) of the following form: given a (time) interval $[t_0, t_f]$ and interior points $t_0 < t_1 < \cdots < t_{f-1} < t_f$, we seek position and velocity variables p, v, and parameters q which minimize the least-squares function

$$\|r_1(p(t_0), v(t_0), \ldots, p(t_f), v(t_f), q)\|_2^2 = \min$$

while satisfying the DAE (2) with multipoint boundary conditions

$$r_2(p(t_0), v(t_0), \ldots, p(t_f), v(t_f), q) = 0.$$

Solving the MPBVP. Following the *boundary value problem approach* [3] the multipoint optimization boundary value problem is discretized and transformed into a finite dimensional optimization problem, which is then solved employing a *generalized Gauß–Newton* method [3] in our case. For discretization we use multiple shooting, which is based on the solution of initial value problems on the subintervals of an appropriately chosen mesh

$$\pi : t_0 = \tau_1 < \ldots < \tau_{m-1} < \tau_m = t_f ,$$

which needs not coincide with the points of measurement. The state variables on every subinterval are replaced by the computed solution $p^\pi(t; \tau_j, p_j, v_j)$, $v^\pi(t; \tau_j, p_j, v_j)$ of the initial value problems

$$\dot{p}^\pi = v^\pi$$
$$A(p^\pi, q) \begin{pmatrix} \dot{v}^\pi \\ \lambda^\pi \end{pmatrix} = \begin{pmatrix} f(t, p^\pi, v^\pi, q) \\ \gamma(p^\pi, v^\pi, q) \end{pmatrix} \tag{4}$$
$$p^\pi(\tau_j) = p_j \quad ; \quad v^\pi(\tau_j) = v_j$$

Thus, the MPBVP is transformed to the finite dimensional nonlinear least–squares problem

$$\|r_1(p_1, v_1, \ldots, p_m, v_m, q)\|_2^2 = \min$$
$$\text{s.t.} \quad p^\pi(\tau_{j+1}; p_j, v_j) - p_{j+1} = 0$$
$$v^\pi(\tau_{j+1}; p_j, v_j) - v_{j+1} = 0 \quad j = 1, \ldots, m-1 \quad \text{(continuity conditions)}$$
$$r_2(p_1, v_1, \ldots, p_m, v_m) = 0 \quad \text{(boundary conditions)}$$

In [7] it is pointed out, that it is advantageous to use the known invariants of the index reduced MBS-system in order to stabilize the IVP solution (see below) as well as the solution process of the nonlinear system of equations above. There a multistage least squares approach is proposed, which exploits this invariant structure. Since the requirements of satisfying the DAE and the invariants formally overdetermine the solution of the BVP, a ranking is introduced, which favors the invariants by replacing the continuity conditions with:

$$\left\| \begin{pmatrix} p^\pi(\tau_{j+1}; p_j, v_j) - p_{j+1} \\ v^\pi(\tau_{j+1}; p_j, v_j) - v_{j+1} \end{pmatrix} \right\|^2 = \min_{p_{j+1}, v_{j+1}}$$
$$\text{s.t.} \quad g_p(p_{j+1}, q_{j+1}) = 0$$
$$g_v(p_{j+1}, v_{j+1}, q) = 0$$

Thus the continuity conditions are only satisfied in the kernel of the invariants. For theoretical and implementational details see [7].

IVP solution for descriptor models In the present context of parameter estimation of multibody descriptor models a core task is to solve IVP (4) on the multiple shooting subintervals. Firstly, this requires the correct and efficient treatment of discontinuities as described earlier. In contrast to numerical integration methods for (2), which must avoid drift from the invariant manifolds given by the constraints (3), in the optimization context one must allow for *inconsistent* initial values. This is because the initial values on the subintervals are iteratively changed by the generalized Gauß–Newton method, so that the second requirement for the numerical solution of (4) is to avoid *additional* drift from the invariant manifolds. This makes it necessary to project onto the *relaxed constraints*

$$\hat{g}_p(t, p, q, t_0, p_0, v_0) \quad := \quad g_p(p, q) - (t - t_0)g_{v,0} - g_{p,0} = 0$$
$$\hat{g}_v(t, p, v, q, t_0, p_0, v_0) \quad := \quad g_v(p, v, q) - g_{v,0}$$

where the index 0 denotes the initial violation of the respective constraints.

A key idea for maintaining the constraints in an efficient manner is to apply a *sequential projection* ([2, 1]). This idea is arrived at by noting that the position constraint does not depend on velocity, so that as a first step one may compute p^* "closest" (in a suitably chosen norm) to the value p_{int} obtained during integration with $\hat{g}_p(t, p^*, q, t_0, p_0, v_0) = 0$. Subsequently, we can find v^* such that $\hat{g}_v(t, p^*, v^*, q, t_0, p_0, v_0) = 0$ without influencing the position constraint. In this second step we only need to solve a linear least–squares problem which does not require any iterations.

Remark 1 By using special linear algebra methods as in MBSSIM, the solution of the linear systems in the integration step can be optimally dovetailed with the projections in the sense that factorizations needed in one process can be reused in the other.

2   Internal Numerical Differentiation for MBS

In order to solve boundary value problems, not only trajectory values are needed but also their derivatives with respect to initial values and parameters. This is the so called Wronskian (or sensitivity matrix), which is defined as

$$\mathcal{W}_{p,v}(t, t_0) := \left(\begin{array}{cc} \mathcal{W}_p^p(t, t_0) & \mathcal{W}_v^p(t, t_0) \\ \mathcal{W}_p^v(t, t_0) & \mathcal{W}_v^v(t, t_0) \end{array} \right), \qquad \mathcal{W}_q(t, t_0) := \left(\begin{array}{c} \mathcal{W}_q^p(t, t_0) \\ \mathcal{W}_q^v(t, t_0) \end{array} \right)$$
$$\text{where} \quad \mathcal{W}_x^y(t, t_0) := \tfrac{\partial}{\partial y_0} x(t; t_0, p_0, v_0), \quad \text{for } x \in \{p, v\} \text{ and } y \in \{p, v, q\}.$$

When computing the derivatives of the solution of a DAE with respect to initial values and possibly parameters, one observes that this solution itself is the result of an intricate adaptive discretization algorithm in the course of the integration. The principle of Internal Numerical Differentiation (IND) is to compute the derivative only of the discretization scheme approximating the DAE initial value problem. This approximate scheme is generated by an adaptive procedure. The resulting adaptive components of the numerical integration procedure, however, are not differentiated (in contrast to a brute force application of automatic differentiation). Since intermediate coefficients, matrices, decompositions etc. are used both for the computation of the solution and its derivatives, high computational savings may be gained. Details for various integration methods are given, e.g., in [3].

2.1   IND FOR DESCRIPTOR MODELS

In order to obtain the relevant gradient information in an efficient way, we make use of
the fact that our original index one DAE (2) has the same Wronskian as

$$\dot{\tilde{p}} = \tilde{v}$$

$$\mathcal{A}(p,q) \begin{pmatrix} \dot{\tilde{v}} \\ \lambda \end{pmatrix} = \begin{pmatrix} f(t,\tilde{p},\tilde{v},q) \\ \gamma(\tilde{p},\tilde{v},q) \end{pmatrix} + [\mathcal{A}(p,q) - \mathcal{A}(\tilde{p},q)] \begin{pmatrix} \dot{v} \\ \lambda \end{pmatrix} \qquad (5)$$

$$\tilde{p}(t_0) = p_0 \quad ; \quad \tilde{v}(t_0) = v_0$$

where $(p, v, \lambda)^T$ denotes the solution of (2) with the same initial values [7].

Thus, e.g., for a Runge-Kutta or polynomial extrapolation method one may obtain the
following algorithm for the computation of the Wronskian $\mathcal{W}_{p,v}$ (and similarly for $\mathcal{W}_q$)
by a finite difference approximation in integration step k:

1. compute the discretized nominal trajectory $y_{k+1} := (p_{k+1}, v_{k+1})$ solving IVP (2),

2. compute a varied step $\eta_{k+1}^{(i)}$ from $\eta_k^{(i)} := y_k + \delta_i e_i$ solving (5) using the same
   discretization scheme and replacing y_{k+1} by the results of 1. in all stages.

3. calculate $\mathcal{W}_{p,v}^{k+1} e_i = (\eta_{k+1}^{(i)} - y_{k+1})/\delta_i$.

This offers the advantage that repeated factorizations of the matrix $\mathcal{A}$ are avoided.

2.2   IND FOR DAE WITH INVARIANTS

As pointed out in [7], the exact Wronskian also satisfies constraints. These are the
derivatives with respect to initial values and parameters of the relaxed constraints, so that
on position level we find:

$$\frac{\partial \hat{g}_p}{\partial(p_0, v_0, q)} = G \left[\mathcal{W}_p^p \mid \mathcal{W}_v^p \mid \mathcal{W}_q^p \right]$$

$$- \left[(t - t_0)H_0 + G_0 \mid (t - t_0)G_0 \mid (t - t_0)H_{q,0} + G_{q,0} - G_q \right] \equiv 0$$

where $H := \frac{\partial}{\partial p_0} g_v$, $G_q := \frac{\partial}{\partial q} g_p$, and $H_q := \frac{\partial}{\partial q} g_v$. Again, the quantities with the index
0 denote the initial values of the quantities without this index. On velocity level we have:

$$\frac{\partial \hat{g}_v}{\partial(p_0, v_0, q)} = G \left[\mathcal{W}_p^v \mid \mathcal{W}_v^v \mid \mathcal{W}_q^v \right] + H \left[\mathcal{W}_p^p \mid \mathcal{W}_v^p \mid \mathcal{W}_q^p \right] - \left[H_0 \mid G_0 \mid H_{q,0} - H_q \right] \equiv 0,$$

In order to generalize the principle of IND to projection onto invariants, one again performs
projection sequentially, which is possible for the same reasons as above. It should be noted
that the Wronskians are projected at the same time as the nominal solution and with exactly
the same system matrix. In the light of remark 1 and due to the use of (5) to compute
the varied steps, IND does not require any extra matrix factorizations compared to the
solution of the IVP!

2.3   UPDATES IN THE PRESENCE OF STATE EVENTS

Suppose at the *switching point* $t_s \in (t_j, t_{j+1})$, which is implicitly defined by the condition

$$s(t_s, p(t_s), v(t_s), q) = 0 \tag{6}$$

there is a jump in the state variables given by the jump function $\left(\frac{d^p}{d^v}\right)$:

$$\left(\begin{array}{c} p(t_s^+) \\ v(t_s^+) \end{array} \right) = \left(\begin{array}{c} p(t_s^-) \\ v(t_s^-) \end{array} \right) + \left(\begin{array}{c} d^p(t_s^-, p(t_s^-), v(t_s^-), q) \\ d^v(t_s^-, p(t_s^-), v(t_s^-), q) \end{array} \right)$$

and/or some change of the model equations. Extending results given in [5] for ODE to multibody systems, the local Wronskians $\mathcal{W}_{p,v}(t_{j+1}, t_j)$ and $\mathcal{W}_q(t_{j+1}, t_j)$ can be calculated using the well–known properties of the Wronskians and applying the implicit function theorem to (6) from which we find:

$$s_t \frac{\partial t_s}{\partial(p_j, v_j, q)} + (s_p, s_v) \left(\begin{array}{c} \dot{p} \\ \dot{v} \end{array} \right) \frac{\partial t_s}{\partial(p_j, v_j, q)} + (s_p, s_v) \left[\mathcal{W}_{p,v} \mid \mathcal{W}_q \right] (t_s^-, t_j) + [0 \mid s_q] = 0$$

where the subscripts on s denote its derivatives with respect to t, p, v, and q. Furthermore, if we denote by $\left(\frac{\dot{p}}{\dot{v}}\right)^-$ resp. $\left(\frac{\dot{p}}{\dot{v}}\right)^+$ the left resp. right limit of $\left(\frac{\dot{p}}{\dot{v}}\right)$ at t_s, we can define the time derivative $\dot{s}$ of s and the Jacobian $\mathcal{D}_{p,v}$ of the jump function

$$\dot{s} := s_t(t_s^-) + (s_p(t_s^-), s_v(t_s^-)) \left(\begin{array}{c} \dot{p} \\ \dot{v} \end{array} \right)^- \quad \text{and} \quad \mathcal{D}_{p,v} := \left(\begin{array}{cc} d_p^p & d_v^p \\ d_p^v & d_v^v \end{array} \right).$$

By defining the *updates* at t_s

$$\mathcal{U}_{p,v} \;\; := \;\; \left(\left(\begin{array}{c} \dot{p} \\ \dot{v} \end{array} \right)^+ - \left(\begin{array}{c} \dot{p} \\ \dot{v} \end{array} \right)^- - \left(\begin{array}{c} d_t^p \\ d_t^v \end{array} \right) - \mathcal{D}_{p,v} \left(\begin{array}{c} \dot{p} \\ \dot{v} \end{array} \right)^- \right) \frac{(s_p, s_v)}{\dot{s}} + I + \mathcal{D}_{p,v}$$

$$\mathcal{U}_q \;\; := \;\; \left(\left(\begin{array}{c} \dot{p} \\ \dot{v} \end{array} \right)^+ - \left(\begin{array}{c} \dot{p} \\ \dot{v} \end{array} \right)^- - \left(\begin{array}{c} d_t^p \\ d_t^v \end{array} \right) - \mathcal{D}_{p,v} \left(\begin{array}{c} \dot{p} \\ \dot{v} \end{array} \right)^- \right) \frac{s_q}{\dot{s}} + \left(\begin{array}{c} d_q^p \\ d_q^v \end{array} \right)$$

we arrive at the following update formulas for the Wronskians at switching points:

$$\mathcal{W}_{p,v}(t_{j+1}, t_j) \;\; = \;\; \mathcal{W}_{p,v}(t_{j+1}, t_s^+) \, \mathcal{U}_{p,v} \mathcal{W}_{p,v}(t_s^-, t_j)$$

$$\mathcal{W}_q(t_{j+1}, t_j) \;\; = \;\; \mathcal{W}_{p,v}(t_{j+1}, t_s^+) \left(\mathcal{U}_{p,v} \mathcal{W}_q(t_s^-, t_j) + \mathcal{U}_q \right) + \mathcal{W}_q(t_{j+1}, t_s^+)$$

Thus, if there are no jumps in the state variables and if the model equations are continuous or if the switch only depends on time, the update formulas just reflect the chain rule of differentiation.

3 Numerical Results

3.1 PARAMETER ESTIMATION FOR A QUICK–RETURN MECHANISM

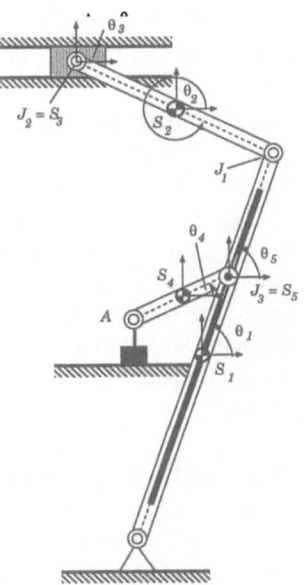

This model of a cutting tool consists of five bodies, the three arms (bodies 1,2, and 4) and the two sliders (bodies 3 and 5), a detailed description of which can be found in [4], where the values of the design parameters are given as:

# of body	1	2	3	4	5
mass (kg)	100	30	50	1000	5
m.o.i $(kg \cdot m^2)$	100	10	1.5	2000	–
lengths (m)	4	1.9	–	1.5	–

During the cutting stroke of the mechanism, a counterforce of 200,000 N acts while the center of mass of the cutting tool (body 5) moves from $x = 1.2m$ to $x = -4.0m$. In all other circumstances no force apart from a reaction force acts directly on the slider. This abruptly changes the equations of motion at the time instances when the slider reaches the positions above.

Our aim is to estimate the mass of body 4 and the lengths of bodies 1 and 2. We use an enhanced version [7] of the parameter estimation code PARFIT [3] together with the integrator presented in this paper. The Gauß–Newton iteration is stopped when a scaled norm of the inrements drops below 10^{-3}. Starting with the values of $m_4 = 800\,kg$, $l_1 = 5\,m$, and $l_2 = 2.3\,m$ and using 20 equally spaced multiple shooting nodes, we obtain the following estimations for the parameters and 95% confidence intervals:

noise	none	2.5%	5%
# iterations	4	4	6
m_4	$.10000E + 04 \pm 0\%$	$.10075E + 04 \pm 8\%$	$.95254E + 03 \pm 19\%$
l_1	$.40000E + 01 \pm 0\%$	$.39975E + 01 \pm 0\%$	$.40600E + 01 \pm 2\%$
l_2	$.19000E + 01 \pm 0\%$	$.18999E + 01 \pm 1\%$	$.18606E + 01 \pm 3\%$

It can be seen that in all cases the tool developed in this paper enabled PARFIT to find values of the parameters as well as confidence intervals that reproduce the original data satisfactorily, even when there is 5% of noise in the data.

3.2 CONCLUSIONS AND FURTHER WORK

A method for the fast and reliable generation of derivative information in the process of solving MPBVP for MBS descriptor models has been presented. One key factor for its efficiency is the application of sequential projection onto the constraints for the nominal trajectory as well as the Wronskians. In connection with a suitable reformulation of the model DAE for the calculation of the varied steps, it is thus possible to avoid any matrix factorizations other than those needed in the calculation of the nominal trajectory. Furthermore, the method described allows the correct treatment of discontinuous effects in optimization problems, which is a prerequisite for its suitability in MBS applications. This is achieved by updating the sensitivity matrices according to the type of discontinuity at the switching points, which are in turn detected by a state–event handler. This approach makes it imperative that the model equations are implemented in a way that reflects the discrete–continuous structure of the model as described in [2].

However, in order to make this approach acceptable in industrial applications, e.g. in mechatronical simulations in vehicle dynamics, the most important classes of discontinuities must be identified and the common practice of their implementation (which generally does not allow to obtain the derivative information necessary for optimization purposes) must be analyzed. Employing techniques of automatic differentiation where appropriate, a preprocessing tool must then be provided that will turn the common way of programming state–events into the one mentioned above. Such a tool in connection with the method for generating the relevant gradient information presented in this paper would then allow to use the proven methods of mathematical optimization in industrial areas like vehicle dynamics, where currently design parameters are mostly hand–tuned.

Acknowledgement.   The authors are indebted to Prof. H.G. Bock and Dr. J.P. Schlöder for many valuable suggestions.

References

[1] T. ALISHENAS. Zur numerischen Behandlung, Stabilisierung durch Projektion und Modellierung mechanischer Systeme mit Nebenbedingungen und Invarianten. PhD thesis, Institut für Numerische Analysis und Informatik, Königliche Technische Hochschule, S-100 44 Stockholm, Schweden, 3/ 1992.

[2] T. ANDRZEJEWSKI, H.G. BOCK, E. EICH, AND R. VON SCHWERIN. Recent advances in the numerical integration of multibody systems. In W. SCHIEHLEN, editor, *Advanced Multibody System Dynamics, Simulation and Software Tools*, pages 127–151. Kluwer Academic Publishers, Dordrecht, Boston, London, 1993.

[3] H.G. BOCK. Randwertproblemmethoden zur Parameteridentifizierung in Systemen nichtlinearer Differentialgleichungen. *Bonner Mathematische Schriften* **183**, 1987. Bonn.

[4] E. J. HAUG. Computer Aided Kinematics and Dynamics of Mechanical Systems — Volume I: Basic Methods Allyn and Bacon, Boston

[5] P. KRÄMER–EIS. Ein Mehrzielverfahren zur numerischen Berechnung optimaler Feedback–Steuerungen bei beschränkten nichtlinearen Steuerungsproblemen *Bonner Mathematische Schriften* **166**, 1985. Bonn.

[6] J.P. SCHLÖDER. Numerische Methoden zur Behandlung hochdimensionaler Aufgaben der Parameteridentifizierung. *Bonner Mathematische Schriften* **187**, 1988. Bonn.

[7] V.H. SCHULZ, H.G. BOCK, AND M.C. STEINBACH. Exploiting invariants in the numerical solution of multipoint boundary value problems in DAE. Technical Report Preprint 93-69, IWR, University of Heidelberg, 1993, submitted to SIAM J. Sci. Comp.

[8] R. VON SCHWERIN and M. WINCKLER, A Guide to the Integrator Library MBSSIM, Version 1.00, IWR-Preprint, University of Heidelberg, 1994

SOME INVERSE PROBLEMS IN TOPOLOGY DESIGN OF MATERIALS AND MECHANISMS

O. SIGMUND

Department of Solid Mechanics
Technical University of Denmark
DK–2800 Lyngby, Denmark

Abstract. This paper presents an alternative approach to design of mechanisms. The approach is based on a transition from an elastic ground structure to a mechanism using topology optimization methods originally developed for optimal design of elastic truss structures. The approach is demonstrated by examples of design of extreme materials with prescribed elastic properties and by examples of design of mechanisms with prescribed output motions. Both applications can be classified as inverse problems: find the topology that has a prescribed output motion.

1. Introduction

A major problem in design of mechanisms is the problem of finding a mechanism topology which can produce a desired output movement subject to a given input movement. The problem is usually solved based on experience [e.g. Erdman and Sandor, (1991)] or by building up a library consisting of many different mechanism topologies from which a candidate mechanism is selected and modified to produce the desired movement [i.e. Hansen, (1994)]. This approach to mechanism synthesis can be difficult especially for 3–D systems due to the huge variation of possible mechanism topologies. This paper will propose a method for <u>automated</u> synthesis of mechanism topologies and due to an elasticity formulation of the design problem it is expected that the resulting mechanisms also will be efficient with respect to power consumption and reaction forces.

The idea behind the proposed mechanism design method comes from work in topology optimization of elastic truss structures [e.g. Pedersen, (1970), Bendsøe, Ben–Tal and Zowe, (1994) or Sankaranarayanan, Haftka and Kapania, (1993)] and from recent work in design of materials with prescribed elastic properties [Sigmund (1994)[a,b] and (1995)].

In topology optimization of elastic truss structures a commonly considered design problem is to find the stiffest possible truss structure within a given design domain, called the ground structure. The truss is subjected to one or more load cases and the total weight of the structure has an upper limit. Solving such truss design problems – especially one load case problems – often results in unstable designs which are stiff in the direction of the applied load but will collapse subject to all other loads. Such "mechanism modes" can be prevented in various ways. Sankaranarayanan, Haftka and Kapania, (1993) add a dummy

277

D. Bestle and W. Schielen (eds.), IUTAM Symposium on Optimization of Mechanical Systems, 277–284.
© 1996 *Kluwer Academic Publishers.*

load perpendicular to the actual design load, Bendsøe, Ben–Tal and Zowe, (1994) suggest the use of several load cases. In these papers methods to prevent the mechanism mode problem are suggested. In this paper, however, we will use the originally undesirable problem in topology optimization of trusses to our advantage in the design of mechanisms.

The mechanism design procedure will be formulated in section 3 but before that, we will show how truss like mechanisms were encountered in a research project on design of material microstructures published in Sigmund (1994)[a,b] and (1995).

2. Material design

Behaviour of a linear elastic material is governed by Hooke's law

$$\{\sigma\} = [E]\{\varepsilon\} \quad \text{or} \quad \begin{Bmatrix} \sigma_{11} \\ \sigma_{22} \\ \sqrt{2}\,\sigma_{12} \end{Bmatrix} = \begin{bmatrix} E_{11} & E_{12} & \sqrt{2}\,E_{13} \\ E_{12} & E_{22} & \sqrt{2}\,E_{23} \\ \sqrt{2}\,E_{13} & \sqrt{2}\,E_{23} & 2E_{33} \end{bmatrix} \begin{Bmatrix} \varepsilon_{11} \\ \varepsilon_{22} \\ \sqrt{2}\,\varepsilon_{12} \end{Bmatrix} \quad (1)$$

where $[E]$ is the symmetric constitutive matrix and $\{\sigma\}$ and $\{\varepsilon\}$ are the stress and strain vectors respectively. For thermodynamical reasons, the constitutive matrix is restricted to be positive semidefinite. For reasons explained in the previously mentioned papers by the author, there is a great interest in designing materials with elastic properties ranging over the entire scale compatible with thermodynamics or summarized shortly: being able to construct materials with any positive semidefinite constitutive matrix can cause great structural benefits. The material design problem has been solved by defining it as a topology optimization problem of a periodic base cell. The goal of the topology optimization problem is to find the lightest and simplest material microstructure with given elastic properties and the optimization problem is solved in an iterative procedure described in detail in afore mentioned papers. The material design problem has been solved for three different types of microstructure discretizations, namely truss–, frame– and continuum–like microstructures. Here we will concentrate on the truss discretization.

Figure 1 shows a variety of material microstructures with various constitutive properties designed by the numerical procedure. The design domain is the topology of the base cell defined as the smallest repetitive unit of a periodic material and as a starting guess for the optimization problem, the 120 bar groundstructure in figure 1a was used. The design variables are the 120 cross sectional areas of the bar elements. The elastic properties of the base cell are found by a homogenization procedure where the cell is loaded in a horizontal, a vertical and a shear test case and subjected to periodic boundary conditions. The 120 bar ground structure was found to be the best starting guess. Simpler starting guesses do not have enough degrees of freedom to model complicated microstructural behaviour and larger ground structures do not result in markedly different or better solutions. The optimization problem is solved by a modified version of the optimality criteria algorithm proposed in Zhou and Rozvany, (1993).

Materials with extreme elastic properties can be defined as materials for which the constitutive matrix $[E]$ defined in (1) has one or more zero eigenvalues. Specifying constitutive matrices with all eigenvalues greater than zero, we get stable microstructural topologies as seen in figure 1b and c which show isotropic materials with Poisson's ratio 0.3 and 1/3, respectively. Studying figure 1c, we notice that the microstructure consists of small square frames with diagonal stiffeners and it is easy to see that the structure is stable

Starting guess (ground structure with 120 possible bar elements)

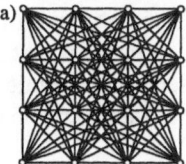

Non–extreme materials

$\nu = 0.3$

$$\begin{bmatrix} 1 & .3 & 0 \\ .3 & 1 & 0 \\ 0 & 0 & .7 \end{bmatrix}$$

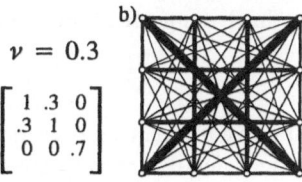

$\nu = 0.33$

$$\begin{bmatrix} 1 & .33 & 0 \\ .33 & 1 & 0 \\ 0 & 0 & .67 \end{bmatrix}$$

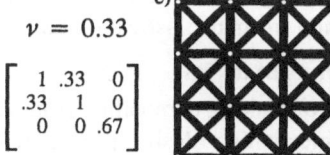

Extreme **isotrotropic** materials

$\nu = 1$

$$\begin{bmatrix} 1 & 1 & 0 \\ 1 & 1 & 0 \\ 0 & 0 & 0 \end{bmatrix}$$

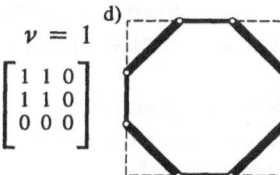

$\nu = -1$

$$\begin{bmatrix} 1 & -1 & 0 \\ -1 & 1 & 0 \\ 0 & 0 & 2 \end{bmatrix}$$

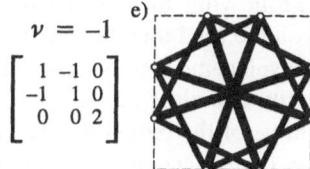

Extreme **orthotropic** materials

$$\begin{bmatrix} 1 & 0 & 0 \\ 0 & 0 & 0 \\ 0 & 0 & 0 \end{bmatrix}$$

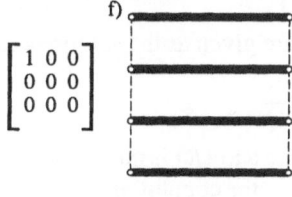

$45°$

$$\frac{1}{4}\begin{bmatrix} 1 & 1 & \sqrt{2} \\ 1 & 1 & \sqrt{2} \\ \sqrt{2} & \sqrt{2} & 2 \end{bmatrix}$$

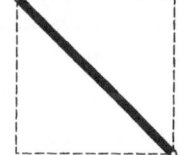

$$\begin{bmatrix} 1 & 0 & 0 \\ 0 & 1 & 0 \\ 0 & 0 & 0 \end{bmatrix}$$

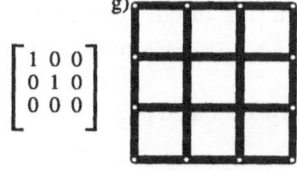

$45°$

$$\frac{1}{2}\begin{bmatrix} 1 & 1 & 0 \\ 1 & 1 & 0 \\ 0 & 0 & 2 \end{bmatrix}$$

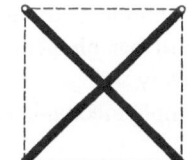

$$\begin{bmatrix} 0 & 0 & 0 \\ 0 & 0 & 0 \\ 0 & 0 & 2 \end{bmatrix}$$

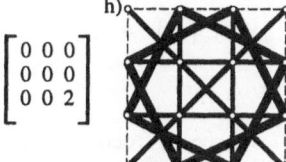

$45°$

$$\begin{bmatrix} 1 & -1 & 0 \\ -1 & 1 & 0 \\ 0 & 0 & 0 \end{bmatrix}$$

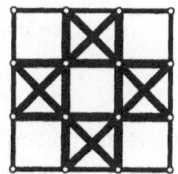

Figure 1. Extreme mechanism like materials

subject to any load. On the other hand, if we specify constitutive matrices with one or two zero eigenvalues, we get mechanism like microstructures as seen in figure 1d–h. Take for example figure 1e, where the constitutive matrix of a material with Poisson's ratio equal to –1 was specified. Such a material has the unusual behaviour that it expands vertically when pulled horizontally. Examining figure 1e it is seen that the microstructure is indeed a mechanism which is composed of two stiff quadratic frames free to rotate on top of each other.

Summarizing the material design problem, we can say that even though we start out from an elastically stable ground structure (figure 1a), we can end up with a mechanism by prescribing extreme elastic behaviour and thus we have suggested a method to make a transition from an elastic structure to a mechanism.

3. Mechanism design

The idea of making a transition from an elastic structure to a mechanism is used in the following mechanism design procedure. The computational details are very similar to the material design procedure developed in Sigmund (1994)[a,b] and (1995) and briefly summarized in the previous section.

In the following, the mechanism design problem will be defined as a problem of minimizing volume (and thereby hopefully complexity) of an elastic truss structure subject to multiple displacement constraints and load cases.

Basically one class of mechanisms can be synthesized by prescribing the output displacement u_O of the output degree of freedom O when the truss structure is subject to an input force or displacement u_F at at the input degree of freedom F. The output displacement u_O can be found by using Betti's reciprocal theorem and is written as

$$u_O = \{\overline{D}\}^T [S]\{D\} \tag{2}$$

where the nodal displacement vectors $\{D\}$ and $\{\overline{D}\}$ are given as the solutions to the finite element problems

$$[S]\{D\} = \{R\} \quad \text{and} \quad [S]\{\overline{D}\} = \{\overline{R}\} \tag{3}$$

where $[S]$ is the stiffness matrix of the full truss structure and $\{R\}$ is the physical load vector containing the input force and $\{\overline{R}\}$ is a virtual load vector containing a unit dummy load applied at the specified output degree of freedom O.

The volume of a truss structure can be written as a sum of bar volumes $x^e\, l^e$ over the number of bar elements N, where x^e and l^e are bar e's cross–sectional area and length respectively.

A simple mechanism optimization problem can now be stated as

$$
\begin{aligned}
Minimize : V &= \sum_{e=1}^{N} l^e\, x^e \\
Subject\ to : \{\overline{D}\}^T [S]\{D\} &= \sum_{e=1}^{N} \{\overline{d^e}\,\}^T [s^e]\,\{d^e\} = u_O^* \\
and : 0 &< x_{min} \leq x^e \quad , \quad e = 1,\ldots,N
\end{aligned}
\tag{4}
$$

where u_O^* is the prescribed displacement at the output degree of freedom O and x_{min} is a lower limit on the design variables. The optimization problem (4) can be solved by many different optimization methods but here we have used a steepest descent algorithm which is very simple to implement but requires many design iterations (several thousand for problems with many design variables).

A simple mechanism design example which was solved using the proposed method is shown in figure 2a–d. In figure 2a, the design area is specified as a quadratic area simply supported at the lower and upper left corners. A linear actuator is connected to the mid-point of the left edge and causes a horizontal displacement of the 3 by 3 node 36 bar ground structure shown in figure 2b. The design problem consist in finding a mechanism that produces an output motion in the opposite direction of the input motion. This is done by prescribing the output motion of the center node of the right edge to be −1 times the actuator stroke, i.e. $u_O^* = -u_F$. The resulting optimal truss topology is shown in figure 2c. The unactuated structure is shown in grey and the actuated structure in black, and it is clearly seen that we have obtained a mechanism with the prescribed motion. It should be noted, that unimportant bars are <u>not</u> removed from the ground structure during the optimization procedure but by choosing the lower limit on the design variables x_{min} in (4) to be very small (10^{-7} times the maximum bar area), the unimportant bars will have no structural significance. Bars with minimum cross–sectional areas are not shown in the graphical representations of the the resulting mechanisms in order to make interpretation easier.

The mechanism in figure 2c has a grave disadvantage. If the output point O is loaded in the vertical direction, the structure will collapse as seen in figure 2d. To prevent this, we can add an extra load case to the optimization problem. The optimization problem (4) is modified to account for several load cases in the following way

$$Minimize : V = \sum_{e=1}^{N} l^e\, x^e$$

$$Subject\ to : \{\overline{D}_I\}^T[S]\{D_I\} = \sum_{e=1}^{N} \{\overline{d_I^e}\}^T[s^e]\,\{d_I^e\} = u_I^* \,,\; I = 1,\ldots,NC \tag{5}$$

$$and : 0 < x_{min} \le x^e \quad,\; e = 1,\ldots,N$$

where NC is the number of displacement constraints or load cases.

The mechanism design problem form figure 2a is reoptimized with an extra load case in the vertical direction of the output node and the allowed displacement u_2^* in the vertical direction is constrained to be 0.1 times the horizontal mechanism mode displacement u_1^*. The resulting topology is shown in figure 2e and we see that the mechanism has been "reinforced" in a way that makes it stable subject to vertical forces as expected.

Choosing a larger ground structure makes it possible to "amplify" the input stroke of the actuator, for example, the displacement at the output degree of freedom can be made four times bigger than the actuator stroke when a 5 by 5 node 300 element ground structure is chosen as the starting guess. This example will not be illustrated here due to the limited amount of space. Instead, three small design examples of "grabber mechanisms" with more complex output displacements are shown in figure 3. Starting guess for all three examples is the 5 by 5 node 300 element ground structure and in design example a, the two grabber nodes were prescribed to move towards each other when the actuator expands. In design example b, the upper grabber was specified to move downwards while the lower grabber should remain stationary. Finally in example c, the upper and lower grabbers were specified to move to the right and to the left, respectively. In all three examples, extra load cases were applied to prevent unstable modes as in the previous example.

4. Conclusion

A simple method for design of mechanism topologies has been proposed. From the design examples it is seen that the method can produce mechanism topologies with complex output motions. Basically the method corresponds to well known compliance minimization methods from structural optimization of trusses but in this paper we allow the formation of "unstable" structures which characterize mechanism behaviour. The formulation of the mechanism design problem as a structural optimization assures that the resulting mechanisms are optimally stiff subject to static loads. This feature indicates that the resulting mechanisms also will be efficient with respect to power consumption and reaction forces. The latter observation is supported by the observation that the resulting mechanisms are composed of bodies connected to each other at near right angles which indicates mechanical efficiency of mechanisms.

At the current state of the developments, the method has a weak point

☞ The finite element analysis of the truss structure assumes infinitesimal displacements and therefore some of the mechanisms might "lock" subject to larger displacements. Furthermore, the prescribed movements of the output points are only tangent directions with respect to the unloaded structure.

To prevent this problem, a large displacement finite element model should be used. This will make the analysis more time consuming but assure that the developed structures are mechanisms indeed and furthermore it will be possible to prescribe more complicated output movements given by arcs or multiple prescribed positions.

5. Acknowledgements

The work presented in this paper received support from Denmark's Technical Research Council (Programme of Research on Computer–Aided Design).

References

Bendsøe, M.P., Ben–Tal, A. and Zowe, J. (1994): "Optimization methods for truss geometry and topology design", *Struct. Optim.*, 7, pp. 141–159.

Erdman, A.G. and Sandor, G.N. (1991): *Mechanism Design, Analysis and Synthesis, Volume 1*, Prentice Hall, London.

Hansen, M.R. (1994): "A multi level approach to synthesis of planar mechanisms", *J. Nonlinear Dynamics*, (to appear).

Pedersen, P. (1970): "On the minimum mass layout of trusses", *AGARD conf. proc. No. 36*, Symposium on structural optimization, AGARD–CP–36–70.

Sankaranarayanan, S., Haftka, R.T. and Kapania, R.K. (1993): Truss topology optimization with stress and displacement constraints", in "Topology Design of Structures" (Eds. M.P. Bendsøe and C.A. Mota Soares), Kluwer, Dordrecht, Holland, pp. 71–78.

Sigmund, O. (1995): "Tailoring materials with prescribed elastic properties". DCAMM Report #480, March 1994. 23 p. (To appear in Mech. Materials).

Sigmund, O. (1994)[a]: "Materials with prescribed constitutive parameters: an inverse homogenization problem". Int. J. Solids Structures, Vol. 31, No. 17, 1994, pp. 2313–2329.

Sigmund, O. (1994)[b]: "Design of Material Structures using Topology Optimization." DCAMM Special Report No. 69, Ph.D.–thesis, Technical University of Denmark, December 1994, 95 p. + 22 p. (appendix).

Zhou, M. and Rozvany, G.I.N. (1993): "DCOC: an Optimality Criteria Method for Large Systems, Part I: Theory", *Struct. Opt.* 5, pp. 12–25.

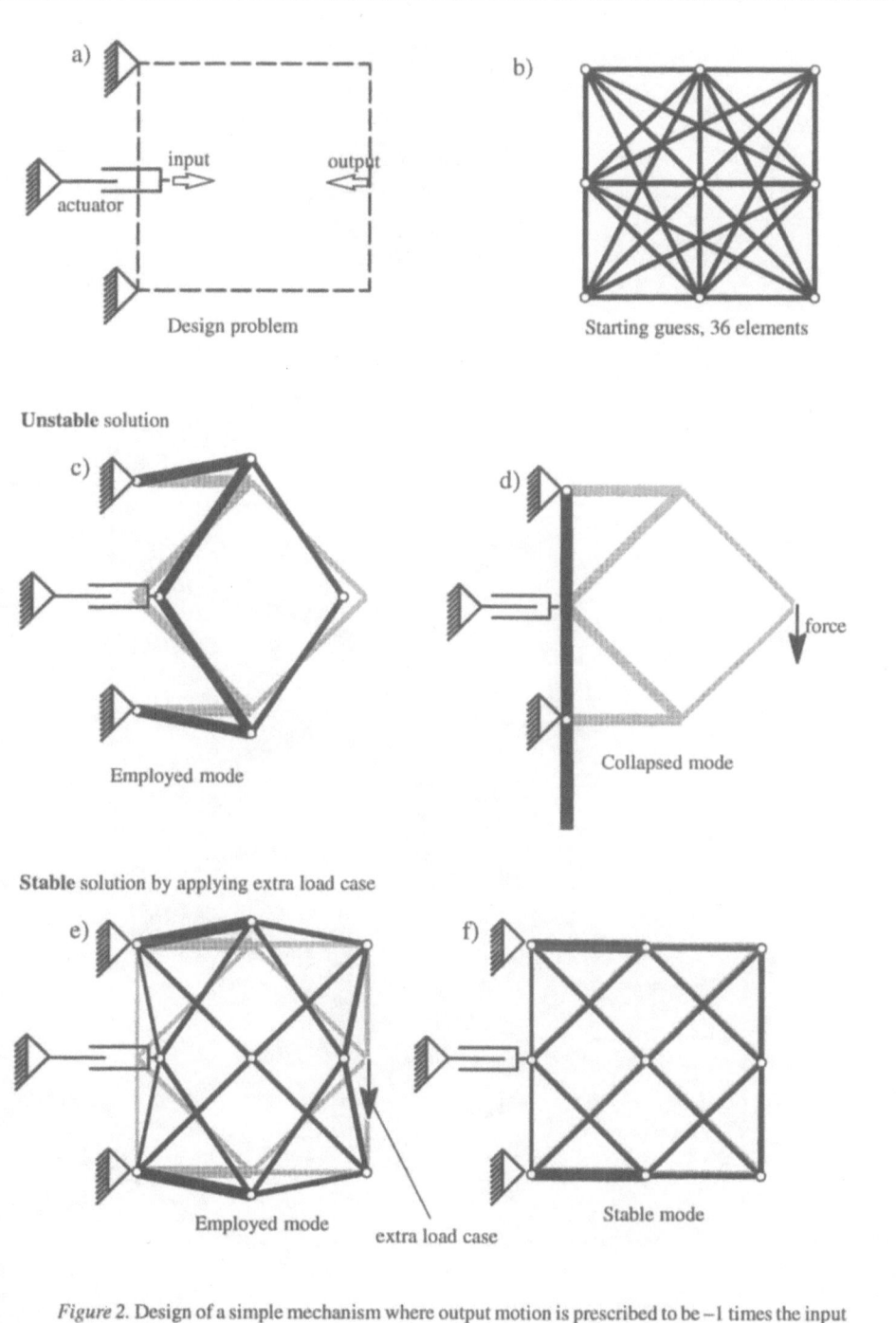

Figure 2. Design of a simple mechanism where output motion is prescribed to be −1 times the input motion of the actuator

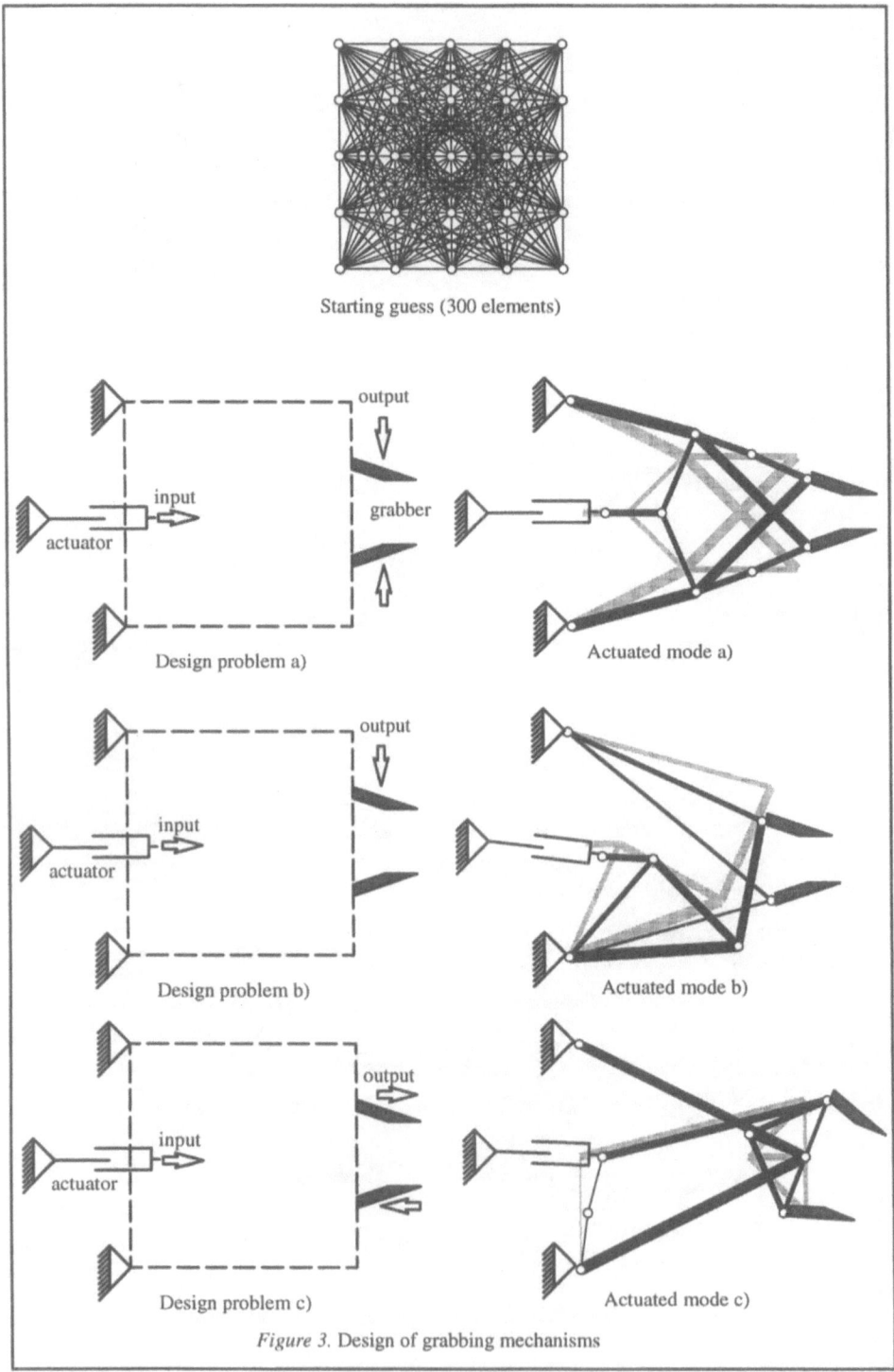

Figure 3. Design of grabbing mechanisms

THE PRACTICAL APPLICATION OF A DYNAMIC SEARCH-TRAJECTORY METHOD FOR CONSTRAINED GLOBAL OPTIMIZATION

J.A. SNYMAN and K.A. GEERTHSEN
Department of Mechanical and Aeronautical Engineering,
University of Pretoria
Pretoria, 0002, South Africa

1. Introduction

The problem of globally optimizing a real valued function is inherently intractable in that no practically useful characterization of the global optimum is available [1]. Nevertheless, the need in practice to find a relative low local minimum has resulted in considerable research over the last decade to develop algorithms that attempt to find such a low minimum (see the survey on global optimization of Törn and Zilinkas [2]). Two distinct approaches to global optimization have been identified, namely *deterministic* and *stochastic*. The methods in the first class implicitly search all of the function domain and are thus guaranteed to find the global optimum. These algorithms are forced to deal with severely restricted classes of functions and are often computationally infeasible because the number of computations required increases exponentially with the dimension of the feasible space. To overcome the inherent difficulties of the guaranteed-accuracy algorithms, much research effort has been devoted to algorithms in which a stochastic element is introduced. This way the deterministic guarantee is relaxed into a *confidence measure*. A general stochastic algorithm for global *unconstrained* optimization consists of three major steps: a sampling step, an unconstrained optimization step, and a check of some stopping criterion. This paper is concerned with the modification of such an algorithm so as to extend its application to *non-convex constrained global optimization*.

The algorithm to be modified is the multi-start trajectory method of Snyman and Fatti [3]. This algorithm is philosophically similar to the simulated annealing technique (see survey by Schoen [1]), in the sense that it also corresponds to the simulation of a physical process, in particular the dynamic trajectory of a particle under the influence of a potential force. The trajectory allows the particle to skip over locally stable configurations as occurs in simulated annealing. The method also has similarities to the trajectory method of Aluffi-Pentini et al [4] in which paths of a stochastic-differential equation are computed. The unconstrained algorithm has been used by Kam and Snyman [5] to investigate the lamination arrangements

D. Bestle and W. Schielen (eds.), IUTAM Symposium on Optimization of Mechanical Systems, 285–292.
© 1996 *Kluwer Academic Publishers.*

of laminated composite plates where many local optima occur.

2. The unconstrained global minimization algorithm

The general global optimization problem may be formulated as follows. Given a real valued objective function $f(\boldsymbol{x})$ defined on the set $\boldsymbol{x} \in D$ in $\Re^n$, find the point $\boldsymbol{x}^*$ and the corresponding function value f^* such that

$$f^* = f(\boldsymbol{x}^*) = \text{ minimum } \{f(\boldsymbol{x})|\boldsymbol{x} \in D\} \qquad (1)$$

if such a point $\boldsymbol{x}^*$ exists. If the objective function and/or the feasible domain D are non-convex, then there may be many local minima which are not global. If D corresponds to all $\Re^n$ the optimization problem is *unconstrained*. Even in this case simple bounds are imposed in practice so that effectively an "almost unconstrained" problem is considered with D corresponding to the hyper box defined by

$$D = \{\boldsymbol{x}|\boldsymbol{\ell} \leq \boldsymbol{x} \leq \boldsymbol{u}\} \qquad (2)$$

where $\boldsymbol{\ell}$ and $\boldsymbol{u}$ are n-vectors defining the respective lower and upper bounds on $\boldsymbol{x}$. D is sometimes also referred to as the domain of interest. The essentials of the original Snyman and Fatti (SF) algorithm [3] using dynamic search-trajectories for unconstrained global minimization will now be discussed.

2.1 DYNAMIC TRAJECTORIES

In the SF-algorithm successive sample points $\boldsymbol{x}^j, j = 1, 2, ...,$ are selected at random from the box D defined by (2). For *each* sample point $\boldsymbol{x}^j$, a sequence of trajectories T^i, $i = 1, 2, ...,$ is computed by numerically solving the successive initial value problems:

$$\ddot{\boldsymbol{x}}(t) = -\boldsymbol{\nabla}f(\boldsymbol{x}(t)); \quad \boldsymbol{x}(0) = \boldsymbol{x}_0^i ; \quad \dot{\boldsymbol{x}}(0) = \dot{\boldsymbol{x}}_0^i \qquad (3)$$

Trajectory T^i is terminated when $\boldsymbol{x}(t)$ reaches a point where $f(\boldsymbol{x}(t))$ is arbitrarily close to the value $f(\boldsymbol{x}_0^i)$ while moving "uphill", or more precisely, if $\boldsymbol{x}(t)$ satisfies the conditions

$$f(\boldsymbol{x}(t)) > f(\boldsymbol{x}_0^i) - \epsilon_1 \text{ and } \dot{\boldsymbol{x}}(t)^T\boldsymbol{\nabla}f(\boldsymbol{x}(t)) > 0 \qquad (4)$$

where ϵ_1 is an arbitrary small positive value. An argument is presented in [3] to show that provided the level set $\{\boldsymbol{x}|f(\boldsymbol{x}) \leq f(\boldsymbol{x}_0^i)\}$ is bounded and $\boldsymbol{\nabla}f(\boldsymbol{x}_0^i) \neq \boldsymbol{0}$, then conditions (4) above will be satisfied at some finite point in time.

Each computed step along trajectory T^i is monitored so that at termination the point x^i_m at which the minimum value was achieved is recorded together with the associated velocity $\dot{x}^i_m$ and function value f^i_m. The values of x^i_m and $\dot{x}^i_m$ are used to determine the initial values for the next trajectory T^{i+1}. From a comparison of the minimum values the best point x^i_b, for the current j over all trajectories to date is also recorded. In more detail the minimization procedure for a given sample point x^j, in computing the sequence x^i_b, $i = 1, 2, ...$, is as follows.

2.2 MINIMIZATION PROCEDURE (MP)

1. For given sample point x^j, set $x^1_0 := x^j$ and compute T^1 subject to $\dot{x}^1_0 := 0$; record $x^1_m, \dot{x}^1_m$ and f^1_m ; set $x^1_b := x^1_m$ and $i := 2$,

2. compute trajectory T^i with $x^i_0 := \frac{1}{2}\left(x^{i-1}_0 + x^{i-1}_b\right)$ and $\dot{x}^i_0 := \frac{1}{2}\dot{x}^{i-1}_m$; record $x^i_m, \dot{x}^i_m$ and f^i_m,

3. if $f^i_m < f(x^{i-1}_b)$ then $x^i_b := x^i_m$; else $x^i_b := x^{i-1}_b$,

4. set $i := i + 1$ and go to 2.

In the original paper [3] an argument is presented to indicate that under normal conditions on the continuity of f and its derivatives, x^i_b will converge to a local minimum. Procedure MP, for a given j, is accordingly terminated if $\|\nabla f(x^i_b)\| \leq \epsilon_2$ for some small prescribed positive value ϵ_2, and x^i_b is taken as the local minimizer x^j_f with $f^j_f := f(x^j_f)$. In the presence of many local minima, the probability of convergence to a relative low local minimum is increased because, with a small value of ϵ_1, it is likely that the particle will move through a trough associated with a relative high local minimum, and move over a ridge to record a lower function value at a point beyond.

2.3 GLOBAL STOPPING CRITERION

The above procedure requires a termination rule for deciding when to end the sampling and to take the current overall minimum function value $\tilde{f}$, i.e.

$$\tilde{f} = \text{minimum} \left\{ f^j_f, \text{ over all } j \text{ to date} \right\} \qquad (5)$$

as the global minimum value f^*. The stopping criterion adopted in [3] is now briefly discussed.

Define the *region of convergence* of the above method for a local minimum $\hat{x}$ as the set of all points x which, used as starting points for the above MP-procedure, converges to $\hat{x}$. One may reasonably expect that in the case where the *regions of attractions* (for the usual gradient-descent methods) of the local minima are more or less equal, that the region of convergence of the global minimum will be relatively increased. Let R_k denote the region of convergence for the above MP-procedure of local minimum $\hat{x}^k$ and let α_k be the associated probability that a sample point be selected in R_k. The region of convergence and the associated probability for the global minimum x^* are denoted by R^* and α^* respectively. The following basic assumption, which is probably true for many functions of practical interest, is now made.

A. *Basic assumption:* $\alpha^* \geq \alpha_k$ for all local minima $\hat{x}^k$. The following theorem may be proved.

B. *Theorem* [3]: Let r be the number of sample points falling within the region of convergence of the current overall minimum $\tilde{f}$ after $\tilde{n}$ points have been sampled. Then under assumption A and a statistically non-informative prior distribution the probability that $\tilde{f}$ corresponds to f^* may be obtained from

$$Pr\left[\tilde{f} = f^*\right] \geq q(\tilde{n}, r) = 1 - \frac{(\tilde{n}+1)!(2\tilde{n}-r)!}{(2\tilde{n}+1)!(\tilde{n}-r)!} \tag{6}$$

On the basis of this theorem the *stopping rule* adopted was: STOP when $Pr\left[\tilde{f} = f^*\right] \geq q^*$, where q^* is some prescribed desired confidence level, typically chosen as 0.99.

3. The constrained global minimization method

Consider now problem (1) with x constrained to the more general feasible set defined by

$$D = \{x | h_i(x) = 0, \quad i \in I; \quad g_j(x) \leq 0, \quad j \in J\} \tag{7}$$

where $h_i(x)$ and $g_j(x)$ represent general equality and inequality constraint functions. Here no restrictions are placed on the convexity of the objective and constraint functions. Until fairly recently relatively few methods have been proposed to deal with such general non-linear constrained problems with multiple minima [2]. It is now proposed that the global optimum of the constrained problem be sought by applying the unconstrained SF-algorithm to an exterior penalty function formulation of the constrained problem. This is done in two stages. In the *first stage*, a rough estimate to the global minimum is obtained by applying the SF-algorithm to a penalty function with moderate values for the penalty parameters. Since an exterior penalty function is used convergence to the constrained optimum, as the penalties are increased, will occur from the infeasible region. The approximate minimum, obtained with the moderate penalty parameter values, may therefore be used to identify the active set of inequality constraints. In *stage 2* a more exact constrained minimum is obtained by seeking the solution to the set of active constraint equations (including the prescribed equality constraint equations), in the neighbourhood of the approximate minimum determined in stage 1. On the assumption that the least-squares problem, defined by the set of active constraints (see least-squares function (10) below) is convex in the neighbourhood of the solution, the SF-algorithm with a single starting point corresponding to the approximate solution, is applied to the least-squares problem to yield a feasible point. This point, to which the trajectories in stage 2 converge, is taken as the constrained global minimum x^*.

If the active constraints are consistent and independent and their number m_a is equal to n, then convergence is to a unique minimum in the neighbourhood of the starting point. If m_a is less than n then with the feasible starting point assumed to be close to the true global optimum, convergence should at least occur to a feasible point very near to the true constrained optimum. Alternatively, and more accurately for the latter case, the Lagrange conditions for all active constraints may be added to the least-squares problem (10) below. A more formal presentation of the proposed procedure for constrained global optimization (SFCON) follows.

STAGE 1: Apply the SF-algorithm to the penalty function

$$p(x) = f(x) + \sum_{i \in I} \sigma_i h_i^2(x) + \sum_{j \in J} \rho_j g_j^2(x) \tag{8}$$

where $\sigma_i = \mu$, with μ some suitable large positive penalty parameter, and $\rho_j = \mu$ if $g_j(x) > 0$, otherwise $\rho_j = 0$. The computed global minimum of (8) is denoted by $x^*(\mu)$ and is assumed to be a rough estimate to x^* for moderate values of μ.

STAGE 2: Identify the set of all active constraints $I_a(\mu)$ at $x^*(\mu)$:

$$I_a(\mu) = \{I \cup J_a(\mu)\}; \quad J_a(\mu) = \{j | g_j(x^*(\mu)) > 0, \quad j \in J\} \tag{9}$$

Apply the SF-algorithm to the minimization of the least-squares function

$$\bar{p}(x) = \sum_{i \in I} h_i^2(x) + \sum_{j \in J_a(\mu)} g_j^2(x) \tag{10}$$

where the single starting point $x^1 := x^*(\mu)$ is used. A simple illustration of this procedure applied to the one-dimensional problem: $\min f(x)$ such that $g(x) = x - a \leq 0$ is depicted in Figure 1.

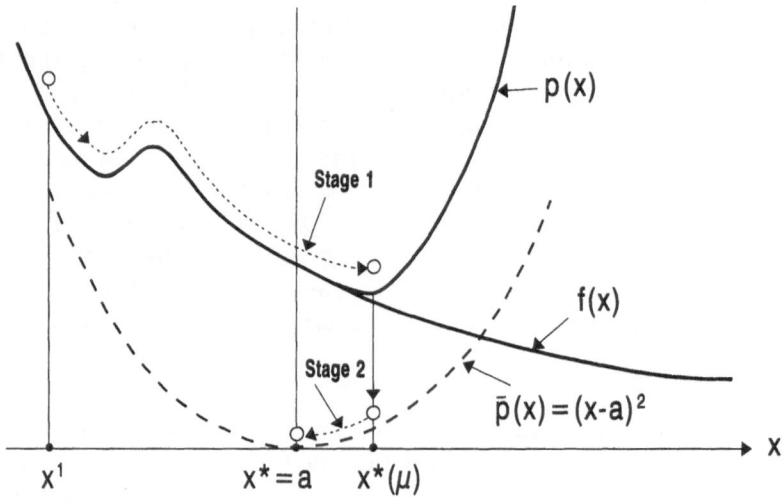

Figure 1. Simple illustration of the SFCON two-stage global optimization procedure.

Moderate values for the penalty parameter is required in order to ensure that the penalty function is sufficiently smooth to allow for the computation of approximately energy conserving trajectories. If μ is too large the problem becomes ill-conditioned in the sense that the trajectories become erratic. On the other hand μ should be large enough to confidently identify the active set of inequality constraints at x^*.

4. Test problems and numerical results

To assess the performance of the SFCON-algorithm we use some selected problems from the collection of constrained test problems of Floudas and Pardalos [6]. The problems selected are listed in Table 1. The number of variables is denoted by n, while m_e and m_i indicate the number of equality and inequality constraints respectively. For each problem the second column indicates the corresponding paragraph (§) in reference [6] in which the particular problem is described. The lower and upper bounds of the region of interest are respectively denoted by the components ℓ_i and u_i. Problems 1-7 are all non-convex quadratic programming test problems with multiple local minima for which the global minima are known. Since SFCON does not specifically utilize information regarding the class of problem being solved, one may expect that if the algorithm works well for these problems, it should also do well when applied to more general classes of problems. Problem 8 is a non-convex quadratically constrained problem. Problems 9 and 10 arise in heat exhanger network synthesis and are relatively large non-convex problems of practical importance. The computations were performed on a Persetal PS8/90-3 IBM plug compatible computer. Table 2 gives a summary of the results. m_a is the total number of active (equality and violated inequality) constraints at $x^*(\mu)$, the global minimum of (8). The ratio $r/\tilde{n}$ indicates that convergence to the computed global minimum was obtained r times for $\tilde{n}$ sample points. The last two columns respectively indicate the total number of times (NFE) the set of objective and constraint functions were evaluated, and the total computer CPU time required to solve each problem. In all cases the value of $\bar{p}(x^*)$ was less than 10^{-13}.

5. Conclusion

The SF-algorithm for unconstrained global optimization has successfully been applied to constrained global optimization problems. The constrained optimum is obtained through a two-stage application of the SF-algorithm to penalty function formulations of the constrained problem. The constrained procedure SFCON successfully solved all the test problems to which it was applied. The results establish it as a robust method of high accuracy. Although the number of function evaluations necessary to obtain a high probability computed global optimum is relatively large, the CPU times required for solution were not excessive for the problems considered here.

TABLE 1. Details of the test problems used

Prob. no	§ in [6]	n	m_e	m_i	i	ℓ_i	u_i
1	2.1	5	0	11		0	1
2	2.2	6	0	13		0	1
3	2.3	13	0	32		0	1
4	2.4	6	0	14	i=1-3	0	0
					i=4-6	0	3
5	2.5	10	0	31		0	1
6	2.7	20	0	30		-50	50
7	2.8	24	10	24		-50	50
8	3.1	8	0	22		0	1
9	7.2	16	13	28		0	1
10	7.3	27	19	42	i=1-15	0	45
					i=16-27	0	400

TABLE 2. Numerical Results

Prob. no.	μ	$p(x^*(\mu))$	m_a	$r/\tilde{n}$	$f(x^*)$	NFE	CPU (sec)
1	10^4	-17.3	5	6/95	-17.00000	37 301	3.21
2	10^2	-307.0	6	10/10	-213.0000	5 156	0.46
3	10	-17.9	13	5/12	-15.0000	2 943	1.19
4	10^3	-11.0	6	9/10	-11.0000^a	6 200	0.76
5	10^4	-270.8	10	10/10	-268.0146^b	14 463	4.53
6-case 1	10^4	-395.2	20	6/36	-394.7506	121 150	65.6
6-case 2	10^3	-891.7	20	6/30	-884.7506	131 548	70.3
6-case 3	10^5	-9817.7	20	6/19	-8695.0122	68 995	37.4
6-case 4	10^3	-759.0	20	6/88	-754.7506	139 833	74.9
6-case 5	10^4	-4209.7	20	6/36	-4150.4101^c	150 878	83.0
7	10^5	15609.0	25	5/5	15639.0000	106 502	72.5
8	10^2	7025.9	6	5/11	7049.3318^d	162 014	29.7
9	10^7	56815.4	16	5/9	56825.828	252 032	115.6
10	10^5	44531.3	27	5/6	45376.157^e	327 595	338.3

a. 11.005 ; b. -266.39 ; c. -4105.2779 ; d. 7049.25; e. 46266.00 in [6]; the other results are the same as the best known solutions listed in [6].

References

1. Schoen, F.: Stochastic techniques for global optimization: A survey of recent advances, *Journal of Global Optimization* **1** (1991), 207-228.

2. Törn, A. and Zilinkas, A.: *Global Optimization: Lecture Notes in Computer Science*, No.350, Springer-Verlag, Berlin Heidelberg, 1989.

3. Snyman, J.A. and Fatti, L.P.: A multi-start global minimization algorithm with dynamic search trajectories, *Journal of Optimization Theory and Applications* **54** (1987), 121-141.

4. Aluffi-Pentini, F., Parisi, V. and Zirilli, F.: Global optimization and stochastic differential equations, *Journal of Optimization Theory and Applications* **47** (1985), 1-16.

5. Kam, T.Y. and Snyman, J.A.: Optimal design of laminated composite plates using a global optimization technique, Composite Structures **19** (1991) 351-370.

6. Floudas, C.A. and Pardalos, P.M.: *A Collection of Test Problems for Constrained Global Optimization Algorithms: Lecture Notes in Computer Science*, No.445, Springer-Verlag, Berlin Heidelberg, 1990.

MULTICRITERIA OPTIMAL DESIGN OF A DYNAMICALLY RESPONSIVE SAFETY CHAIR

W. Stadler
Professor of Mechanical Engineering
San Francisco State University
1600 Holloway Avenue
San Francisco, California U.S. 94110

A. Johnson
Engineering Consultant
Sundial Engineering
3073 Bateman Street
Berkeley, California, 94705

Abstract

The object of our investigation is a chair whose seat is designed to move on a circular track. It thus is capable of responding to an occupant to provide ergonomic comfort or it may serve to "catch" an occupant when the chair experiences sudden deceleration in safety related instances. We provide a relatively complete dynamic analysis of both chair and occupant for the latter case. This analysis then forms the basis for a multicriteria optimal design problem with the deployment time of the chair and the head angle and relative kinetic energy of the head at that time as criteria and with the track radius as the design variable.

1. Introduction

Chairs have long been designed for their function with one, three and four legs or with wheels; designing them for comfort appears to be a more recent innovation. A still more recent approach is the design with occupant safety in mind.

Mr. Serber of American Ergonomics has a long record in designing more comfortable and sensible office chairs, his most recent model incorporating a seat capable of rotating on a circular track. The initial motivation was the comfort and physiological well-being of the user based on an analysis of human body behavior; the use of the same basic chair as a safety device for "catching" the occupant during sudden vehicular stops then became an obvious possibility. Boeing company has opted for use of a similarly designed airplane passenger seat as part of its compliance with FAA Regulation No. 25.562 concerning dynamic conditions during emergency landings. Mr. Serber received U.S. Patent No. 5,244,252 dated September 14, 1993, for his invention.

He approached the first author with a request for a dynamic analysis of the seat motion in 1991.[1] The problem seemed both interesting and workable and this paper presents some of the dynamic and optimization results obtained to date. The initial steps of the analysis consisted of the formulation of benchmark problems meant to provide a

---

[1] The authors gratefully acknowledge the partial financial support of American Ergonomics Corporation.

D. Bestle and W. Schielen (eds.), IUTAM Symposium on Optimization of Mechanical Systems, 293–301.
© 1996 *Kluwer Academic Publishers.*

least upper bound on the seat deployment time. The final model consisted of the three link plane kinematic chain shown in Figure 1 which also includes all of the kinematic and kinetic terminology as used for the present analysis. Once the dynamical model had been established, suitable design variables and criteria became apparent. The variation of the system dynamics with the radius R of the seat deployment track made it an obvious design variable. The choice of criteria was equally straightforward. The chair deployment time, the maximum forward motion of the head segment and its kinetic energy all are clear candidates for minimization. It seemed obvious that there could be no single choice of R minimizing all of those simultaneously and the problem therefore was posed as a multicriteria problem where the required comparisons are treated in an organized fashion. The optimal design problem thus is a multicriteria parameter optimization problem.

2. Problem Formulation (Dynamics)

An indication of the simplified analysis was that a reasonably accurate representation of the combined motion of seat and dummy could be attained by considering a plane kinematic chain consisting of three elastically hinged rigid bodies as shown in Figure 1. Based on that terminology, the governing system equations were then derived in accordance with Euler's laws of motion. The resultant three simultaneous equations with the q_i as dependent variables were then investigated numerically with respect to design changes and criteria. Obviously, the equations also lend themselves to parameter studies including reactive forces and moments at the joints, different mass distributions and corresponding changes in mass center locations, moments of inertia and others.

The raw data for the proportionment and mass distribution for the dummy were taken from NASA studies concerning the human anatomy as cited in [1]. These data were then reapportioned, to some extent, to fit the dummy configuration used here. In particular, the data were adjusted to facilitate the use of the Instantaneous Center of Lumbar Flexion (ICLF), introduced by Serber in [2] as the lower hingepoint location between thigh and torso, instead of the hip joint which normally fills this role.

The actual derivation of the governing equations is standard. We shall discuss some of the terminology only to explain some of the basic assumptions underlying our derivation. Four reference frames are shown in Figure 1, each embedded in one of the rigid body links:

$(Oxyz)_0$ -   a decelerating (a_0) reference frame assumed to be imbedded in a vehicle of considerably greater mass than that of the remaining rigid bodies; the origin O_0 serves as the center of rotation for the seat;

$(Oxyz)_1$ -   imbedded in the seat with origin O_1 a distance R from O_0 (note that R, our design variable, is not the radius of the circular track but may easily be related to it)[2]; the mass of this link consists of that of the seat and thigh

---

[2] The radius of the circular track is related to the design parameter R in the following manner:

$$R_{Track} = \sqrt{R^2 + 0.25L^2}$$ where L is the length of the seat, 43 cm.

segment combined both moving together as one rigid body in pendular motion about O_0;

$(Oxyz)_2$ - imbedded in the torso with origin O_2 at the ICLF;

$(Oxyz)_3$ - imbedded in the dummy's head with origin O_3 at the hinge between torso and head.

Throughout, we take counterclockwise rotation to be positive with θ_i, $i = 1, 2, 3$ denoting the rotation of the i-th frame relative to the preceding frame; e.g. θ_3 denotes the rotation of $(Oxyz)_3$ relative to $(Oxyz)_2$. The only external forces acting on the system are the weights W_i of the individual links and the normal forces N_i exerted by the rollers on the circular seat guide; we neglect what amounts to being rolling friction. Each hinge connection is modeled with the usual internal reactions F_i and with internal resisting moments $M_i = k_i \theta_i \hat{\mathbf{k}}$, $i = 1, 2$, proportional to the relative rotation between adjacent reference frames. Finally, we adhere to the following kinematic terminology:

$r_i = x_i \hat{\mathbf{i}}_i + y_i \hat{\mathbf{j}}_i$ $i = 1, 2, 3$, the position vectors of the centers of mass within the i-th reference frame.

$c_{ij} = a_{ij} \hat{\mathbf{i}}_i + b_{ij} \hat{\mathbf{j}}_i$ $i = 1, 2, j = i + 1$, the location of the origin O_j with respect to the preceding origin O_i; an exception is $c_{01} = -R \hat{\mathbf{j}}_1$, $R > 0$.

$a_0 = -a_0 \hat{\mathbf{i}}_0$ $a_0 > 0$, the deceleration of the translating reference frame $(Oxyz)_0$.

All of the remaining notation is standard, with r denoting the motion, v the velocity and a the acceleration, for example.

Accordingly, we now formulate the equations of motion. We provide some of the steps in the derivation to allow the reader to check our results as well as facilitating any possibly subsequent calculations of internal forces and moments. The basic application of Euler's laws of linear momentum and moment of momentum yields

$\mathcal{B}_1$: $N_1 + N_2 + N_3 + F_1 + W_1 = m_1 a_1$ and

$$\left(c_{12} - r_1\right) \times F_1 - \left(r_1 + c_{01}\right) \times \left(N_1 + N_2 + N_3\right) = I_{G_1} \ddot{\theta}_1 \hat{\mathbf{k}} - k_2 \theta_2 \hat{\mathbf{k}}$$

$\mathcal{B}_2$: $-F_1 + F_2 + W_2 = m_2 a_2$ and

$$r_2 \times F_1 + \left(c_{23} - r_2\right) \times F_2 = I_{G_2}\left(\ddot{\theta}_1 + \ddot{\theta}_2\right)\hat{\mathbf{k}} + k_2 \theta_2 \hat{\mathbf{k}} - k_3 \theta_3 \hat{\mathbf{k}}$$

$\mathcal{B}_3$: $W_3 - F_2 = m_3 a_3$ and $r_3 \times F_2 = I_{G_3}\left(\ddot{\theta}_1 + \ddot{\theta}_2 + \ddot{\theta}_3\right)\hat{\mathbf{k}} + k_3 \theta_3 \hat{\mathbf{k}}$

The elimination of the internal reactions leads to the following preliminary form of the three moment of momentum equations

$\mathcal{B}_1$: $\left(c_{01} + c_{12}\right) \times \left(W_1 + W_2 + W_3 - m_1 a_1 - m_2 a_2 - m_3 a_3\right) + \left(r_1 - c_{12}\right) \times \left(W_1 - m_1 a_1\right)$

$$= \left(I_{G_1} \ddot{\theta}_1 - k_2 \theta_2\right)\hat{\mathbf{k}}$$

$\mathcal{B}_2$: $r_2 \times \left(W_2 - m_2 a_2\right) + c_{23} \times \left(W_3 - m_3 a_3\right) = \left(I_{G_2}\left(\ddot{\theta}_1 + \ddot{\theta}_2\right) + k_2 \theta_2 - k_3 \theta_3\right)\hat{\mathbf{k}}$

$\mathcal{B}_3$: $r_3 \times \left(W_3 - m_3 a_3\right) = \left(I_{G_3}\left(\ddot{\theta}_1 + \ddot{\theta}_2 + \ddot{\theta}_3\right) + k_3 \theta_3\right)\hat{\mathbf{k}}$

As usual, the system kinematics are the messiest part of the derivation. The substitution of the accelerations and the execution of the cross-products result in the final set of simultaneous equations

$$
\begin{bmatrix} M_{11} & M_{12} & M_{13} \\ M_{21} & M_{22} & M_{23} \\ M_{31} & M_{32} & M_{33} \end{bmatrix}\begin{pmatrix} \ddot{\theta}_1 \\ \ddot{\theta}_2 \\ \ddot{\theta}_3 \end{pmatrix} = -\begin{bmatrix} 0 & N_{12} & N_{13} \\ N_{21} & 0 & N_{23} \\ N_{31} & N_{32} & 0 \end{bmatrix}\begin{pmatrix} \dot{\theta}_1^2 \\ \left(\dot{\theta}_1 + \dot{\theta}_2\right)^2 \\ \left(\dot{\theta}_1 + \dot{\theta}_2 + \dot{\theta}_3\right)^2 \end{pmatrix}
$$

$$
-\begin{bmatrix} P_{11} & 0 & 0 \\ 0 & P_{22} & 0 \\ 0 & 0 & P_{33} \end{bmatrix}\begin{pmatrix} \sin\theta_1 \\ \sin(\theta_1 + \theta_2) \\ \sin(\theta_1 + \theta_2 + \theta_3) \end{pmatrix}
$$

$$
-\begin{bmatrix} Q_{11} & 0 & 0 \\ 0 & Q_{22} & 0 \\ 0 & 0 & Q_{33} \end{bmatrix}\begin{pmatrix} \cos\theta_1 \\ \cos(\theta_1 + \theta_2) \\ \cos(\theta_1 + \theta_2 + \theta_3) \end{pmatrix} - \begin{bmatrix} 0 & R_{12} & 0 \\ 0 & R_{22} & R_{23} \\ 0 & 0 & R_{33} \end{bmatrix}\begin{pmatrix} \theta_1 \\ \theta_2 \\ \theta_3 \end{pmatrix}
$$

These equations are to be solved subject to the initial conditions

$$
\theta_1(0) = \theta_2(0) = \theta_3(0) = 0, \quad \text{and} \quad \dot{\theta}_1(0) = \dot{\theta}_2(0) = \dot{\theta}_3(0) = 0
$$

essentially assuming that seat and mannequin all are in an upright position, and that vehicle and occupant are in a uniform state of motion at the instant deceleration is applied. The matrix entries have the following form:

$$
\begin{aligned}
M_{11} &= M_1 + M_2 \sin\theta_2 + M_3 \cos\theta_2 + M_4 \sin(\theta_2 + \theta_3) + M_5 \cos(\theta_2 + \theta_3) \\
M_{12} &= M_2 \sin\theta_2 + M_3 \cos\theta_2 + M_4 \sin(\theta_2 + \theta_3) + M_5 \cos(\theta_2 + \theta_3) \\
M_{13} &= M_4 \sin(\theta_2 + \theta_3) + M_5 \cos(\theta_2 + \theta_3) \\
M_{21} &= N_1 + N_2 \sin\theta_2 + N_3 \cos\theta_2 + N_4 \sin\theta_3 + N_5 \cos\theta_3 \\
M_{22} &= N_1 + N_4 \sin\theta_3 + N_5 \cos\theta_3 \qquad\qquad M_{23} = N_4 \sin\theta_3 + N_5 \cos\theta_3 \\
M_{31} &= P_1 + P_2 \sin\theta_3 + P_3 \cos\theta_3 + P_4 \sin(\theta_2 + \theta_3) + P_5 \cos(\theta_2 + \theta_3) \\
M_{32} &= P_1 + P_2 \sin\theta_3 + P_3 \cos\theta_3 \qquad\qquad M_{33} = P_1
\end{aligned}
$$

$$
\begin{aligned}
N_{11} &= 0, \qquad N_{22} = 0, \qquad N_{33} = 0 \\
N_{12} &= -M_3 \sin\theta_2 + M_2 \cos\theta_2 \qquad\qquad N_{13} = -M_5 \sin(\theta_2 + \theta_3) + M_4 \cos(\theta_2 + \theta_3) \\
N_{21} &= N_3 \sin\theta_2 - N_2 \cos\theta_2 \qquad\qquad N_{23} = -N_5 \sin\theta_3 + N_4 \cos\theta_3 \\
N_{31} &= P_5 \sin(\theta_2 + \theta_3) - P_4 \cos(\theta_2 + \theta_3) \qquad N_{32} = P_3 \sin\theta_3 - P_2 \cos\theta_3
\end{aligned}
$$

with relatively sparsely populated remaining matrices

$$
\begin{aligned}
P_{11} &= M_6, \quad Q_{11} = M_7, \quad R_{12} = M_8 \\
P_{22} &= N_6, \quad Q_{22} = N_7, \quad R_{22} = N_8, \quad R_{23} = N_9
\end{aligned}
$$

$$P_{33} = P_6, \quad Q_{33} = P_7, \quad R_{33} = P_8$$

with

$$M_1 = \frac{I_{G_1}}{m_1} + \left(\frac{m_2}{m_1} + \frac{m_3}{m_1}\right)\left(a_{12}^2 + (R - b_{12})^2\right) + x_1^2 + (y_1 - R)^2$$

$$M_2 = -\frac{m_2}{m_1}\left(a_{12}y_2 + (R - b_{12})x_2\right) - \frac{m_3}{m_1}\left(a_{12}b_{23} + a_{23}(R - b_{12})\right)$$

$$M_3 = \frac{m_2}{m_1}\left(a_{12}x_2 - (R - b_{12})y_2\right) + \frac{m_3}{m_1}\left(a_{23}a_{12} - b_{23}(R - b_{12})\right)$$

$$M_4 = -\frac{m_3}{m_1}\left(y_3 a_{12} + x_3(R - b_{12})\right)$$

$$M_5 = \frac{m_3}{m_1}\left(x_3 a_{12} - y_3(R - b_{12})\right)$$

$$M_6 = \left(1 + \frac{m_2}{m_1} + \frac{m_3}{m_1}\right)\left(a_0 a_{12} + g(R - b_{12})\right) + a_0(x_1 - a_{12}) - g(y_1 - b_{12})$$

$$M_7 = \left(1 + \frac{m_2}{m_1} + \frac{m_3}{m_1}\right)\left(a_{12}g - a_0(R - b_{12})\right) + g(x_1 - a_{12}) + a_0(y_1 - b_{12})$$

$$M_8 = -\frac{k_2}{m_1}$$

and, finally,

$$N_1 = \frac{I_{G_2}}{m_2} + x_2^2 + y_2^2 + \frac{m_3}{m_2}\left(a_{23}^2 + b_{23}^2\right) \qquad P_1 = \frac{I_{G_3}}{m_3} + \left(x_3^2 + y_3^2\right)$$

$$N_2 = \frac{m_1}{m_2}M_2 \qquad\qquad\qquad\qquad P_2 = \frac{m_2}{m_3}N_4$$

$$N_3 = \frac{m_1}{m_2}M_3 \qquad\qquad\qquad\qquad P_3 = \frac{m_2}{m_3}N_5$$

$$N_4 = \frac{m_3}{m_2}\left(b_{23}x_3 - a_{23}y_3\right) \qquad\qquad P_4 = \frac{m_1}{m_3}M_4$$

$$N_5 = \frac{m_3}{m_2}\left(a_{23}x_3 + b_{23}y_3\right) \qquad\qquad P_5 = \frac{m_1}{m_3}M_5$$

$$N_6 = a_0 x_2 - g y_2 + \frac{m_3}{m_2}\left(a_0 a_{23} - g b_{23}\right) \qquad P_6 = x_3 a_0 - y_3 g$$

$$N_7 = a_0 y_2 - g x_2 + \frac{m_3}{m_2}\left(a_0 b_{23} + g a_{23}\right) \qquad P_7 = x_3 g + y_3 a_0$$

$$N_8 = \frac{k_2}{m_2} \quad N_9 = -\frac{k_3}{m_2} \qquad\qquad\qquad P_8 = -\frac{m_2}{m_3}N_9$$

These equations were integrated using a standard Runge-Kutta four section algorithm with a fixed time increment of 0.02 milliseconds.

3. Problem Formulation (Optimization)

Our overall optimal design framework is that of multicriteria optimization with Edgeworth-Pareto optimality as the basic optimality concept. In brief, suppose we have N generally conflicting and noncommensurate criteria $g_i(d)$ to be minimized simultaneously on some decision set $\mathcal{D}$. There rarely exits a single decision $\bar{d} \in \mathcal{D}$ which accomplishes this task and a compromise is required. The basic compromising decision concept is described in the following definition.

Definition. *Edgeworth-Pareto (EP)-Optimality.* A decision $d^* \in \mathcal{D}$ is EP-optimal iff
$$g(d) \le g(d^*) \Rightarrow g(d) = g(d^*)$$
for every d^* - comparable $d \in \mathcal{D}$.

Here $g(d) = (g_1(d),...,g_N(d))$ and the inequality "$\le$" between vectors is componentwise, the result being a partial order on R^N; d_1 is comparable to d_2 iff $g(d_1)$ is comparable to $g(d_2)$.

This mathematically precise definition sometimes seems conceptually obscure. From a practical point of view, a design cannot be EP-optimal if we can deviate from it and manage to decrease all of the criteria; conversely, if a design is EP-optimal, then a deviation from it will result in the increase of at least one criterion value. The fundamentals of such a multicriteria approach may be found in Chapter 1 of [3].

Here, we shall formulate a parameter optimization problem with $R_0 \le R \le R_1$ as our design parameter. We have chosen the relatively obvious criteria:

$g_1(R)$ the time t_1 (in milliseconds) for the chair to reach full deployment, $\theta_1(t_1) = 30°$;
$g_2(R)$ $\theta_3(t_1)$, the angle of the head segment at time t_1;
$g_3(R)$ the relative kinetic energy of the head segment at time t_1.

Thus, we take the relative kinetic energy at the time of full deployment to be given by

$$T = \frac{1}{2}m_3 v_{G_3}{}^2 + \frac{1}{2}I_{G_3}\left(\dot{\theta}_1 + \dot{\theta}_2 + \dot{\theta}_3\right)^2$$

where v_{G_3} excludes the velocity v_0 of the translating reference frame $(Oxyz)_0$. The final results are represented graphically in terms of plots of the criteria versus R and the bicriteria graph g_3 versus g_1.

4. The Numerical Results

All of the raw data for the mass distribution and moments of inertia of the standard dummy stem from [1]. These studies investigated total body centers of mass and moments of inertia for different postures as well as varying body sizes.

We make use of a model based on 50th percentile seated male data. This construction resulted in a mannequin having a mass distribution very close to that observed by NASA researchers.

The NASA data do not incorporate the ICLF as a reference location; they take the hip joint to be the primary mid-body hinge. Thus, some adjustment of their data was required to account for the ICLF as a body reference location. Specifically, 5% of the body mass was taken from the measured torso and transferred to the the thigh. The adjusted data finally led to the use of the numerical values listed in Table 1. The use of the ICLF as a body hinge is appropriate in the study of aircraft seating while a pelvic link joining torso and thigh is the preferred model for automobiles.

TABLE 1. Numerical values for the Model Parameters

$m_1 = 35.1$ kg	$I_{G1} = 32797$ kg-cm^2			
$m_2 = 43.8$ kg	$I_{G2} = 24202$ kg-cm^2			
$m_3 = 6.5$ kg	$I_{G3} = 12782$ kg-cm^2			
$x_1 = 14.8$ cm	$y_1 = 3.8$ cm	$a_{12} = -12.7$ cm	$b_{12} = 23.7$ cm	$k_2 = 0$
$x_2 = 0$ cm	$y_2 = 11.9$ cm	$a_{23} = -5$ cm	$b_{23} = 35.8$ cm	$k_3 = 0$
$x_3 = 5$ cm	$y_3 = 7.5$ cm	$a_0 = 19620$ cm/sec^2		

The results of the numerical investigation are presented in Figures 2 - 4. We have chosen to present the results three different ways - by graphing the criteria versus R, versus each other in so-called criteria space and by presenting a time sequence of the seat-dummy deployment.

Figure 2 is a plot of the criteria versus the design parameter R. The criteria obviously are not commensurate. However, from the viewpoint of multicriteria optimization only their increasing or decreasing behavior and the corresponding ranges of R are of interest. As implied by EP-optimality (and by common sense) the need for compromise arises when different criteria are both increasing and decreasing within the same range of R. For example, until the kinetic energy reaches a minimum at R = 31.5 all of the criteria are monotonically decreasing; that is, increasing R continues to provide improvement for all. Thus, our first design conclusion: R ≥ 31.5. Similarly, the deployment time g_1 decreases until R = 42 for which $g_1(R) = 53$ ms, and then increases. Thus, for R ≥ 42, both the energy and the time are increasing with increasing R. It follows that for R ≥ 42, *decreasing* R will improve both criteria. Conclusion: 31.5 ≤ R ≤ 42 is desirable. Only in this range is a compromise required between the energy and the time. Deliberations with respect to θ_3 are easily included since $g_2(R)$ is a monotonically decreasing function of R - the larger the R the smaller the θ_3.

Within the present problem the crucial trade-off occurs between g_1 and g_3 and we thus also depict their relationship in criteria space, as a plot of g_1 versus g_3 (Figure 3). The heavily drawn curve segment shows the EP-optimal criteria values for the bicriteria problem involving g_1 and g_3. Finally, we use Figure 4 to depict the dummy-seat motion as a function of time in the range $0 \leq t \leq t_1$. (for R = 35).

5. Conclusions

The deadline oriented design process which prevails in the corporate world generally leaves no margin for a design optimization. The first design that works is implemented and only subsequently is it tweaked here and there to improve performance in isolated segments of the design. Optimization should be an intrinsic part of the design effort.

Here, the greatest effort by far concerned the establishment and analysis of the dynamical model; only a comparatively small additional effort was required for the formulation and completion of an optimal design aspect. Multicriteria optimization represents an organized approach to the reach of compromise between conflicting criteria. It separates the decision space into two disjoint regions, one where decisions ought not to be made and one within which any final decision should be made. The latter set, the set of EP-optimal decisions usually consists of an infinity of desirable selections and there are just as many justifications for making it. Our final choice of radius should be made within the range $31.5 \leq R \leq 42$. We may then introduce additional design constraints or use some rule of thumb to arrive at a final selection R*.

6. References

1. Webb Associates (eds.), (1978) *NASA Reference Anthropometric Source Book,* VVI Publication 1024, National Aeronautics and Space Administration, Technical Information Service, Washington, D.C., Appendix D.
2. Serber, H (1994) *Hard Facts about Soft Machines*, Chapter 31, R. Lueder (ed.), Taylor & Francis, London.
3. Stadler, W. (1988) Fundamentals of Multicriteria Optimization, Chapter 1 in W. Stadler (ed.), *Multicriteria Optimization in Engineering and the Sciences*, Plenum Press, New York, NY, 1-26.

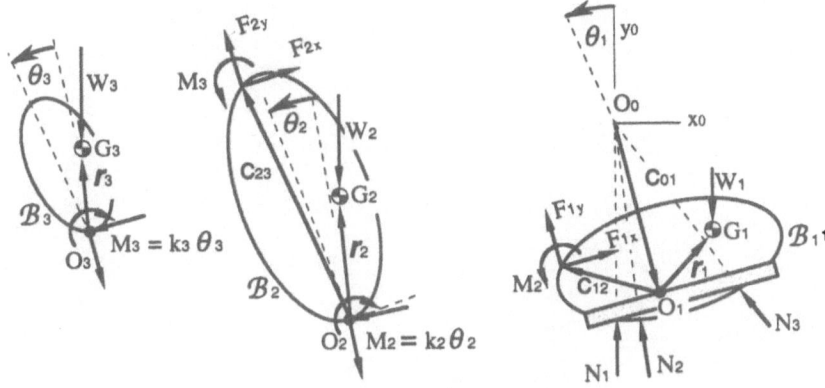

Figure 1.
Kinematic and kinetic terminology for seat and mannequin.

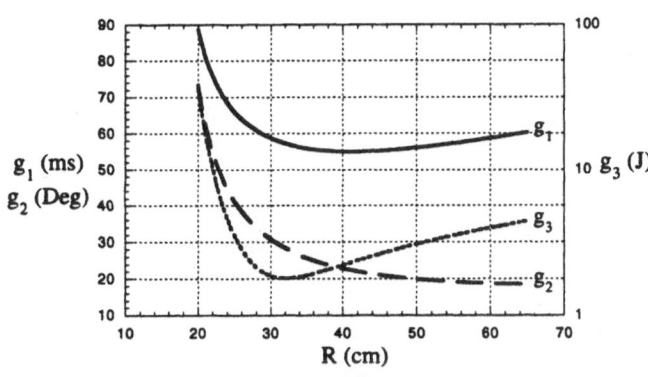

Figure 2.
Criteria versus design parameter R
(note the logarithmic scale for g_3)

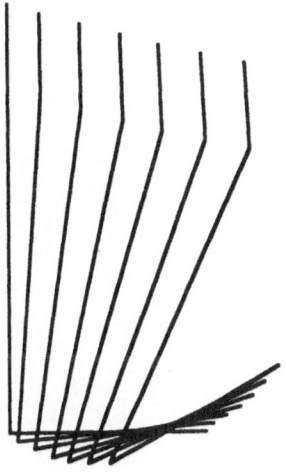

Figure 4.
Time history of seat and
dummy motion for R = 35.

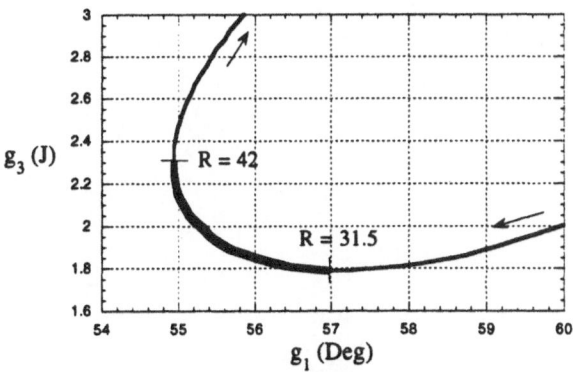

Figure 3.
Criteria space: g_3 versus g_1.

MULTICRITERIA DESIGN OF MACHINES

R. STATNIKOV and J.MATUSOV
The Russian Academy of Sciences
Mechanical Engineering Research Institute,
101830, Moscow Centre, Russia, 4 Griboyedov str.,

1. Abstract

Engineering optimization problems, design problems among them, are multicriteria in their essence. When designing machines, mechanisms, structures, one has to deal with numerous contradictory criteria: material demands, strength, longevity, vibration and noise levels, energy-intensiveness, production capacity, efficiency, overall dimensions, economic characteristics.

But the most important thing is as follows. The experience gained in solving engineering optimization and optimal design problems shows that the specialist cannot state them correctly. Unfortunately, the known optimization methods offer only slender assistance to him in this case. However, it should be noted that some interesting approaches were given in [1].

For correct statement and solving of engineering optimization problems a METHOD OF PARAMETER SPACE INVESTIGATION (PSI Method) has been developed [2].

The method has no analogs. It takes into account the specific features of these problems most fully; it allows for the first time the determination of optimal solutions with any number of performance criteria under consideration.

The method also makes it possible to solve such important engineering problems as engineering development of the pilot models of object (machines, mechanisms, structures) and vector (multicriteria) identification problems.

The PSI Method has become one of the basic working tools for choosing optimal parameters in many fields of the economy in automotive industry, machine-tool industry, agricultural industry, aircraft construction, shipbuilding. It has displayed its high effectiveness in all these fields without exception.

The method is widely used for optimization of complex finite element models [3,4].

2. Formulation of Multicriteria Optimization Problems

The problems statement is valid for the majority of optimization problems in engineering.

D. Bestle and W. Schielen (eds.), IUTAM Symposium on Optimization of Mechanical Systems, 303–307.
© *1996 Kluwer Academic Publishers.*

Assume that there is a mathematical model of an object (e.g., a system of equations) that enables one to calculate all characteristics of the object.

Assume that the system studied depends on r design variables $a_1,...,a_r$, which are considered as a point $\alpha = (a_1,...,a_r)$ in r-dimensional space. Usually, the mentioned system includes α.

In general, in order to state the multicriteria optimization problem correctly, we should take into account three types of constraints: design-variable, functional, and criteria. The design-variable constraints have the form

$$a_j^* \leq a_j \leq a_j^{**}, j = 1, 2, ..., r. \tag{1}$$

In mechanical systems, variables a_j may be represented by rigidities, moments of inertia, masses, damping coefficients, dimensions, etc.

The functional constraints may be written as

$$c_l^* \leq f_l(\alpha) \leq c_l^{**}, l = 1, 2, ..., l. \tag{2}$$

The functional dependencies $f_l(\alpha)$ may be represented by functionals dependent on integral curves of the considered differential equations, or by functions of α; c_l^* and c_l^{**} are the standardization constraints (e.g., admissible tensions in the elements of structures, the railway gage width, etc.).

The particular performance criteria $\Phi_\nu(a)$, $\nu = 1, 2, ..., k$ are the values to be, *caeteris paribus*, made extremal. For simplicity, we will assume that they are to be minimized.

Evidently, the constraints (1) define a parallelepiped Π in r-dimensional parameter space.

To avoid the situation when, from the designer's viewpoint, the values of certain criteria become intolerably high, criterion constraints have to be introduced:

$$\Phi_\nu(a) \leq \Phi_\nu^{**}, \nu = 1, 2, ..., k, \tag{3}$$

where Φ_ν^{**} is the worst value of the criterion $\Phi_\nu(\alpha)$, which the designer can accept.

Criteria and functional constraints differ in that the values of Φ_ν^{**} are determined when solving the problem and are repeatedly revised (made stronger or weaker). The constraints (1)-(3) define the feasible solution set D, i.e., the set of solutions α^i that satisfy these constraints.

Let us now state one of the basic problems in multicriteria optimization. The objective is to find a set $P \subset D$, such that

$$\Phi(P) = \min_{\alpha \in D} \Phi(\alpha), \tag{4}$$

where $\Phi(\alpha) = (\Phi_1(\alpha), ..., \Phi_k(\alpha))$ is the vector of criteria and P is called the Edgeworth-Pareto optimal set (EP-set).

It should be noted that $\Phi(\alpha) < \Phi(\beta)$ if and only if $\Phi_\nu(\alpha) \leq \Phi_\nu(\beta)$, $\nu = 1, 2, ..., k$, and at least for $\nu_0 \in \{1, k\}$ $\Phi_\nu(\alpha) < \Phi_\nu(\beta)$. After solving the problem, the vector α^0, which is the preferable one in EP-set P, can be determined.

Consider the principal peculiarities of engineering problem optimization. Usually, a designer states the problem and then tries to solve it by applying various methods available. However, although traditional, this approach cannot be applied to the problem class under consideration.

Indeed, as stated above, in problems with contradictory criteria, the designer cannot correctly formulate the criteria constraints Φ_ν^{**} before the problem is solved. This is also valid for soft functional constraints, i.e., those where c_l^*, c_l^{**} should be set at the designer's will. In addition, in many problems, setting the design-variable constraints α_j^* and α_j^{**} presents considerable difficulties for the designer. In other words, if the traditional approach is applied, the designer is unable to state the problem correctly.

It follows that finding the feasible solution set is the basis for stating the optimization problem correctly. However, this set cannot be correctly defined using the methods employed in the well-established program packages, because the problems of defining design-variable, functional, and criteria constraints are disregarded. These constraints are assumed to be fixed prior to solving the problem. To state the optimization problem correctly, the Method of Parameter Space Investigation (PSI Method) has been developed.

3. PSI Method

The PSI Method is based on the exploration of the search domain, defined by inequalities (1), using uniformly distributed points. For variable vectors α' corresponding to these points, the values of functional constraints are calculated. If these constraints are satisfied, the values of performance criteria $\Phi_\nu(\alpha^i)$, $\nu = 1, 2, ..., k$ are calculated. The correct definition of design-variable, functional, and criteria constraints is performed by a special algorithm developed in [2].

The search is performed in *three steps*. In the *first step*, the test tables are obtained for each criterion. They contain the values of $\Phi_\nu(\alpha^1), ..., \Phi_\nu(\alpha^N)$ in ascending order (assuming that all criteria are to be minimized). In the *second step*, the designer performs a preselection of the criteria constraints Φ_ν^{**}, $\nu = 1, 2, ..., k$. Φ_ν^{**} are the maximal values of the criteria $\Phi_\nu(\alpha)$ that guarantee an acceptable functioning of the object. If Φ_ν^{**} is not taken as maximal, many interesting solutions may be lost, because of contradictory criteria. As a rule, a designer may define some $\Phi_\nu(\bar{\alpha})$ that is definitely feasible, as the Φ_ν^{**}. If the maximal values are taken for Φ_ν^{**}, one should switch to the *third step*.

In the *third step*, the problem solvability is verified. The vectors α^i that simultaneously satisfy all constraints $\Phi_\nu(\alpha^i) \leq \Phi_\nu^{**}$, $\nu = 1, 2, ..., k$ are determined. If the set of these vectors is nonempty, it is possible to construct the feasible solution set. If not, either the values Φ_ν^{**} should be corrected or the *first step* should be run once more, with more points, so that the tables of the *second step* are larger.

The process repeats until D becomes nonempty. Then, the EP-set P is constructed according to [2].

Let us consider the case when the designer meets with difficulties in defining the maximal Φ_ν^{**}. Usually, the values of $\Phi_\nu(\alpha)$ within the interval $\Phi_\nu(\bar{\alpha}) \leq \Phi_\nu(\alpha) \leq \tilde{\Phi}_\nu^{**}$ are indefinite in terms of feasibility. Here, $\tilde{\Phi}_\nu^{**}$ is the value of the νth criterion, for which the values $\Phi_\nu(\alpha) > \tilde{\Phi}_\nu^{**}$ are *a fortiori* inadmissible. In other words, the designer is not sure whether $\Phi_\nu(\alpha) > \Phi_\nu(\bar{\alpha})$ are feasible or not. In this situation, as before, one should pass to the *third step* and construct the admissible set D and the EP- set P under the constraints $\Phi_\nu^{**} = \Phi_\nu(\bar{\alpha})$. Then, one should obtain the set $\tilde{D}$ under the constraints $\tilde{\Phi}_\nu^{**}$, $\nu = 1, 2, ..., k$ and the corresponding EP-set $\tilde{P}$. Then, $\Phi(P)$ and $\Phi(\tilde{P})$ are compared. If vectors in $\Phi(\tilde{P})$ have no considerable advantage over the vectors in $\Phi(P)$, the values $\Phi_\nu^{**} = \Phi_\nu(\bar{\alpha})$ may be taken as criteria constraints. Otherwise, if the advantage is considerable, the criteria values may be taken equal to $\tilde{\Phi}_\nu^{**}$. One should verify the obtained optimal solution for its admissibility. If the designer cannot verify the feasibility of the solution, then the former $\Phi_\nu^{**} = \Phi_\nu(\bar{\alpha})$ are taken as the criteria constraints. (All possible values of $\Phi_\nu(\bar{\alpha})$ and Φ_ν^{**} fit into this scheme.) The PSI Method permits us to find the opportunity for solving the important problem of constructing the feasible set and EP-set with a given accuracy [2].

4. Conclusion

The PSI Method enables a designer to correctly state the problems of engineering optimization. It is important that alternative solutions can be obtained and that the number of criteria used is virtually unlimited.

The following conclusions can be made. In order to solve the problems of optimization in engineering, including optimal design, all general-purpose packages should be complemented by a program module intended for the statement and solution of the above-discussed problems. The MOVI (Multicriteria Optimization and Vector Identification) software package, which implements the PSI Method, can play this role.

Presented investigation have been carried out under support of Russian Fund for Fundamental Research (Grant no. 95-01-03341).

5. References

1. Stadler, W. Fundamentals of Multicriteria Optimization, in W.Stadler(ed), *Multicriteria Optimization in Engineering and in the Sciences*. New York, London: Plenum Press, 1988.

2. Statnikov, R., and Matusov, J. *Multicriteria Optimization and Engineering*. New York: CHAPMAN and HALL, 1995.

3. Bondarenko, M.I., Nazemkin, A.Y., Pozhalostin, A.A., Statnikov, R.B., and Shenfel'd, V.S. Construction of Consistent Solutions in Multicriteria Problems of Optimization of Large Systems, *Physics-Doklady, Vol. 39, No. 4, 1994, p.p. 274-279. Translated from Doklady Akademii Nauk, Vol. 335, No. 6, 1994, p.p. 719-724.*

4. Statnikov, R.B., and Matusov I.B. General-Purpose Finite Element Programs in Search for Optimal Solutions. *Physics- Doklady, Vol. 39, No. 6, 1994, p.p. 441-443. Translated from Doklady Akademii Nauk, Vol. 336, No. 4., 1994, p.p. 481-484.*

References

OPTIMUM DESIGN OF 3-D TRUSS STRUCTURE CONSIDERING CONTROL WITH STRESS CONSTRAINT AT INITIAL DEFORMED STATE

Y. TADA AND S. MITA
Department of Computer and Systems Engineering,
Faculty of Engineering, Kobe University,
Rokkodai, Nada, Kobe 657 Japan

Abstract

As a simultaneous optimization problem of structure and control, cross sectional areas of a 3-D truss structure are optimized considering the minimization of the structural weight and its control cost under the stress constraint at the initial deformed state. A quadratic performance index as in the optimal regulator theory is adopted as a control cost, and it is minimized with structural weight fixed at a constant value in order to obtain a Pareto optimum solution of the two-objective problem. The optimization is carried out by the energy-ratio method and the stress constraint is treated by the stress-ratio method.

1. Introduction

Large scale structures such as a space station are desired to be designed so that they may have a least weight considering the transportation cost to space. However, they have several problems; Since the stiffness of a light-weight structure is comparatively low, vibration is apt to occur and the vibration once induced does not readily decay because of its low damping. For such a vibration problem, it is well known that the active control is effective for vibration control as discussed in [1]. In the field of large space structure, for the satisfaction of the severe specification, the necessity of the simultaneous optimization of structural system and control one is recognized because of close relationship between them, and several papers such as [2] have been published on it. In this paper, as a treatment of the simultaneous optimization of structure and control, we formulate an optimization problem of a three-dimensional truss structure aiming to minimize a control cost under the constraint on the structural weight and under the stress constraint at the initial deformed state of the system. As the control cost, we adopt an integral

309

D. Bestle and W. Schielen (eds.), IUTAM Symposium on Optimization of Mechanical Systems, 309–316.
© 1996 *Kluwer Academic Publishers.*

of a quadratic form of state variables and control inputs when the truss is controlled according to the optimal regulator theory. The design variables are cross sectional areas of truss members.

2. Vibration Control

2.1. EQUATION OF MOTION

In this paper, we treat a three-dimensional truss with n members as the object of control and optimization. By the use of Finite Element Method, the equation of motion of the truss is expressed as

$$\mathbf{M}\ddot{\mathbf{q}}(t) + \mathbf{D}\dot{\mathbf{q}}(t) + \mathbf{K}\mathbf{q}(t) = \mathbf{L}\mathbf{u}(t) , \tag{1}$$

where $\mathbf{M}$, $\mathbf{D}$ and $\mathbf{K}$ are the mass, damping and stiffness matrices, respectively, and $\mathbf{q}$ is the vector of nodal displacements. $\mathbf{u}$ is the vector of control inputs and $\mathbf{L}$ is the matrix which is determined by the location of control inputs. Usually, Eq.(1) is transformed into the form of state equation as

$$\dot{\mathbf{x}}(t) = \mathbf{A}\mathbf{x}(t) + \mathbf{B}\mathbf{u}(t) , \tag{2}$$

where the state variables are defined by

$$\mathbf{x} = [\mathbf{q}^T , \dot{\mathbf{q}}^T]^T \tag{3}$$

and the coefficient matrices $\mathbf{A}$ and $\mathbf{B}$ are given by

$$\mathbf{A} = \begin{bmatrix} \mathbf{0} & \mathbf{I} \\ -\mathbf{M}^{-1}\mathbf{K} & -\mathbf{M}^{-1}\mathbf{D} \end{bmatrix} , \qquad \mathbf{B} = \begin{bmatrix} \mathbf{0} \\ \mathbf{M}^{-1}\mathbf{L} \end{bmatrix} . \tag{4}$$

2.2. OPTIMAL REGULATOR THEORY

In the system represented by Eq.(2), we assume that the state variables are described by setting its equilibrium state

$$\mathbf{x} = \mathbf{0} , \qquad \mathbf{u} = \mathbf{0} , \tag{5}$$

as the reference state. Now, we consider the case that the system deviated from its equilibrium state. If we assume that the time begins to be counted at the moment when the state deviates from the equilibrium point, then the initial state of the system is written by

$$\mathbf{x}(0) = \mathbf{x}_o . \tag{6}$$

The control is required to bring the system back to its equilibrium state as quickly as possible and by small energy. One of the index for evaluation of the control cost is an integral of quadratic form of state variables and inputs given by

$$J_c(\mathbf{x}_o , \mathbf{u}) = (1/2) \int_0^{\infty} [\mathbf{x}^T\mathbf{Q}\mathbf{x} + \mathbf{u}^T\mathbf{R}\mathbf{u}]dt \tag{7}$$

where $\mathbf{Q}$ is a semi-positive definite symmetric matrix and $\mathbf{R}$ is a positive definite symmetric matrix. The solution of the optimal regulator problem, that is, the input which minimizes the functional J_c is represented by the state feedback as

$$\mathbf{u}^*(\mathbf{t}) = -\mathbf{R}^{-1}\mathbf{B}^T\mathbf{P}\mathbf{x}(t) \ , \tag{8}$$

where $\mathbf{P}$ is a matrix which is a solution of the Riccati equation

$$\mathbf{A}^T\mathbf{P} + \mathbf{P}\mathbf{A} - \mathbf{P}\mathbf{B}\mathbf{R}^{-1}\mathbf{B}^T\mathbf{P} + \mathbf{Q} = \mathbf{0} \ . \tag{9}$$

When the system is controlled by this input, the minimum value of the index function is given by

$$J_c^*(\mathbf{x}_o) = J_c(\mathbf{x}_o, \mathbf{u}^*) = (1/2)\mathbf{x}_o^T\mathbf{P}\mathbf{x}_o \ . \tag{10}$$

3.   Problem of Optimum Design

In this study, we consider a problem which searches for cross sectional areas of truss members that minimize a control cost under the constraint of structural weight constancy and the stress constraint, which is a treatment of optimization of structure considering control. As a control cost, we adopt the quadratic performance index when the truss is controlled according to the optimal regulator theory. Then, the present optimization problem is formulated by

$$\underset{\mathbf{a}}{\text{Minimize}} \quad J(\mathbf{a}) = (1/2)\mathbf{x}_o^T\mathbf{P}\mathbf{x}_o \tag{11}$$

$$\text{subject to} \ \ W(\mathbf{a}) = \sum_{i=1}^{n} \rho_i a_i l_i = C \tag{12}$$

$$|s_i| \leq s_a \quad (i = 1, \cdots, n) \ , \tag{13}$$

where $\mathbf{x}_o$ is the initial state of the system and $\mathbf{P}$ is the matrix which is the solution of the Riccati equation in the optimal regulator theory. ρ_i , a_i and l_i are the density, cross sectional area and length of the i -th member, respectively, and C is the specified value for the weight. s_i is the stress induced in the i -th member when the truss is subjected to a static load $\mathbf{p}_o$ as undermentioned, and s_a is the allowable stress. The treatment of the above problem is one method for obtaining the Pareto optima of a two-objective problem of the structural and control systems. As the initial state of the system, we adopt the deformed state $\mathbf{q}_o$ by a static load $\mathbf{p}_o$ and zero velocity, that is,

$$\mathbf{x}_o = [\mathbf{q}_o^T \ , \mathbf{0}]^T \tag{14}$$

$$\mathbf{K}(\mathbf{a})\mathbf{q}_o = \mathbf{p}_o \ . \tag{15}$$

4. Optimization Scheme

4.1. METHOD OF OPTIMALITY CRITERIA

First, we consider the problem without stress constraint. The problem, Eqs.(11) and (12), is described by the Lagrangian function as

$$L(\mathbf{a}, g) = J(\mathbf{a}) + g[W(\mathbf{a}) - C] \longrightarrow \text{ stationary} , \tag{16}$$

where g is the Lagrangian multiplier. The differentiation of Eq.(16) with respect to each variable gives the optimality conditions as

$$\partial J/\partial a_i + g \partial W/\partial a_i = 0 \quad (i = 1, \cdots, n) \tag{17}$$

in addition to the weight constraint, Eq.(12). If we define new variables g_i

$$g_i = -(\partial J/\partial a_i)/(\partial W/\partial a_i) \quad (i = 1, \cdots, n) , \tag{18}$$

Eq.(17) is reduced to

$$g_i - g = 0 \quad (i = 1, \cdots, n) . \tag{19}$$

Then, we solve Eqs.(19) and (12) by the energy-ratio method as in [3]. That is, we reform the cross sectional areas by the algorithm

$$a_i^* = a_i^t + \eta(g_i - \bar{g})/\bar{g} \quad (i = 1, \cdots, n) \tag{20}$$
$$a_i^{t+1} = a_i^* C/W(\mathbf{a}^*) \quad (i = 1, \cdots, n) , \tag{21}$$

where $\bar{g}$ is the average of n g_i 's, the superscript "t " and "$t+1$ " represent the iteration number and η is a positive number for controlling the computation. This procedure is repeated until the condition

$$\underset{i}{\text{Max}} \quad |(g_i - \bar{g})/\bar{g}| \leq e \tag{22}$$

is satisfied for a small constant e .

When cross sectional areas are restricted to their lower limits due to the stress constraint, we use

$$a_i^{t+1} = a_i^*\{C - \sum \rho_j \bar{a}_j l_j\}/\{W(\mathbf{a}^*) - \sum \rho_j \bar{a}_j l_j\} \tag{23}$$

instead of Eq.(21), where $\sum$ means the summation about members whose cross sectional areas are restricted to the lower limits due to the stress constraint.

4.2. STRESS CONSTRAINT

In order to incorporate the stress constraint in the optimization procedure, we transform the stress constraint into the size constraint as in [4], by assuming no stress redistribution. That is, Eq.(13) is reduced to

$$a_i \geq \bar{a}_i = |N_i|/s_a \quad (i = 1, \cdots, n) , \tag{24}$$

where N_i is the internal force of the i -th member at the t -th iteration. By introducing this explicit constraint on the design variables, a_i , we can treat the stress constraint easily in the optimization process.

4.3. SPECIFIED VALUE FOR WEIGHT

When the given specified value for the structural weight is small, the cross sectional areas of some members do not satisfy the optimality condition for the control cost, Eq.(19), but are determined by the limited value from the stress constraint, Eq.(24). Moreover, if the given specified value for the structural weight is too small, we can not at all obtain feasible solutions which satisfy the stress constraint. This means that when we solve the two-criteria problem by the constraint method, the assumed constraint value of the weight is not included in the region where the solutions exist. Then, in order to ensure hitting the region of the Parerto optimality, we manipulate the constraint value of the structural weight by the following two methods.

method A   In this method, for the weight specification we set a value which is large enough for all members to satisfy the optimality condition, Eq.(19). In order to select such a value, we renew the specification value by the following method;

$$C^{(t+1)} = (k^{(t)})^r C^{(t)} , \qquad (25)$$

where $C^{(t)}$ is the specified value at the t -th outer cycle and $k^{(t)}$ is determined by

$$k^{(t)} = \underset{i}{Max} \, (\bar{a}_i^{(t)}/a_i^{(t)}) , \qquad (26)$$

according to the stress-ratio method and r is a constant for the control of optimization process. Then, the optimization calculation is restarted from new initial values

$$\mathbf{a}^{0,(t+1)} = (k^{(t)})^r \mathbf{a}^{(t)} . \qquad (27)$$

method B   In order to make much of the structural weight, the constraint value is kept at the initial value $C^{(1)}$. If a_j which is calculated by the algorithm, Eq.(20), is less than the lower limit $\bar{a}_j$, then a_j is set as $a_j = \bar{a}_j$ and the j -th cross sectional area is removed from the set of design variables. This ordinary method is called method B. In this paper, we start the design process from the state where all members have a uniform cross sectional area as initial values, that is, $a_i^{0,(1)} = 0.2 \, (i = 1, \cdots, n)$, and then, its weight is adopted as the initial specified value $C^{(1)}$.

In the case that we consider the stress constraint after the reformation by the control cost J through Eqs.(20) and (21), when we adopt $\bar{a}_j$ as the cross sectional area of the j -th member, if we do not reform the sizes again, the weight of the

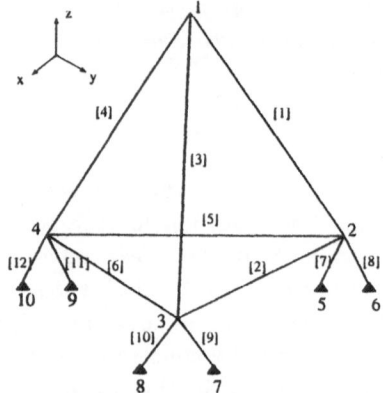

Figure 1: 3-dimensional truss

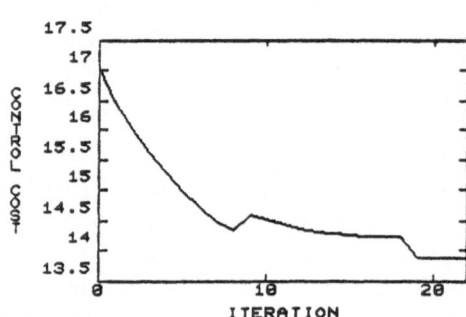

Figure 2: Behavior of control cost in no-stress-constraint design

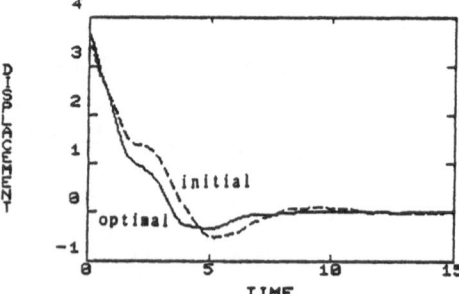

Figure 3: Transient responses in no-stress-constraint design

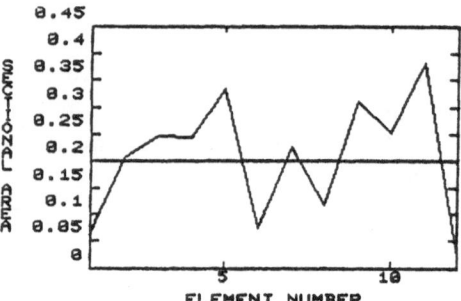

Figure 4: Optimum cross sectional areas by no-stress-constraint design

truss increases slightly. Then, we can adopt the weight which is calculated using these new sizes as the new specification value for the weight constraint. This method is a revised method of method B, and is called method B'.

5. Design Example

As a numerical example, we optimize the truss with 12 members as shown in Fig.1. As to the control system, it is assumed that actuators are located at the node 1 in x , y and z directions and sensors can know displacements and velocities at all nodes. The weighting matrices in Eq.(7) are selected as

$$\mathbf{Q} = \begin{bmatrix} \mathbf{K} & \mathbf{0} \\ \mathbf{0} & \mathbf{M} \end{bmatrix}, \quad \mathbf{R} = \mathbf{I}. \tag{28}$$

We consider the case where initial static loads with magnitude 2 are applied at the node 1 from x and y directions. We start the optimization from the initial

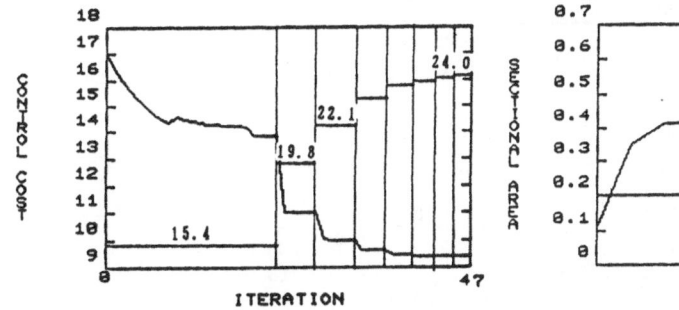

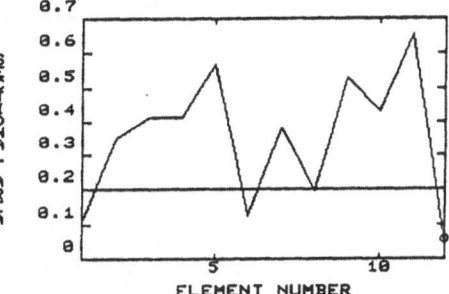

Figure 5: Behaviors of control cost and structural weight in method A

Figure 6: Optimum cross sectional areas by method A

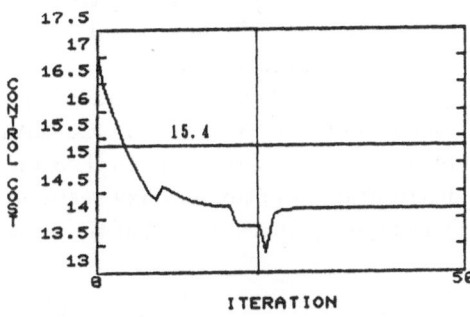

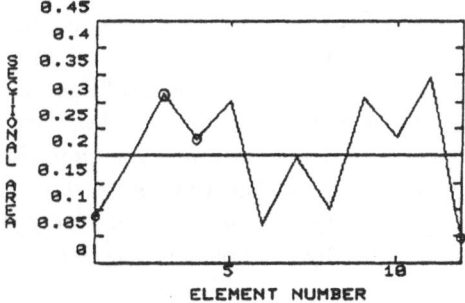

Figure 7: Behaviors of control cost and structural weight in method B

Figure 8: Optimum cross sectional areas by method B

design in which all members have the same cross sectional area, $a_i^{0,(1)} = 0.2$, and its weight is set as the initial specified value, $C^{(1)} = 15.4$.

No-stress-constraint   First, we solve the problem under the specified value $C^{(1)}$ without stress constraint. Figure 2 shows the behavior of the control cost J in the optimization process. It is observed that the computational process is stable. Figure 3 shows the examples of the transient responses of the initial design and optimized design. In the optimized truss, the vibration goes down with comparatively short time. Figure 4 shows the distribution of cross sectional area obtained. The horizontal line in the figure represents the initial value of the cross sectional area, that is, $a_i = 0.2$.

Stress-constraint   Second, by using the result of no-stress-constraint problem as the initial design, we solve the problem with stress constraint and under the specified value $C^{(1)}$, calling the result "B". Moreover, adopting the result of no-stress-constraint problem as initial guess, we solve the problem by changing the specified value according to Eq.(25). In the optimized state of the latter, all

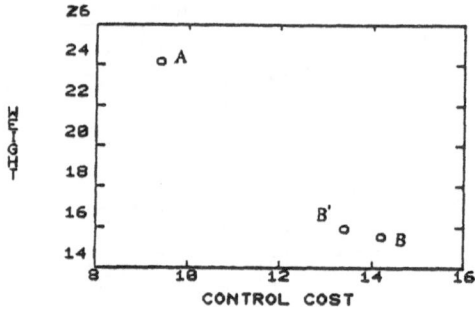

Figure 9: Pareto optimality

members satisfy both conditions of stress and optimality w.r.t. the control cost but having a relatively large volume, calling the result "A". Figures 5 and 7 show the behaviors of the control cost J in the optimization process in the cases A and B, respectively. The horizontal lines accompanied by numbers in the figures show the value of the specified weight. The obtained solutions "A" and "B" are both Pareto optima of the multicriteria problem of structural weight and control cost as shown in Fig.9. Figures 6 and 8 show the distribution of optimum cross sectional areas in designs A and B, respectively. In these figures, circles show that the areas are limited by the stress constraint.

6. Conclusion

In this paper, we solved the optimization problem with two objectives, control cost and structural weight, under the stress constraint by two methods. Method A attaches importance to the minimization of the control cost and method B respects the keeping of the structural weight. Both methods were effective in optimization and we could obtain Pareto optima for two objectives.

References

1. Seto, K., Yoshida, K. and Nonami, K. (Eds.) (1992) Proc. 1st Int'l Conference on Motion and Vibration Control, Jpn. Soc. Mec. Eng.

2. Salama, M., Garba, J., Demsetz, L. and Upwadia, F. (1988) Simultaneous Optimization of Controlled Structures, Computational Mechanics 3, 275-282.

3. Tada, Y., Matsumoto, R. and Nagai, M. (1992) Optimum Structural Design Considering Vibration Control, JSME International J. (Ser.III) 35, 413-420.

4. Tada, Y. and Minami, E. (1993) Optimum Design of 3-D Truss Structure and Its Effect on Control, JSME International J. (Ser.A) 36, 90-96.

APPLICATION OF CONTINUUM SENSITIVITY ANALYSIS AND OPTIMIZATION TO AUTOMOBILE STRUCTURES

S. WANG*, K.K. CHOI*, and H.T. KULKARNI**

* Center for Computer-Aided Design, University of Iowa
Iowa City, Iowa 52242, U.S.A.

** Advanced Vehicle Technology, Ford Motor Company
P.O. 2053, Dearborn, Michigan 48121-2053, U.S.A.

Abstract

Continuum element sensitivity analysis (CONTESA) and system optimization (SYSOPT) for Noise, Vibration, and Harshness (NVH) have been developed and applied to automobile structures for sizing, topology, and configuration design using Mindlin plate and Timoshenko beam theories. The topology optimization has been developed using the density approach, sequential linear programming, and the adjoint variable method. CONTESA has been tested using various vehicle models. Optimized vehicles using CONTESA and SYSOPT are manufactured to validate the simulation-based design methodology.

1. Introduction

Automobile design engineers have pursued the design of lightweight cars which are also quiet, durable, and safe. Manufacturing thinner gauge, lightweight bodies often results in problems involving vibration, durability, and ultimately passenger safety. In this paper, CONTESA and SYSOPT have been developed and applied to automobile structures for sizing, topology, and configuration design using Mindlin plate and Timoshenko beam theories [1]. In CONTESA, isoparametric bilinear plate and linear beam elements are used and deficiencies in bending energy has been accounted for [2]. To handle complex vehicle built-up structures, the warped plate element and general cross-sectional beam element with offset effect are included in CONTESA [1].

For sizing design sensitivity analysis (DSA), the thickness of the plate element and geometric sectional properties of the beam element such as cross-sectional area, nonprincipal moments of inertia, and torsional rigidity are considered as design variables. A concept vehicle model is studied and accurate sensitivity results are obtained. Sensitivity of a detailed vehicle model is computed by CONTESA and optimized by SYSOPT, and the NVH index is significantly improved compared to that of the original vehicle model.

Changing topology of structural layout significantly affect on structural responses such as compliance and eigenvalue. Recently Bendsoe et al. [3-4] presented topology optimization using the homogenization method. The homogenization method uses microscopic composite material properties and the optimality criteria. Yang and Chuang [5] presented the topology optimization of compliance and eigenvalues using the density approach [6], sequential linear programming, and the adjoint variable method. The

D. Bestle and W. Schielen (eds.), IUTAM Symposium on Optimization of Mechanical Systems, 317–324.
© 1996 Kluwer Academic Publishers.

homogenization method allows to work with larger design space than the density approach. However, the homogenization method with optimality criteria cannot be used for the optimization problem of multiple constraints. In this paper, the topology optimization of NVH has been developed using the density approach, sequential linear programming, and the adjoint variable method. The density approach is employed due to isotropic materials used in NVH analysis and multiple NVH constraints. Young's modulus and material density become functions of design variables, i.e., compactness. A general nonlinear optimization is formulated and solved for multiple NVH constraints. The package tray of a vehicle is studied to obtain the best structure layout that minimizes the vibration at the seat track at a given frequency range for a particular NVH load case.

The configuration design variation of a design component can be characterized by changes in the domain shape and orientation of the design component. Twu and Choi [7] developed a continuum configuration DSA method for static and eigenvalue using Batoz element [8], Wang and Choi [9] for transient responses, and Shim [10] for NVH. Due to the use of cubic shape function, the above methods are applicable only to special cases. In this paper, the Mindlin plate theory allows use of linear design velocity which can be applied to broad built-up structural problems [11]. A simplified vehicle model is studied to validate configuration design sensitivity results.

2. Sizing Design Sensitivity Analysis of NVH

The variational equation of structural-acoustic system can be obtained as [12]

$$b_u(z,\bar{z}) - \iint_{\Gamma^{as}} p\bar{z}^{*T} n\,d\Gamma + d(p,\bar{p}) - \omega^2 \iint_{\Gamma^{as}} \bar{p}^* z^T n\,d\Gamma = \ell_u(\bar{z}) \tag{1}$$

which must hold for all kinematically admissible virtual states $\{\bar{z}^*, \bar{p}^*\} \in Q$ where Q is a complex vector space

$$Q = \left\{(z,p) \in Z \otimes P \mid f_p = pn \text{ and } \nabla p^T n = \omega^2 \rho_o z^T n, x \in \Gamma^{as} \equiv \Omega^s \right\} \tag{2}$$

and

$$Z = \left\{ z \in \left[H^2(\Omega^s) \right]^3 \mid Gz = 0, x \in \Gamma^s \right\} \\ P = \left\{ p \in H^1(\Omega^a) \mid \nabla p^T n = 0, x \in \Gamma^{ar} \right\} \tag{3}$$

and H^1 and H^2 are complex Sobolev spaces of orders one and two, respectively [13]. In Eqs. (1)-(3), z and p are structural displacement and acoustic pressure, respectively, and '*' denotes complex conjugate. In Eq. (1), the sesquilinear forms $b_u(\bullet,\bullet)$ and $d(\bullet,\bullet)$, and semilinear form $\ell_u(\bullet)$ are defined, using complex L_2-inner product $(\bullet,\bullet)$ on a complex function space, as

$$b_u(z,\bar{z}) \equiv (D_u z,\bar{z}) = -\iint_{\Omega^s} \omega^2 m\bar{z}^{*T} z\,d\Omega + i\omega c_u(z,\bar{z}) + a_u(z,\bar{z}) \tag{4}$$

where

$$c_u(z,\bar{z}) \equiv \iint_{\Omega^s} \bar{z}^{*T} C_u z\,d\Omega \text{ and } a_u(z,\bar{z}) \equiv \iint_{\Omega^s} \bar{z}^{*T} A_u z\,d\Omega \tag{5}$$

$$d(\mathbf{p},\bar{\mathbf{p}}) \equiv (\mathbf{B}\mathbf{p},\bar{\mathbf{p}}) = \iiint_{\Omega^a} \left(-\frac{\omega^2}{\beta}\mathbf{p}\bar{\mathbf{p}}^* + \frac{1}{\rho_o}\nabla\mathbf{p}^T\nabla\bar{\mathbf{p}}^* \right) d\Omega \tag{6}$$

and

$$\ell_u(\bar{\mathbf{z}}) \equiv \iint_{\Omega^s} \mathbf{f}^T\bar{\mathbf{z}}^* \, d\Omega \tag{7}$$

In Eq. (5), C_u is the linear differential operator that corresponds to the damping of the structure and A_u is the fourth order symmetric partial differential operator for the structure. In Eq. (6), ω is forced frequency, β and ρ_o are bulk modulus and mass density of acoustic medium, respectively.

Equation (1) is solved using finite element analysis (FEA) codes such as MSC/NASTRAN, ABAQUS, and MOTRAN. The acoustic pressure $\mathbf{p}(x)$ and the structural displacement $\mathbf{z}(x)$ are approximated using shape functions as [10]

$$\begin{bmatrix} \left[-\omega^2\mathbf{M}_{ss} + i\omega\mathbf{C}_{ss} + \mathbf{K}_{ss} \right] & \left[\mathbf{K}_{sf} \right] \\ \left[-\omega^2\mathbf{M}_{fs} \right] & \left[-\omega^2\mathbf{M}_{ff} + \mathbf{K}_{ff} \right] \end{bmatrix} \begin{bmatrix} \mathbf{z} \\ \mathbf{p} \end{bmatrix} = \begin{bmatrix} \mathbf{f} \\ \mathbf{0} \end{bmatrix} \tag{8}$$

Direct and modal frequency FEA methods are used to solve the coupled Eq. (8). In the direct frequency FEA method, Eq. (8) is directly solved as a linear algebraic equation with complex variables. Even though the method is straightforward in application and gives very accurate solutions, it requires a large amount of computational time for repeated analyses of a large system due to several frequencies and different loading conditions. The modal frequency FEA method is an efficient method for a large size coupled system. In this method, a finite number of modes of the structure and acoustic medium are obtained and used to diagonalize the mass and stiffness submatrices, even though the off-diagonal submatrices in Eq. (8) cannot be diagonalized in this process since the modes are not orthogonal with respect to the off-diagonal submatrices.

Harmonic performance measures of the structural-acoustic system can be expressed in terms of complex phasors of the structural displacement and the acoustic pressure. For the adjoint variable method, first consider the pressure at a point $\hat{x}$ in the acoustic medium enclosed by the structure under harmonic excitation

$$\psi_p = \iiint_{\Omega^a} \hat{\delta}(x-\hat{x})\mathbf{p}d\Omega \tag{9}$$

The first variation of Eq. (9) is

$$\psi_p{}' = \iiint_{\Omega^a} \hat{\delta}(x-\hat{x})\mathbf{p}'d\Omega \tag{10}$$

Using the adjoint variable method, design sensitivity expression can be obtained as ,

$$\psi_p{}' = \ell'_{\delta u}(\boldsymbol{\lambda}) - b'_{\delta u}(\mathbf{z},\boldsymbol{\lambda}) = \iint_{\Omega^s} \mathbf{f}_u{}^T\boldsymbol{\lambda}^*\delta u + \iint_{\Omega^s} \omega^2 m_u\boldsymbol{\lambda}^{*T}\mathbf{z}\delta u \, d\Omega - i\omega c'_{\delta u}(\mathbf{z},\boldsymbol{\lambda}) - a'_{\delta u}(\mathbf{z},\boldsymbol{\lambda}) \tag{11}$$

where adjoint response is obtained from adjoint equation

$$b_u(\bar{\boldsymbol{\lambda}},\boldsymbol{\lambda}) - \iint_{\Gamma^{as}} \bar{\boldsymbol{\eta}}\boldsymbol{\lambda}^{*T}\mathbf{n}\,d\Gamma + d(\bar{\boldsymbol{\eta}},\boldsymbol{\eta}) - \omega^2\iint_{\Gamma^{as}} \boldsymbol{\eta}^*\bar{\boldsymbol{\lambda}}^T\mathbf{n}\,d\Gamma = \iint_{\Omega^a} \hat{\delta}(x-\hat{x})\bar{\boldsymbol{\eta}}d\Omega \tag{12}$$

which must hold for all kinematically admissible virtual states $\{\overline{\lambda}, \overline{\eta}\} \in Q$ where λ and η are adjoint structural displacement and acoustic pressure, respectively.

The structural displacement at a point $\hat{x}$ such as the vibration amplitude at a seat of the passenger vehicle can be written as

$$\psi_{z_i} = \iiint_{\Omega^s} \hat{\delta}(x - \hat{x}) z_i \, d\Omega, \quad i = 1, 2, 3 \tag{13}$$

The same design sensitivity expression of Eq. (11) can be used with a different adjoint equation for this performance measure defined as

$$b_u(\overline{\lambda}, \lambda) - \iint_{\Gamma^{su}} \overline{\eta} \lambda^{*^T} n \, d\Gamma + d(\overline{\eta}, \eta) - \omega^2 \iint_{\Gamma^{su}} \eta^* \overline{\lambda}^T n \, d\Gamma = \iint_{\Omega^s} \hat{\delta}(x - \hat{x}) \overline{\lambda}_i \, d\Omega \tag{14}$$

which must hold for all kinematically admissible virtual states $\{\overline{\lambda}, \overline{\eta}\} \in Q$. It is noted that, since sizing design variable u is defined only on the structural part, Eq. (11) requires only the structural response λ^* of the adjoint Eqs. (12) or (14) which are the same structural-acoustic system with different adjoint loads. The design sensitivity result of Eq. (11) is also valid for structural systems without the acoustic medium.

3. Topology Optimization

The topology optimization problem of NVH can be written as

Minimize ψ_p or ψ_{z_i}

Subject to $\int_\Omega c(x) \rho \, d\Omega \leq m_o, \quad \psi_p \leq \overline{\psi}_p, \quad \psi_{z_i} \leq \overline{\psi}_{z_i}, \quad 0 \leq c(x) \leq 1$

where ψ_p is acoustic pressure in Eq. (9), ψ_{z_i} is structural displacement in Eq. (13), $\overline{\psi}$ denotes NVH constraint value, $c(x)$ is compactness which is design variable, ρ is density, and m_o is mass limit. In the density approach, Young's modulus and density become functions of compactness. The relationships between them are

$$E_i = c^{n_1} E_o \text{ and } \rho_i = c^{n_2} \rho_o \tag{15}$$

where n_1 and n_2 are exponents, E_i and E_o are intermediate and original Young's modulus, and ρ_i and ρ_o are intermediate and original density, respectively. In this paper, $n_1 = 2$ and $n_2 = 1$ are used empirically. Bigger value of n_1 means more penalty for intermediate values of the compactness and forces the solution to be either 1 or 0. A continuum approach with the adjoint variable method as in Section 2 is used to compute design sensitivities. Sequential linear programming is used to solve the optimization problem of multiple constraints of large structure. LINDO is used in this paper.

4. Configuration Design Sensitivity Analysis

The first configuration variation of Eq. (9) is [10]

$$\psi_p{}' = \iiint_{\Omega^a} \hat{\delta}(x - \hat{x}) \dot{p} d\Omega \tag{16}$$

Using the adjoint variable method, design sensitivity expression can be obtained as,

$$\psi_p{}' = \ell_V'(\lambda) - b_V'(z,\lambda) + \iint_{\Gamma^{au}} p\lambda^{*T}\left[\tilde{V}_\theta n + \nabla^T V_\Omega n\right] d\Gamma - d_V'(p, \eta)$$
$$+ \omega^2 \iint_{\Gamma^{au}} \eta z^{*T}\left[\tilde{V}_\theta n + \nabla^T V_\Omega n\right] d\Gamma \tag{17}$$

where adjoint response is obtained from adjoint equation

$$b_{\Omega^a}(\overline{\lambda}, \lambda) - \iint_{\Gamma^{au}} \overline{\eta}\lambda^{*T} n\, d\Gamma + d_{\Omega^a}(\overline{\eta}, \eta) - \omega^2 \iint_{\Gamma^{au}} \eta^* \overline{\lambda}^T n\, d\Gamma = \iint_{\Omega^a} \hat{\delta}(x - \hat{x}) \overline{\eta} d\Omega \tag{18}$$

which must hold for all kinematically admissible virtual states $\{\overline{\lambda}, \overline{\eta}\} \in Q$. In Eq. (17), configuration design velocity, V is the sum of shape design velocity, V_Ω and orientation design velocity, V_θ where orientation design velocity matrix, $\tilde{V}_\theta$ for the plate design component is

$$\tilde{V}_\theta = \begin{bmatrix} S & 0 \\ 0 & S \end{bmatrix}, \quad S = \begin{bmatrix} 0 & 0 & -V_{3,1} \\ 0 & 0 & -V_{3,2} \\ V_{3,1} & V_{3,2} & 0 \end{bmatrix} \tag{19}$$

In similar sizing design, the same design sensitivity expression of Eq. (17) can be used for the structural displacement at a point $\hat{x}$ with a different adjoint equation as

$$b_{\Omega^a}(\overline{\lambda}, \lambda) - \iint_{\Gamma^{au}} \overline{\eta}\lambda^{*T} n\, d\Gamma + d_{\Omega^a}(\overline{\eta}, \eta) - \omega^2 \iint_{\Gamma^{au}} \eta^* \overline{\lambda}^T n\, d\Gamma = \iint_{\Omega^a} \hat{\delta}(x - \hat{x}) \overline{\lambda}_i\, d\Omega \tag{20}$$

which must hold for all kinematically admissible virtual states $\{\overline{\lambda}, \overline{\eta}\} \in Q$.

5. Numerical Examples

5.1 SIZING DSA AND OPTIMIZATION OF NVH

A concept vehicle model has 5,000 plate elements with 100 properties and 1,300 beam elements with 300 properties [1]. A what-if study is carried out for 26 plate elements with 2 properties and 34 beam elements with 8 properties by adding design sensitivities of elements. MSC/NASTRAN is used for NVH analysis and reanalysis using the direct frequency method. For accurate central finite difference (CFD), a convergence test was carried out and compared with the design sensitivity results obtained. Tables 1 and 2 show design sensitivity of acoustic response at frequencies 40 and 72 Hz of beam and plate, respectively. In Tables 2, prediction1 and prediction2 are obtained with and without the residual bending flexibility, respectively. In Tables 1 and 2, reanalysis is obtained using CFD. Agreement between reanalysis and prediction is used as accuracy checking. Tables 1-2 show that design sensitivities of NVH response are accurate.

Table 1. Design sensitivity of sound pressure level with respect to section properties of beam using direct frequency method

Response	Pertur	Change of Amplitude		
Frequency Amplitude	-bation	CFD	Prediction1	Agree-ment%
Rear Sound 40 Hz 1.0127E+4	1% 0.1% 0.01%	-3.360E+2 -3.350E+2 -3.350E+2	-3.354E+2	100.1%
Rear Sound 72 Hz 4.2342E+3	1% 0.1% 0.01%	6.720E+1 6.700E+1 6.500E+1	6.559E+1	100.9%

Table 2. Design sensitivity of sound pressure level with respect to thickness of shell using direct frequency method

Response	Pertur	Change of Amplitude				
Frequency Amplitude	-bation	CFD	Prediction1	Agree-ment%	Prediction2	Agree-ment%
Rear Sound 40 Hz 1.0127E+4	1% 0.1% 0.01%	2.450E+1 2.500E+1 0	2.420E+1	98.8% 96.8%	2.603E+1	106.2% 104.1%
Rear Sound 72 Hz 4.2342E+3	1% 0.1% 0.01%	-7.946E+2 -1.020E+3 -1.020E+3	-1.021E+3	100.2% 100.1%	-7.341E+2	70.6% 72.0%

A detailed vehicle model has 70,000 elements and a half-million degrees of freedom [1]. Eigenvectors of structure and air, respectively, are computed using MSC/NASTRAN super-elements and NVH responses are calculated using MOTRAN with the modal frequency FEA method. CONTESA is used to reduce the rear seat sound pressure level at 70 and 81 Hz of mostly warped plate elements. Vehicles are optimized using SYSOPT and the responses of original and optimized body models are compared in Figure 1. The numbers from 4 to 10 at the right of Figure 1 are the NVH index with the number 10 being the most quiet. It is shown in Figure 1 that the NVH index is significantly improved with weight reduction. The optimized vehicles are manufactured to validate the simulation-based design methodology.

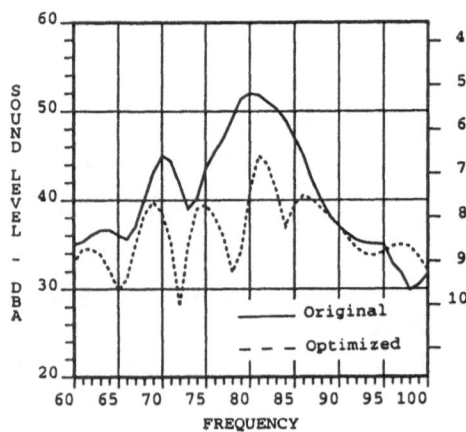

Figure 1. Comparison of NVH index between original and optimized vehicles

5.2 TOPOLOGY OPTIMIZATION OF NVH

The topology optimization of the package tray of a concept vehicle is carried out to reduce peaks at 63 and 70 Hz. The optimization problem is such that minimize the sum of rear seat vibrations of out-of-phase loading at 69, 70, and 71 Hz subjected to 9 vibration constraints. The mass limit is 0.3. Topology obtained at the seventh iteration is given in Figure 2.

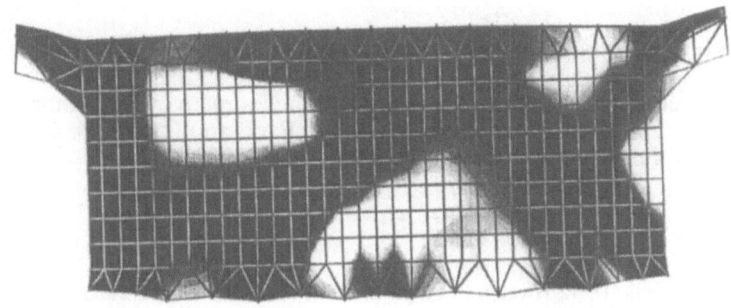

Figure 2. Optimal topology of a package tray

5.3 CONFIGURATION DSA

The configuration DSA is tested using a simplified vehicle [12]. Linear design velocity is computed using the isoparametric mapping method [11]. The configuration sensitivity results are given in Table 3. The shape and configuration design change of a simplified vehicle is shown in Figure 3.

Table 3. Design sensitivity of displacement of a simplified vehicle
with respect to shape and configuration change (z=1.4684 at node 3)

Design Change	perturbation	CFD	Prediction1	Agreement%
Shape	1%	5.72E-5	5.61E-5	98.1%
Configuration	1%	-2.42E-5	-2.49E-5	103.0%

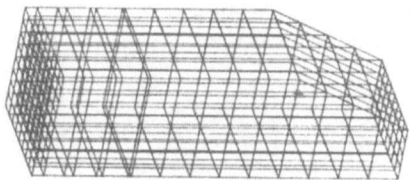 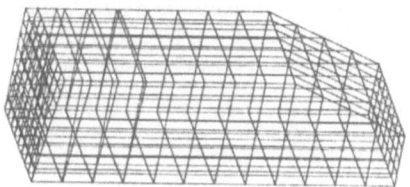

Figure 3. Shape and configuration design change of a simplified vehicle

6.   Conclusions

Continuum DSA and optimization method of NVH have been developed and applied to automobile structures for sizing, topology, and configuration design using Mindlin plate and Timoshenko beam theories. The topology optimization has been developed using the density approach, sequential linear programming, and the adjoint variable method. The sensitivity results have been tested using various vehicle models. Optimized vehicles are manufactured to validate the simulation-based design methodology.

Acknowledgments

This research was supported by Ford Motor Company. The authors would like to acknowledge Drs. M. Adelberg, M. Godse, Y. Park, and C. Soto of Ford Motor Company for their contribution of CONTESA and SYSOPT.

References

1.   Wang, S., Hung, H.H., Hwang, H.Y., Park, Y.H., Choi, K.K. and Kulkarni, H.T.: Design sensitivity analysis of NVH of vehicle body structure, *Proceeding of 5th AIAA Symposium on Multidisciplinary Analysis and Optimization* (1994), 1192-1201.
2.   MacNeal, R.H.: A simple quadrilateral shell element. *Computers and Structures* **8** (1978), 1175-183.
3.   Bendsoe, M.P. and Kikuchi, N.: Generating optimal topologies structural design using a homogenization method *Computer Methods in Applied Mechanics and Engineering* **71** (1988), 197-224.
4.   Olhoff, N., Bendsoe, M.P., and Rasmussen, J.: On CAD-integrated structural topology and design optimization *Computer Methods in Applied Mechanics and Engineering* **89** (1991), 257-279.
5.   Yang, R.J. and Chuang, C.H.: Optimal topology design using linear programming *Computers and Structures* **52** (1994), 265-275.
6.   Mlejnek, H.P.: Some aspects of the genesis of structures *Structural Optimization* **5** (1992), 64-69.
7.   Twu, S.L. and Choi, K.K.: Configuration design sensitivity analysis of built-up structure, Part 1, Theory, *Numerical Methods in Engineering* **35** (1992), 1127-1150.
8.   Batoz, J.L., Bathe, K.J., and Ho, L.W.: A study of three node triangular plate bending elements *International Journal for Numerical Methods in Engineering* **15** (1980), 1771-1812.
9.   Wang, S. and Choi, K.K.: Configuration design sensitivity analysis of transient response *Proceeding of 33rd AIAA Structures, Structural Dynamics and Materials Conference* (1992), 1460-1470.
10. Shim, I.B.: Design sensitivity analysis of dynamic frequency responses of structural-acoustic systems *Ph.D. Thesis*, The University of Iowa, 1993.
11. Choi, K.K. and Chang, K.H.: A study of design velocity field computation for shape optimal design *Finite Elements in Analysis and Design* **15** (1994), 317-341.
12. Choi, K.K., Shim, I.B., and Wang, S.: Design sensitivity analysis of structure-induced noise and vibration *submitted to ASME Journal of Vibration and Acoustics*, 1994.
13. Haug, E.J., Choi, K.K., and Komkov, V.: *Design Sensitivity Analysis of Structural Systems*, Academic Press, Orlando, 1986.

MULTIBODY MODEL-BASED MULTI-OBJECTIVE PARAMETER OPTIMIZATION OF AIRCRAFT LANDING GEARS

H. Wentscher, W. Kortüm
DLR-German Aerospace Research Establishment
Institute for Robotics and System Dynamics
D-82230 Wessling

1. Abstract

Landing gears of commercial passenger transport aircraft play an increasingly important role for safety, comfort and for weight considerations. As a result and in particular for future large aircraft with take-off weights up to 600 metric tons, the contribution of the landing gears to the structural weight would become prohibitive if standard designs would simply be extrapolated.

Thus research has been performed to investigate new concepts for advanced landing gears and optimize their associated design parameters to achieve minimum weight, maximum comfort under strict requirements with respect to safety and even increase lifetime by reducing the loads during landing impact and taxiing. A detailed and verified mechanical model is derived for an existing aircraft (Airbus A300); because this model will be used for simulating the standard design as well as the active optimized landing gear it needs rather careful and detailed modelling, e.g. it requires fuselage and wing flexibility to be taken into account.

The simulation was performed within SIMPACK, DLR's prime multibody computer code [1]. SIMPACK has the major features to allow for arbitrary complex elastic mechanical systems including active components. Using the complete SIMPACK model the concept for a semi-active actuator is considered and realistic model assumptions for its dynamic behavior are made. Even though SIMPACK allows for the portation of the full nonlinear multibody simulation (MBS) model into any of the common design packages like ANDECS, $MATRIX_X$ or MATLAB it is very useful to reduce the model complexity to the necessary effects because of the further, computationally very involved design procedures. Order reduction is performed through physical and engineering reasoning and formal mathematical procedures, both supported by SIMPACK modules.

Results will be presented on using the multi-objective parameter optimization software MOPS (Multi-Objective Programming System) implemented in ANDECS [2]. The design case study concentrates on taxiing of a flexible aircraft where reduction of the so-called "beaming effect" (e.g. dynamic coupling of runway excitation with elastic fuselage eigenmodes) is the major design goal. Comparisons are made showing the achieved improvements.

D. Bestle and W. Schielen (eds.), IUTAM Symposium on Optimization of Mechanical Systems, 325–332.
© *1996 Kluwer Academic Publishers.*

2. Introduction

Aircraft landing gears are designed to meet the various requirements of the ground operating conditions. These can be divided into two main groups:

- landing impact: even severe landings with sink rates up to 3.05 m/s have to be absorbed without damage to the airframe,

- ground roll: unevenness of taxiways and runways has to be absorbed to improve ride quality for pilots and passengers.

The active control requirements for each of this case conflict with the other. In a practical active landing gear, the controller is considered to be switched between touchdown and taxi mode.

The trend of aircraft design of the last decade showed the advent of rather flexible airframes, either due to a stretch of the passenger compartment or to highly optimized fuselage structures. Along with this development came the complaints of pilots and passengers of heavy vibrations during ground roll. It was therefore decided to concentrate the analysis on this interesting aspect of the interaction of runway excitation and elastic airframe response. In recent landing gear research like [3], two general types of active control were considered:

- Fully active control: the oleo force is controlled directly by introducing additional hydraulic or pneumatic flow, using the oleo effectively as an actuator. Depending on the amounts and the pressure of the extra flow, these systems may add extensively to system complexity, weight and stowing volume.

- Active damping (or semi-active) control: here only the oleo damping characteristics are controlled. This could be achieved using valves or variable diameter orifices, thus limiting extra weight and power consumption. Additionally, these systems are not able to introduce energy into the system, rendering this system inherently stable.

Research has shown that semi-active control achieves performance improvements comparable to those of the fully active systems. Due to the advantages of the semi-active control concept in system complexity and extra weight, it is likely to be chosen as the first attempt to be integrated into a production aircraft and therefore was selected for this analysis.

3. System Model Synthesis

3.1 AIRCRAFT MODEL

As aircraft to be modeled the AIRBUS A300 was chosen. It represents a configuration widely used for commercial airliners: two wheel front gear, two four wheel bogie main gears, wide body fuselage, two engines, maximum take off weight 140 metric tons.

The main model elements include the airframe structure (one elastic body), the landing gears (several rigid bodies), the force laws (representing the forces acting on the bodies) and the runway excitation.

3.2 AIRCRAFT STRUCTURE

To include elastic deformation into aircraft motion, the main airframe members (fuselage and the wings) are described as linear EULER-BERNOULLI beams. The assumptions of this theory (length must by far exceed width and height, loads act vertical to datum line) are met as long as changes in the aircraft attitude remain small.

Each member consists of several beam parts with piecewise constant dimensions and mass properties. Engines and other external masses were connected to the appropriate nodes including the full inertial tensor.

The human body is most susceptible in the frequency range between 4 and 8 Hertz [4], so the performance analysis of the semi-active gear was limited to the range between 0.1 and 10 Hertz. To account for aliasing effects, eigenmodes of the airframe up to 50 Hertz were considered. The deformations of the third airframe eigenmode (with its dominating first fuselage bending eigenmode) are shown in Figure 1.

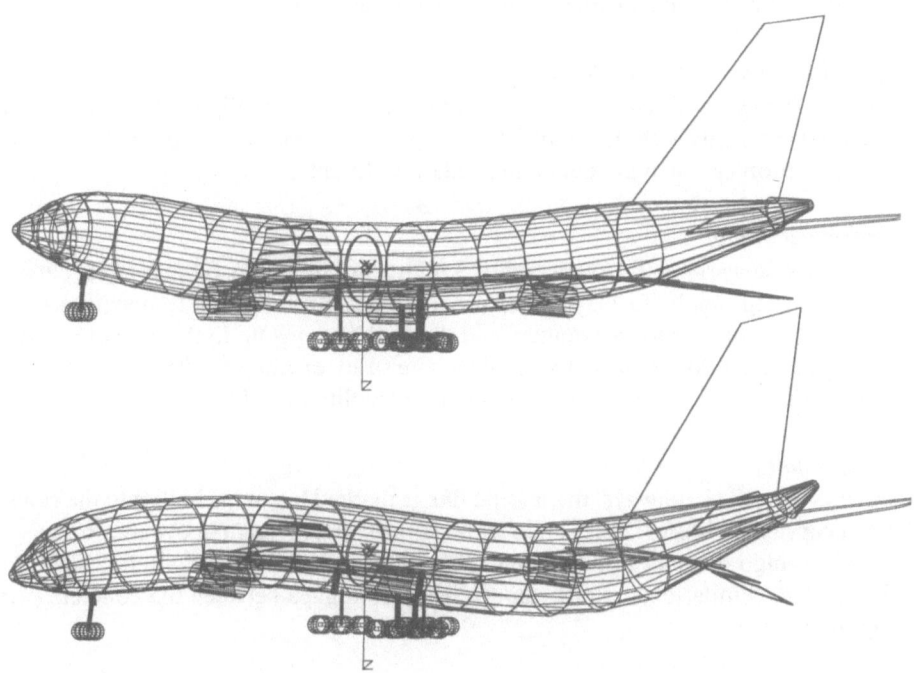

Figure 1: First and Second Fuselage Bending Eigenmode Deformations

3.3 FORCE LAWS

3.3.1 Oleos
Each of the three landing gears of the A300 were reduced to a system consisting of a linear spring/damper system representing the tire, the unsprung mass with inertial tensor, and a nonlinear force element. For the passive gears, this force element incorporates the airspring characteristics found in the A300 oleo-pneumatic shock absorbers (oleos) com-

bined with the piecewise constant damping. This oleo model was tested and verified against drop test results made available by DAIMLER BENZ AEROSPACE AIRBUS.

For the active oleo, the damping part was modified such that the controller was able to govern the force by altering the damping. The gas spring remained unchanged.

3.3.2 Aerodynamic forces

The main aerodynamic forces, like lift, drag and pitch damping, were included as external forces acting on the center of gravity. The influence on the simulation results were found to be negligible, since the governing states (angle of attack, forward speed and pitch rate) remained almost constant during simulation.

3.4 RUNWAY EXCITATION

During normal ground operations a multitude of different excitations are encountered. To represent the majority, three distinct scenarios were selected.

3.4.1 Quasi-Stochastic Runway Input

In order to use measured data for existing runways, [5], for nonlinear analysis, a deterministic runway profile was developed that displays a power spectrum similar to measured runway profiles; the maximum amplitude was 15 cm.

3.4.2 Harmonic Input

The worst case scenario is the excitation of a high amplitude, low frequency eigenmode. Especially for long slender fuselages this resonance has been reported frequently and was found to pose discomfort for passengers and the cockpit crew up to the extend to endanger the aircraft. Therefore this case was taken care of by exciting the first bending eigenmode of the fuselage with a sinusoidal input of an amplitude of 15 cm.

3.4.3 Step Input

On taxiways as well as runways, the middle line is marked by lamps built into the runway that are encapsulated in steel cases and extend approximately 5 cm above the surface. Hit by the gear at high velocities, these center line lights induce high amplitude oscillations into the aircraft. Similar excitations can originate from gaps between the concrete plates forming the runway.

3.5 CONTROL SYSTEM LAYOUT

To determine the number and locations of actuators needed in the three landing gears, two different designs were compared:

1. a semi-active configuration incorporating three active gears with separate controller and damping actuator sets,
2. a semi-active configuration with two standard main landing gears and a damping controlled front gear.

Evaluation showed that the full semi-active landing gear provided no significant improvements versus the semi-active front gear. The main landing gears are located close to the center of gravity of the aircraft and are mounted in the structural highly reinforced area of the wing root. An elastic deformation would not contribute any displacement to the movement of the center of gravity and the main gear, rendering the elastic modes as not being controllable with the main gear.

The primary contributor to normal accelerations along the fuselage is the rigid body heave motion with an associated frequency below 1 Hertz, depending on weight and the gas chamber pressure of the oleos. Other components for vertical acceleration originate solely in airframe eigenmodes which are dominated by the fuselage bending modes. In the frequency range where the human body is most susceptible, i.e. between 4 and 8 Hertz [4], these are the third and sixth airframe eigenmode with their associated frequencies of 3.6 Hertz and 8.2 Hertz, respectively. Passenger discomfort and crew disorientation are primarily due to vertical acceleration, so this data was used to define landing gear performance.

The feedback principle is based on the "sky-hook" damping idea of Karnopp [8], originally developed for semi-active car suspensions. The absolute velocity of the upper part of the oleo is used as feedback signal, resulting in a damper force component that a passive damper, coupling the front fuselage to the ground, would supply.

4. Optimal Design

4.1 MULTI-OBJECTIVE PARAMETER OPTIMIZATION

The design strategy for multi-objective parameter optimization is sketched in Figure 2:

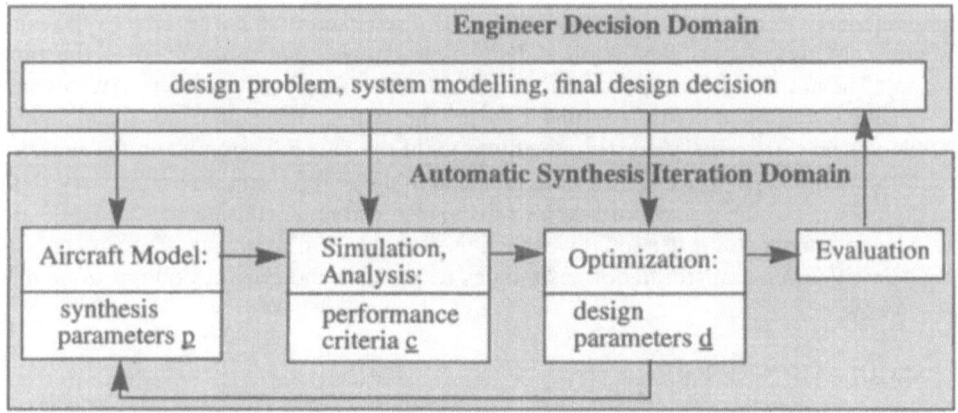

Figure 2: Automatic Synthesis Iteration Loop

the first block in the *automatic synthesis domain* contains the *plant* and the *controller model*. They have to be defined by the *engineer* and may contain linear and nonlinear model descriptions. Also included in this model are *free synthesis parameters,* p, which are subsequently "tuned" to yield the optimal design. For the A300 model, p included the controller parameters, namely the gains.

The *performance criteria* c_i, with respect to which the parameters should be optimized (minimized) are summarized in the criteria vector:

$$c=[c_1,c_2,...,c_i,...]^T. \qquad \text{(EQ 1)}$$

This criteria vector includes sixteen vertical accelerations, measured in the cockpit, the passenger compartment and on the wings and is calculated by time domain integration and *analysis* (second block of Figure 2). In the last block, the actual *optimization* procedure is carried out. The desired "design direction" is defined by a *design parameter* vector $\underline{d}$:

$$d=[d_1,d_2,...,d_i,...]^T, \qquad \text{(EQ 2)}$$

where d_i is a chosen weighting factor corresponding to criterion c_i; d_i are upper limits for the desired stepwise descend of the criteria

$$c_i(p) \le d_i. \qquad \text{(EQ 3)}$$

As a design comparator for detecting "better" designs the maximum function

$$\alpha = \max\{c_i/d_i\} \qquad \text{(EQ 4)}$$

is chosen. The strategy is to find a minimum for α in varying the synthesis parameters p:

$$\alpha^* = \min_p(\alpha(\underline{p})). \qquad \text{(EQ 5)}$$

After a sufficient number of loops in the automatic synthesis iteration domain, an improved design is reached which is Pareto optimal, i.e. no criterion can be further improved without deteriorating others [8].

4.2 EVALUATION SETUP

Model evaluation took place adopting a model-family approach, using three models simultaneously to calculate the responses for all three excitation cases. Due to the non-linearities of the model, simulation and dynamic analysis were restricted to the time domain. The multi-objective performance criteria vector was built up using sixteen vertical accelerations on the fuselage and two on the wings. Peak and root-mean-square (RMS) values were reduced systematically to yield the best compromise in the sense of pareto-optimality. The peak values were included in the criteria vector because very short but high peaks (which possess low RMS values) also had to be eliminated.

The optimized controller setup was then used to perform frequency domain analysis using Fast Fourier Transformation techniques to produce the graphics output to be discussed now.

4.3 SIMULATION RESULTS

In Figure 3, the frequency response of the active front gear is plotted for the frequency range from 1 to 10 Hertz. It was calculated using the cross-correlation function found for various inputs between wheel and cockpit acceleration, [6]. It can be seen that additionally to a broad band amelioration a significant performance gain is achieved in the frequency range of 3 to 4 Hertz, where the first elastic airframe eigenmode is located. Figures 4 to 6 show the power spectral density (PSD) of the cockpit vertical acceleration plotted against frequency, using a semi-logarithmic format.

As already seen from the frequency response, a general performance gain can be achieved for the whole frequency band under consideration with the biggest amelioration

Power Spectral Density

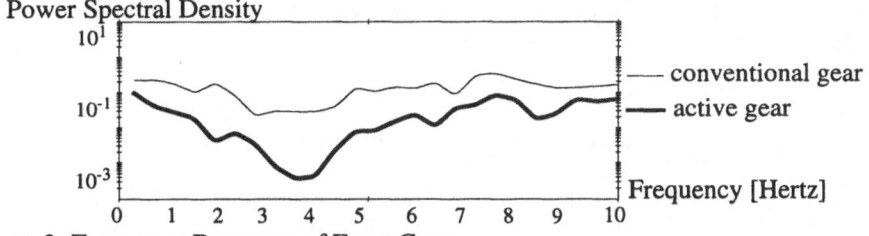

Figure 3: Frequency Response of Front Gear

between 3 and 4 Hertz. This particular behavior, reduction of the elastic body response, can best be seen for the impulse excitation of the center line lights (Figure 4). Taxiing

Power Spectral Density of Cockpit Acceleration [m^2/s^3]

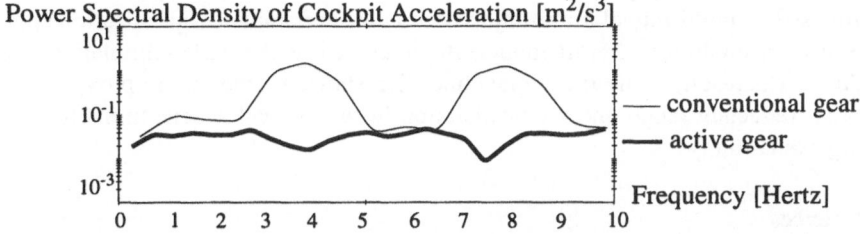

Figure 4: Power Spectral Density of Cockpit Acceleration for Center Line Light Case

with the conventional gear at the critical speed would result in cockpit movement dominated by elastic airframe resonance. The active gear eliminates this movement to a great extend without penalizing at other frequencies.

The system response for the harmonic runway excitation (Figure 5) shows the same

Power Spectral Density of Cockpit Acceleration [m^2/s^3]

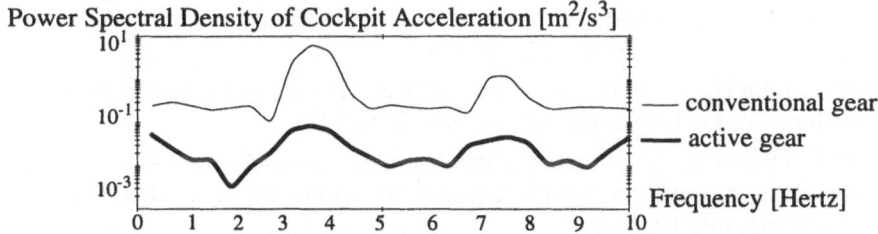

Figure 5: Power Spectral Density of Cockpit Acceleration for Runway 5

pattern. The conventional gear induces airframe vibrations resulting in high acceleration levels for the eigenfrequencies, whereas the optimized semi-active gear augments the overall vibration isolation performance.

Even for the quasi-stochastic runway excitation, representing a broad band of frequencies acting on the front gear, vertical accelerations can be reduced significantly over the whole frequency band (Figure 6).

The results for these three excitations which represent a broad band of possible runway perturbations show robust controller behavior derived with the multi-model multi-objective control parameter optimization approach.

Power Spectral Density of Cockpit Acceleration [m²/s³]

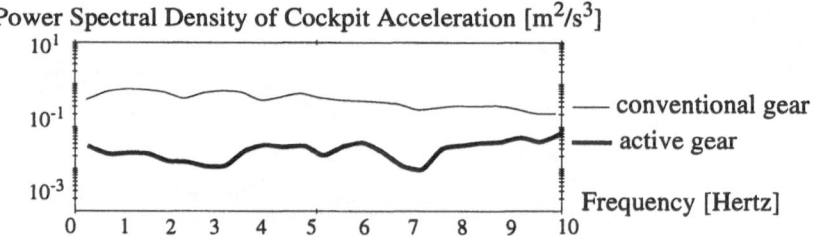

Figure 6: Power Spectral Density of Cockpit Acceleration for Runway 4

5. Summary

For a semi-active landing gear of a commercial aircraft a controller optimization process was proposed. A multi-model multi-objective controller parameter optimization approach was used on a nonlinear aircraft model, implemented in the MBS-Simulation code of SIMPACK. The result, a pareto-optimal controller, showed important improvements over the whole frequency range under consideration in the vertical accelerations for several, differing excitation types.

6. References

[1] W. Kortüm and W. Rulka and A. Eichberger: *Recent Enhancements of SIMPACK and Vehicle Applications*. European Mechanics Colloquium, EUROMECH 320, Prague 1994

[2] G. Grübel, H.-D. Joos, M. Otter: *The ANDECS Design Environment for Control Engineering*. In IFAC World Congress, Sidney, 18-23 July 1993

[3] T. Catt, D. Cowling, A. Shepard: Active Landing Gear Control for Improved Ride Quality During Ground Roll. Proc. of AGARD meeting "Smart Structures for Aircraft and Spacecraft, CP 531, 1993

[4] W. Kortüm and P. Lugner: *Systemdynamik und Regelung von Fahrzeugen*. Springer-Verlag Berlin, Heidelberg, New York 1994

[5] W. E. Thompson: *Measurements and Power Spectra of Runway Roughness at Airports in Countries of the North Atlantic Treaty Organization*. NACA Technical Note 4303, July 1958

[6] H. Schlitt: *Stochastische Vorgänge in linearen und nichtlinearen Regelkreisen*. Vieweg, Braunschweig 1968

[7] D. Karnopp, M. J. Crosby, R. A. Harwood: *Vibration Control Using Semi-Active Force Generators*. ASME Journal of Engineering for Industry, Vol 96, No.2, 1974

[8] H.-D. Joos: *Informationstechnische Behandlung des mehrzieligen optimierungsgestützten regelungstechnischen Entwurfs*, Ph.D. thesis, Universität Stuttgart, Prof. R. Rühle, 1992

MULTICRITERIA OPTIMIZATION AS A TOOL IN THE VEHICLE'S DESIGN PROCESS

J. WIMMER AND J. RAUH
Daimler-Benz AG, Research and Technology
E222, 70322 Stuttgart, Germany
email: wimmer@str.daimler-benz.com

1. Modelling and Simulation in Vehicle Dynamics

Modelling and simulation in the field of vehicle dynamics is a complex inter-disciplinary topic. For the optimization of the vehicle's dynamic behaviour, not only model accuracy but also an efficient implementation of suitable approaches is of great importance.

Practical work on this topic has resulted in an integrated treatment of the mathematical modelling combined with adapted integration schemes. By using such an approach, small and efficient simulators can be applied even for systems which are known to be difficult to handle, e.g. mechanical systems with clearance or dry friction. Furthermore, best results are obtained with very simple one step integrators having constant step size. This approach also seems to be very well suited for applications in optimization. The predictable computing time makes load balancing in multi-processor environments easy to implement, while side effects of system parameter variations due to step size and order control of the integrator are avoided.

1.1. DIFFERENT APPROACHES FOR THE MODELLING OF SUSPENSION SYSTEMS

As an example for different modelling approaches, the vehicle's suspension characteristics will be considered. Generally, suspension models can be assigned to three different categories:

- kinematic models
- combined kinematic and elastokinematic models
- combined kinematic and elastokinematic models including inertia properties

D. Bestle and W. Schielen (eds.), IUTAM Symposium on Optimization of Mechanical Systems, 333–340.
© 1996 *Kluwer Academic Publishers.*

Kinematic models only roughly describe the real suspension behaviour. In the following discussion, suspension models of the second and third category are presented. The models are part of the CASCaDE simulation environment, which will be described in detail in subsection 2.2.

1.1.1. *Partially Nonlinear Model*

In this model the geometric locations of the bushings, which are nonlinearly dependent on the suspension deflection, result from kinematic analysis. The material properties of the bushings are linearized in bushing-fixed coordinate systems. With this partial linearization, the elastic changes in the wheel carrier locations are related to the known forces and moments exerted on the tire by

$$
\begin{bmatrix} \mathbf{F}^{le} \\ \mathbf{L}^{le} \\ \mathbf{F}^{ri} \\ \mathbf{L}^{ri} \end{bmatrix} = C_R^{le/ri} \begin{bmatrix} \Delta \mathbf{r}_{el}^{le} \\ \Delta \mathbf{s}_{el}^{le} \\ \Delta \mathbf{r}_{el}^{ri} \\ \Delta \mathbf{s}_{el}^{ri} \end{bmatrix} ,
\tag{1}
$$

where the coupling of the left and right suspension system by a suspension subframe is also considered. The matrix $C_R^{le/ri}$ is obtained by systematic combination of single stiffness matrices, which describe the material properties of the bushings. Based upon the topology of the suspension system the individual matrices are combined using a spatial form of an electric circuit analogy.

1.1.2. *Nonlinear Polynomial Model*

Another way of modelling the elastokinematic properties of the suspension system is to concentrate on the most important effects. One possible approach is to calculate the changes in the wheel camber and toe as well as the changes in the longitudinal deflection of the wheel center by

$$
\Delta s_{el,z} = \alpha F_x + \beta \hat{F}_x + \gamma F_y + \epsilon \hat{F}_y + \zeta L_y + \eta \hat{L}_y + \theta L_y^2
\tag{2}
$$

$$
\Delta s_{el,x} = \mu F_x + \nu L_x + \xi \hat{L}_x + \rho L_y + \sigma L_y^2
\tag{3}
$$

$$
\Delta r_{el,x} = \phi F_x + \chi L_y + \psi L_y^2
\tag{4}
$$

for both wheel carriers, where $\hat{F}$ and $\hat{L}$ are the forces and moments applied to the opposite tire. The wheel carrier location is obtained from Eqs. (2)-(4) and from the results of the kinematic analysis. The coefficients of the polynomials can be determined by FEM-calculations or from measured data.

1.1.3. *Dynamic Suspension Model*

Investigations of detailed phenomena, which for instance are due to dynamic interactions between the wheel, the suspension system, and the steering system, respectively, require a more detailed suspension model. In such cases the inertia properties of some of the suspension components have to be taken into account as well. Allowing for inertia properties of each component, as is done in general purpose programs, would be highly inefficient. Elastokinematic models of the third category are also available in the CASCaDE environment.

2. Optimization

In the design process of a vehicle, several conflicting criteria have to be considered in order to achieve a high safety standard and ride comfort. In this context, multicriteria optimization is an efficient tool.

2.1. INTRODUCTION TO MULTICRITERIA OPTIMIZATION

The general multicriteria optimization problem can be stated as (Bestle, 1994):

$$\text{opt } \mathbf{f}(\mathbf{p}),$$
$$\mathbf{p} \in \mathcal{P}$$
$$\text{where } \mathcal{P} := \{\mathbf{p} \in \mathbf{R}^h \,|\, \mathbf{g}(\mathbf{p}) = \mathbf{0},\, \mathbf{h}(\mathbf{p}) \leq \mathbf{0},\, \mathbf{p}^l \leq \mathbf{p} \leq \mathbf{p}^u,$$
$$\mathbf{g}: \mathbf{R}^h \to \mathbf{R}^k, \mathbf{h}: \mathbf{R}^h \to \mathbf{R}^m\}. \tag{5}$$

Hence, a solution $\mathbf{p}^\star$ has to be found, which satisfies both the equality and inequality constraints, and optimizes a vector criterion function $\mathbf{f}(\mathbf{p})$, $\mathbf{f}: \mathbf{R}^h \to \mathbf{R}^n$. $\mathbf{f}(\mathbf{p})$ defines a partial order, which allows to separate nonoptimal points. The remaining set

$$\mathcal{P}^P := \{\mathbf{p}^P \in \mathcal{P} \,|\, \nexists \mathbf{p} \in \mathcal{P} : \mathbf{f}(\mathbf{p}) < \mathbf{f}(\mathbf{p}^P)\} \tag{6}$$

contains the so-called Edgeworth-Pareto optimal solutions. At EP-optimal points no further improvement in any of the criterion values can be achieved without worsening at least one other criterion value. In most cases, there exist an infinite number of such solutions, and it is in the responsibility of the user to make a final selection among them.

EP-optimal points can be found by scalarization of the vector optimization problem. Frequently used scalarization approaches are the weighting criteria method, the global criterion method, the weighting min-max-method, and the hierarchical optimization method, respectively. The remaining single criterion optimization problem can be solved by standard optimization algorithms.

2.2. THE OPTIMIZATION ENVIRONMENT OPENMIND

The multicriteria optimization methods, which have been mentioned in section 2.1, are implemented in OpenMind (**Op**timization **En**vironment for **M**ulticriterial **I**mprovement in **N**onlinear **D**ynamics). It has been developed for optimizing the vehicle's overall behaviour. Up to now, each vehicle component has been optimized separately, followed by a dynamic analysis of the entire system. Hence, dynamic interactions between the subsystems of the assembled vehicle could not be considered in the optimization process. In the new environment, variables which directly describe the dynamic behaviour can be applied for the formulation of design criteria.

At present, two simulation packages are implemented in OpenMind. CASCaDE (**C**omputer **A**ided **S**imulation of **Car**, **D**river, and **E**nvironment) is an efficient simulation program, where the vehicle and its driver as well as environmental effects are simulated (Rauh, 1990). The program serves as a platform for the development, the integration, and the test of new methods.

With respect to the suspension characteristics, ECCO (**E**lastokinematics **C**omputation and **O**ptimization) is used as a preprocessing tool in order to gain computational efficiency. Above that, ECCO is applied for the pre-design of the suspension systems. Usually, approximately 1000 elastokinematic data sets have to be calculated during the optimization process of a suspension system. Since the suspension models have been implemented efficiently, the elastokinematic optimization can be performed within a few minutes on a personal computer.

Figure 1 shows the architecture of OpenMind. The user-interface features

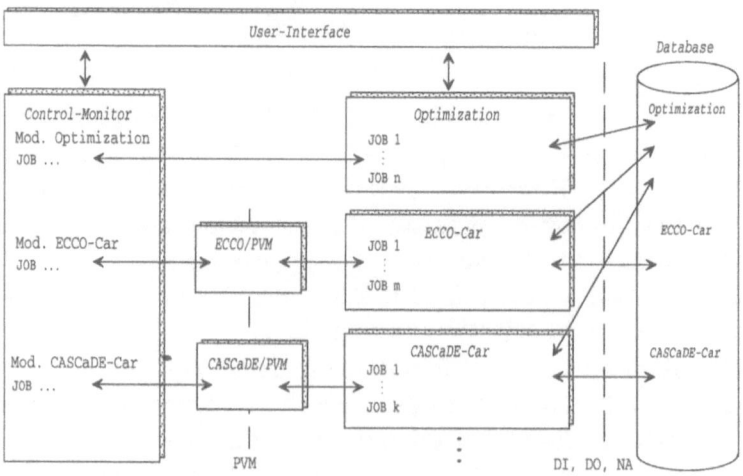

Figure 1.   Architecture of OpenMind

an iterative and interactive optimization process. Several tools for online and offline analysis support the user in making his final selection from among the set of EP-optimal solutions. The job sequence is laid down by a control monitor, which also handles the communication between the independent modules. Further, the data exchange between the program modules and the database is handled by a set of interfacing routines in order to assure data encapsulation. A direct data exchange between the modules is not permitted.

Due to the chosen architecture, the implementation of further modules is straightforward. In addition, parallelization of the optimization process can be easily achieved. For that purpose, a second interface between the control monitor and the program modules has been developed.

3. Application Example

The requirements with respect to the vehicle's dynamic behaviour are partially contradictory. One of the fundamental problems, which renders the chassis and suspension tuning difficult, is used in the optimization example.

3.1. DESIGN CONFLICT BETWEEN DRIVING SAFETY AND RIDE COMFORT

The variation of the design parameters of the suspension system always leads to a trade-off between safety and ride comfort, as shown in figure 2. In

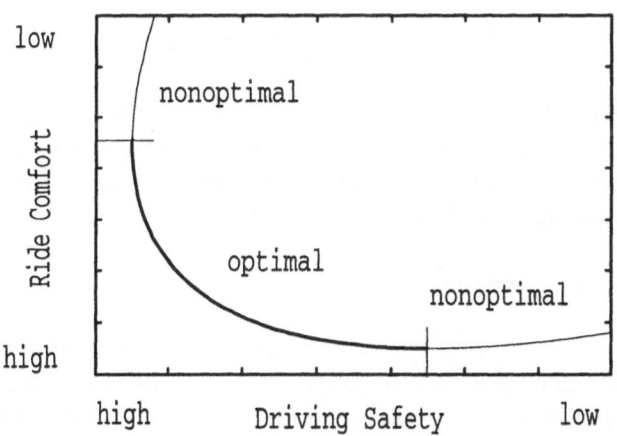

Figure 2.   Design conflict between driving safety and ride comfort

general, the ride comfort is represented by the time integral of the vehicle's vertical acceleration:

$$f_1^c = \frac{1}{T} \int_T \ddot{z}^2(t)\, dt\,. \tag{7}$$

However, Eq. (7) is neither a unique nor a complete criterion. For a refined description, it is useful to concentrate on the direction of influence on the driver, which can be achieved by considering the vehicle's pitch and roll accelerations separately. It is known from experiments that drivers are about three times more sensitive to longitudinal disturbances than to disturbances in lateral direction. In addition, according to the VDI 2057 standard, filters can be defined in which the acceleration is weighted with respect to the appearing frequency spectrum. Thus, a more precise criterion for the vehicle's ride comfort is found to be

$$f_2^c = \frac{1}{T} \int_T (\ddot{z}^F(t))^2\, dt \; ; \; \ddot{z}^F(t) = f_F(\ddot{z}(t), ...)\,, \tag{8}$$

where f_F represents an adequate filter model.

In case of a closed-loop optimization, where a driver model is integrated in the optimization process, a criterion for the driving safety can be defined by the deviation from a given nominal track s_n during critical driving manoeuvres:

$$f_1^s = \frac{1}{T} \int_T |s - s_n|^p\, dt \; ; \; p \geq 2\,. \tag{9}$$

Since experienced drivers might be able to maintain road position even with mistuned vehicles, an additional criterion is formulated, where the angular velocity of the steering wheel input is implied:

$$f_2^s = \frac{1}{T} \int_T |\omega_{st}|^p\, dt \; ; \; p \geq 2\,. \tag{10}$$

Eq. (10) is a measure for the driver's activity, and, as a consequence, for the requirements he has to meet. Due to the importance of the driving safety, it is appropriate to increase the penalty in case of extreme deviations from the nominal values by choosing $p > 2$.

3.2. MODEL DESCRIPTION

In the example, a multibody system vehicle model with 20 degrees of freedom is used. The elastokinematic properties of the suspension systems are described by a nonlinear polynomial model of second order, where the corresponding polynomial coefficients have been determined from measured data. The driver is represented by an algorithmic model, which not only provides elementary operations for maintaining road position, but also allows for a realistic imitation of the driver's behaviour even during critical manoeuvres (Reichelt, 1990). The nominal track for the driver model is provided by an environment model, which also generates road irregularities by stationary stochastic processes (Ammon, 1989).

3.3. CLOSED-LOOP OPTIMIZATION

The point of main effort in optimization is to make the vehicle's overall behaviour predictable for the driver. Therefore, a closed-loop optimization is an important aspect in vehicle design. In the following example, a lane change manoeuvre on a country road with an average road irregularity is performed. The stabilizer stiffness as well as the shock-absorber stiffness serve as design parameters in the optimization process. Further, five criteria are chosen according to Eqs. (8)-(10) in order to consider the vertical acceleration, the pitch and roll accelerations, the deviation from the nominal track, and the steering wheel velocity. Figures 3 and 4 show the optimization result obtained by using the weighting criteria method. In contrast to the simulation results for the intentionally mistuned initial design, where the low stabilizer stiffness hardly allows the driver to maintain road position, the handling characteristics of the optimized vehicle are substantially improved.

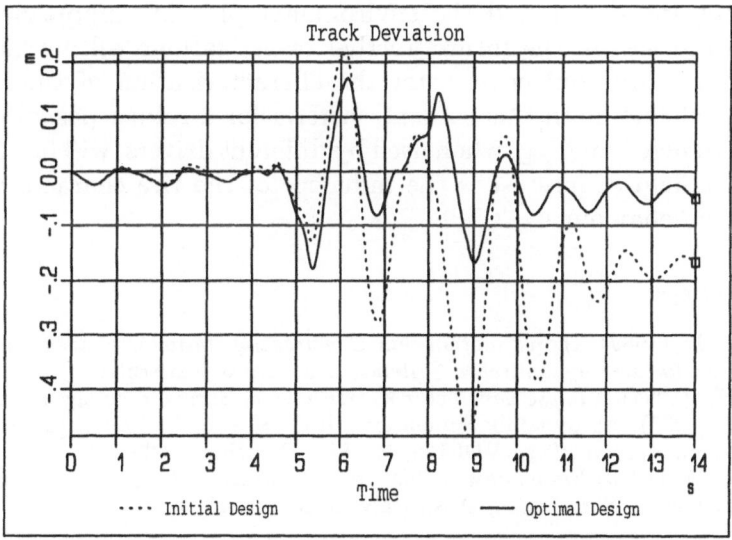

Figure 3. Track deviation versus time

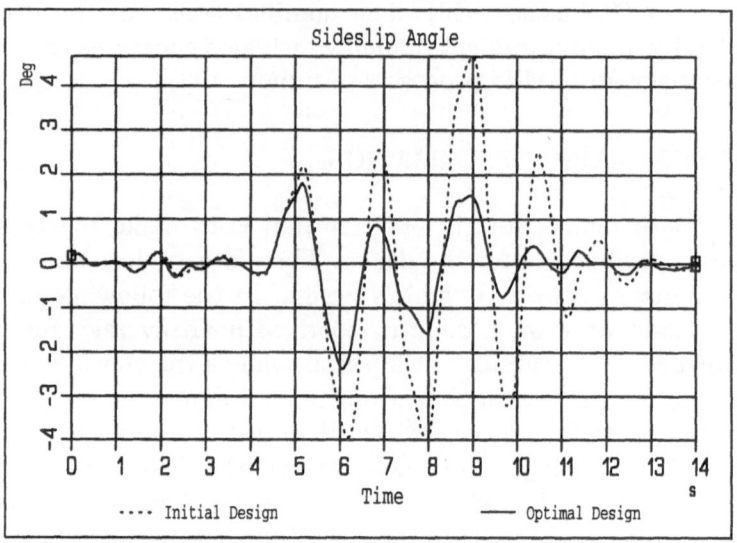

Figure 4.   Sideslip angle versus time

4. Conclusions

With the new environment OpenMind a further step is taken with respect
to a systematical improvement of the vehicle's dynamic behaviour. Apparently, the driver and the environment play an important role in the
design process. In the future, further investigations will be made in order
to define additional criteria for the characterization of the vehicle's dynamic behaviour. In this context, criteria for a robust design with respect
to the vehicle handling, when used by different drivers, will be also included.
Another field of interest is the influence of the tire characteristics on the
vehicle's behaviour.

References

Ammon, D. (1989) *Approximation und Generierung stationärer stochastischer Prozesse
    mittels linearer dynamischer Systeme*, Diss., Univ. Karlsruhe.
Bestle, D. (1994) *Analyse von Mehrkörpersystemen*, Springer Verlag, Berlin Heidelberg.
Rauh, J. (1990) Fahrdynamik-Simulation mit CASCaDE, in *VDI-Tagungsbericht Berech-
    nung im Automobilbau*, VDI-Ber. Nr. 816, Düsseldorf, 599–608.
Reichelt, W. (1990) *Ein adaptives Fahrermodell zur Bewertung der Fahrdynamik von Pkw
    in kritischen Situationen*, Diss., Univ. Braunschweig.

OPTIMIZATION OF HIGH PERFORMANCE RESILIENT MACHINERY FOUNDATIONS ON SHIPS

D. WITTEKIND
Thyssen Nordseewerke GmbH
Am Zungenkai
26725 Emden
Germany

L. GAUL
Institute A of Mechanics
University of Stuttgart
70550 Stuttgart
Germany

1. Introduction

Resilient mounting systems are a common feature on naval ships, research vessels and even merchant ships where the requirement for high comfort goes along with the contradictory requirement for light weight structures to maximize payload.

In extreme cases a double resilient mounting system is employed. It consists of two elastic spring levels with an intermediate mass [1], [2], [3].

This paper reports on research work to minimize noise transmission of a small diesel generator to the hull of a submarine to be able to operate the ship in very quiet conditions. Emphasis is put on the design of the intermediate mass.

2. The Ideal System

The simplest model of a double resilient mounting is the two-degree-of-freedom-system, *figure 1*. Various transfer functions of the system are shown.

Theoretically, the level at the foundation can be predicted if the source level and the transfer mobility (with the attachment to the foundation blocked) are known. If the behavior of the total system is considered and compared to reality it must be taken into account that accelerations can be measured quite easily but forces cannot.

What we are interested in is the true reduction of the force on the foundation. This is shown in *figure 1c*. However, neither the internal force nor the force on the foundation can be measured very easily in practical systems. Besides, this model covers two degrees of freedom only but the full threedimensional system would have 12 rigid body dof. To ensure optimum performance, all natural frequencies of the rigid body modes have to be lower than the second natural frequency in the vertical direction. The latter is determined by the weight of the masses and the spring stiffness and cannot be

341

D. Bestle and W. Schielen (eds.), IUTAM Symposium on Optimization of Mechanical Systems, 341–348.
© 1996 *Kluwer Academic Publishers.*

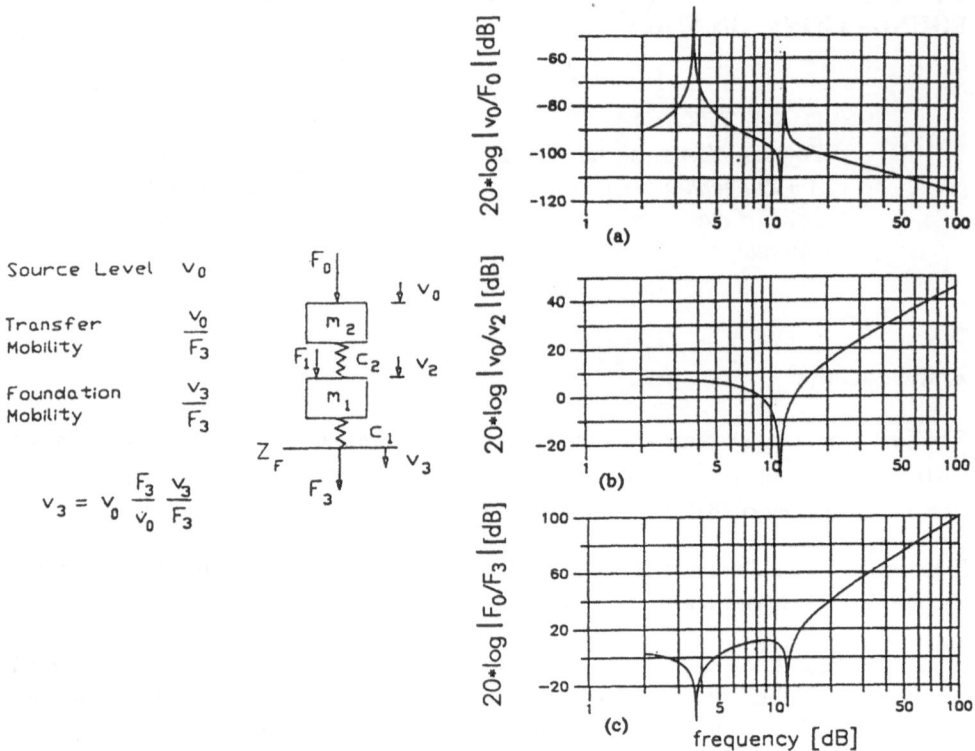

Source Level v_0

Transfer
Mobility $\dfrac{v_0}{F_3}$

Foundation
Mobility $\dfrac{v_3}{F_3}$

$$v_3 = v_0 \,\frac{F_3}{v_0}\,\frac{v_3}{F_3}$$

Figure 1. Two-degree-of-freedom system with main parameters: (a) Mobility of machine, (b) Transmission loss of upper stage, (c) Transmission loss of forces

influenced beyond that. All others are adjustable by spring arrangement and by variations of the mass moments of inertia of the intermediate mass.

All transfer functions taking account of the rigid body modes only have in common that, in the representation of *figure 1* (level versus logarithm of frequency), they will merge into straight lines with sufficient distance above the highest rigid body mode. If the real system does not show this characteristic behavior, this must be due to nonlinearities in the spring or other non-ideal performance of system parts.

One of the main reasons that ideal system behavior is not observed in real installations is the intermediate mass not being rigid. This structure is usually designed as a steel frame. Special designs feature e.g. polymere concrete for the material. The influence of the springs on the dynamic behavior of such a mass has been reported on in [4].

All designs have in common that their lowest-frequency mode shapes are bending modes. In the vicinity of the respective natural frequency, structureborne noise attenuation will be deteriorated. The questions arising are:

- what determines the degree of deterioration?
- how can negative effects be avoided?
- can attenuation be maximized over sufficiently large frequency ranges?

These questions have to be answered keeping in mind that no information about phase relationships of engine support motions are available.

3. Effect of Resonances

The intermediate mass can be visualized as a flexible structure with defined points of excitation (the attachment points of the springs of the upper level) and defined points of response (the attachments of the springs of the lower level). Keeping the attenuation high means maximizing the transfer mobility between these points:

$$Y_{ij}(\omega) = \sum_{k=1}^{n} \frac{i\omega u_{ik} u_{jk}}{\omega^2_{0k} - \omega^2 + 2i\omega\omega_{0k}\zeta_k}$$

Y_{ij}	Transfer mobility
n	number of modes
ω	circular frequency
ω_{0k}	natural frequency of the kth mode
u_{ik}	complex ith element (displacement) of the kth mass normalized mode
ζ_k	damping ratio of the kth mode

The value of this function at a given frequency is determined by the product of the modal components of the excitation and the response points.

If the weight of the intermediate mass is fixed only the stiffness/mass distribution can be varied. This optimization has to be done under the severe space constraints in a submarine, where it is good practise to design the intermediate mass frame into a given space.

With the finite element method the principal behavior of a structure under dynamic load can be investigated. *Figure 2* shows the attenuation of two systems with equal weight

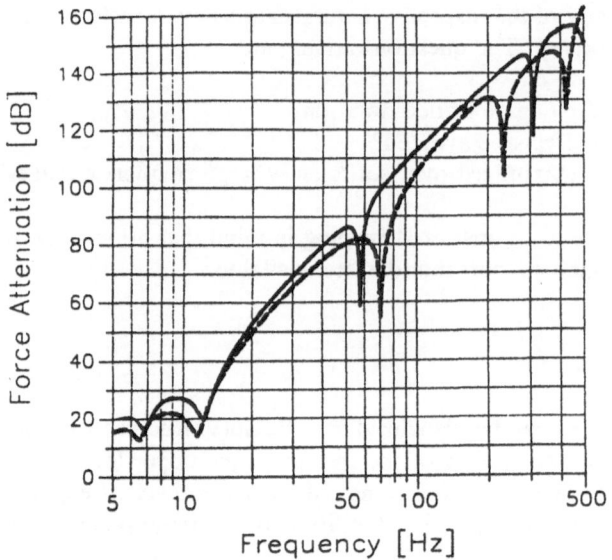

Figure 2. Attenuation of two systems with an intermedate mass of equal weight but different stiffness/mass distribution. Representation as in *figure 1c*

of the intermediate mass but different stiffness/mass distributions. While resonances of the intermediate mass structure can not be avoided, the overall attenuation is affected differently. Damping of the intermediate mass would only influence the direct vicinity of the resonance. The absolute damping values of such systems are almost completely determined by the resilient elements - e.g. rubber springs - and other attachments, [4], [5], [6].

4. Optimization

It is obvious that three conditions have to be met if an eigenmode of the intermediate mass causes a reduction in attenuation:

a)   the exciting frequency is close to a natural frequency of the intermediate mass
b)   neither all excitation nor all response points are located at modal nodes
c)   the excitation is at least partly in phase with the respective components of the mode

The condition a) is necessary but not sufficient. From b) follows that attenuation is kept high if it can be ensured that the exciting points and/or the response points are located on or close to nodes and c) indicates a strong influence of the phase properties of the excitation.

Most emphasis for optimization will be put on b), because it can be calculated with good accuracy with the finite element method. The natural frequencies can be calculated with limited accuracy only. Approximate values are sufficient. A shift of a resonance is possible with simple structural modification, e. g. adding a mass at a mode maximum. It must only be ensured that this potential is larger than the inaccuracy of the calculation. Because phase relationships of the source´s supports are generally not known, c) cannot be adopted. In our calculations, we assume the machine to be rigid which is the worst case in most (not in all) applications.

A possible optimization philosophy is the variation of the stiffness/mass distribution.

Figure 3 shows a structure which allowed variations of mass concentration locations. The representation of attenuation is according to *figure 1b* indicated by the dotted line. It can clearly be seen that only one configuration yields good results in the frequency range shown. After verification by modal analysis and further finite element calculations the following rules were found for proper design of the intermediate mass:

-   mass concentrations must be located at the edges of the structure.
-   almost all modes will show a node in the vicinity of the mass concentrations, consequently, the springs of the upper elastic level or the springs of the lower elastic level have to be mounted in the direct vicinity of the concentrations. At best, both are located close to the nodes.
-   the lowest vibration mode will be even lower in frequency but can easily be arranged in the minima of the excitation.

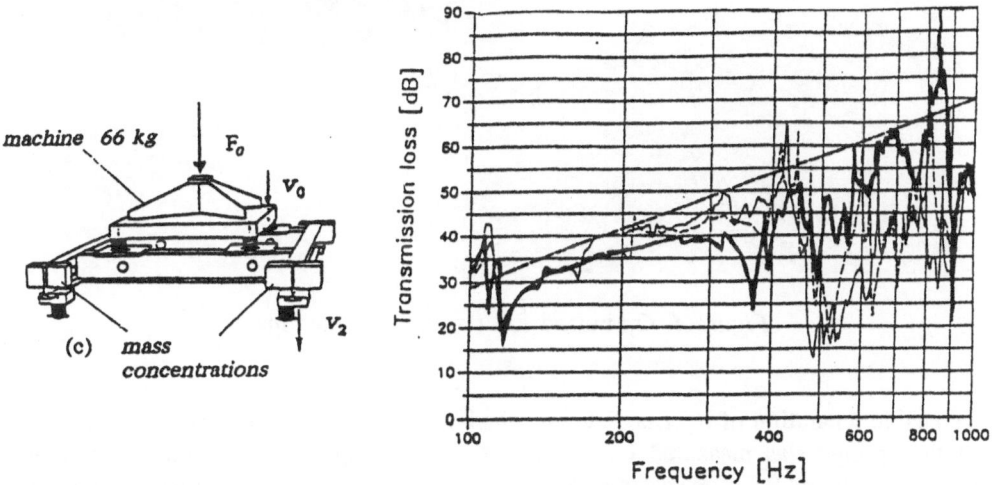

Figure 3. Measured influence of the mass distribution of the intermediate mass. The attenuation to the attachment of the lower springs is highest if the mass is mounted there (heavy line). All other mounting points are worse. Dotted line is the ideal system according to *figure 1b*

5. Full Size Tests

The structure of *figure 3* will not be applicable due to its bulky appearance in a system of true size. Further development of the principles lead to an even simpler structure shown in full scale in *figure 4*. It consists of two transverse beams which serve as mass concentrations in total up to their first natural frequency in bending at around 600 Hz. The lower springs are attached to the ends of these beams. The machine is supported by the much weaker longitudinal beam. Although not necessary, damping has been applied to the logitudinal girder by glueing 4 hollow square shaped beams together with a damping material, which, however, was not optimized for this particular application. The 4 beams are welded to a flange which is screwed to the transverse beam. The total weight of the structure is 500 kg. It serves as an intermediate mass for a 1500 kg motor mounting.

Figure 4. Optimized full size intermediate mass

To enable separation of intermediate mass and spring resonances the dynamic stiffness of the springs was measured on a test bed. The springs selected were stiffer than usually employed for such systems to ensure that structureborne noise transmission will prevail over airborne noise directly exciting the intermediate mass. For further comparison another intermediate mass with equal weight but designed to the conventional criteria was constructed. Although its stiffness was maximized the lowest bending mode was at 258 Hz, while that of the optimized mass was about 100 Hz.

The exciting frequencies considered are 25 Hz harmonics. *Figure 5* compares the ideal system and real systems. The representation is as in *figure 1b* with a linear frequency scale. The results for 25 Hz harmonics are shown. The classical intermediate mass is slightly better at low frequencies but much worse in higher. The optimized mass shows an attenuation minimum at 275 Hz which is due to a longitudinal resonance. Above 500 Hz the intermediate mass is even better than could be expected theoretically. Here, the machine is not rigid any more as assumed for the ideal system. The motion of the supports will be uncorrelated and the transverse beam will be fully effective as a mass concentration.

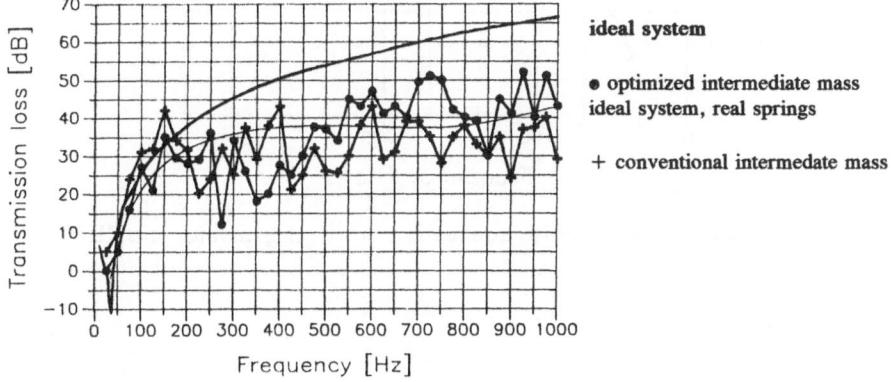

Figure 5. Attenuation of different intermedate mass configurations.

According to these principles a 3000 kg diesel generator was mounted to an optimized intermediate mass of 2400 kg including the acoustic enclosure. The intermediate mass has a lowest natural frequency in bending of around 75 Hz which is lower than the very strong excitation at 100 Hz which goes along with this particular type of diesel. A conventional mass would have had its lowest bending mode at 100 to 110 Hz. There is no damping treatment.

Figure 6 shows the result in a diagram according to *figure 1b*. Almost ideal conditions at dominating excitations are achieved before airborne noise limits the attenuation.

6. Conclusion

Application of very simple analytically tractable models can effectively serve as tools for improving the acoustic behavior of high performance double resilient mounting systems. For a practical non-ideal system the theoretical optimum in structureborne noise attenuation could almost be achieved even with limited knowledge of the properties of the excitation.

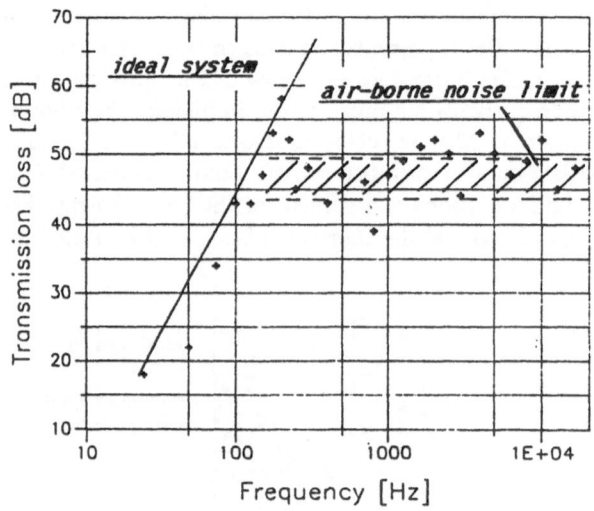

Figure 6. Structureborne noise attenuation of diesel generator installation

7. References

[1]   Wittekind, D. (1992) Körperschalldämmung auf Schiffen durch doppelelastische Lagerungen. Thesis, University of the Federal Armed Forces Hamburg, Bericht aus dem Institut für Mechanik, Heft Januar.

[2]   Wittekind, D. (1992) Optimizing the Intermediate Mass of Double Resilient Mounting Systems, Proceedings of the Underwater Defence Technology Conference, London, 181-186.

[3]   Wittekind, D. (1992) Noise Reduction Measures of a Submarine Closed Cycle Diesel, Proceedings of the International Conference on Submarine Systems, Stockholm, 265-271.

[4]   Geissler, P. (1987) Modal Tests under Different Boundary Conditions for a Polymere Concrete Machinery Foundation, 5th International Modal Analysis Conference, London, Vol. II, 1122-1128.

[5]   Gaul, L., Bohlen, S. (1987) Identification of Nonlinear Joint Models and Implementation in Discretized Structure Models, ASME New York, The Role of Damping in Vibration and Noise Control, 213-219.

[6]   Gaul, L., Chen, C.M. (1993) Modeling of Viscoelastic Elastomer Mounts in Multibody Systems. Schiehlen, W. (ed.) Advanced Multibody System Dynamics, Kluwer Academic Publishers, Dordrecht, 257-276.

MULTIOBJECTIVE OPTIMIZATION FOR INTEGRATED DESIGN OF MACHINE PRODUCTS BASED ON WORKING ENVIRONMENT INFORMATION

M. YOSHIMURA and T. KANEMARU

Department of Precision Engineering
Kyoto University
Sakyo-ku, Kyoto 606-01, Japan
email: yoshimura@prec.kyoto-u.ac.jp

1. Introduction

The working environments of machine products are diversified. The formulation of product design optimization should be fundamentally different according to the specific features of each working environment. Operational accuracy, operational efficiency and operation cost (including cost for controling the operation) are fundamental characteristics considered in the evaluation of machine products performances. For obtaining optimum designs applicable to practical circumstances, integrated evaluation of those characteristics is essential.

In this paper, methodologies for obtaining optimum designs of machine products are proposed based on consideration of the working environment. Here, as an example of machine products, industrial robots used in manufacturing shops are mainly considered. First of all, operational accuracy, operational efficiency and operation cost, which are the fundamental evaluative characteristics of industrial robots, are analyzed. Then, the relationships among these three evaluative characteristics are clarified using a multiobjective optimization strategy [1][2]. Next, the operations of the robots are categorized according to the required levels of operational accuracy and efficiency. Procedures for obtaining the optimum designs of the robots which are most suitable for their assigned operations are constructed. Finally, it is demonstrated by applied examples that diversification of product design is obtained by having a variety of working environments.

2. Product Performances of Industrial Robots and the Evaluative Characteristics

2.1. PRODUCT PERFORMANCES REQUIRED FOR INDUSTRIAL ROBOTS

The fundamental characteristics to be considered in the evaluation of industrial robots performances are as follows: (a) Static accuracy, (b) Operational efficiency, (c) Dynamic accuracy and (d) Operation cost

Those characteristics which are determined by the structural design and the control design are related with each other. For concurrently conducting the structural design and the control design (for example, see the reference [3]), the rela-

349

D. Bestle and W. Schielen (eds.), IUTAM Symposium on Optimization of Mechanical Systems, 349–356.
© 1996 *Kluwer Academic Publishers.*

tionship between these operational performances and the operation cost should be clarified.

Figure 1 shows the structural model having two arms and two driving joints which is popularly used in manufacturing shops.

For simplicity, each arm has a square cross-sectional shape and the cross-sectional widths d_i of arm i (i=1, 2) are the design variables. M_h is the concentrated mass of the hand effector and object. The coordinate system for evaluating the displacement at the hand effector point of a robot is shown in Figure 2.

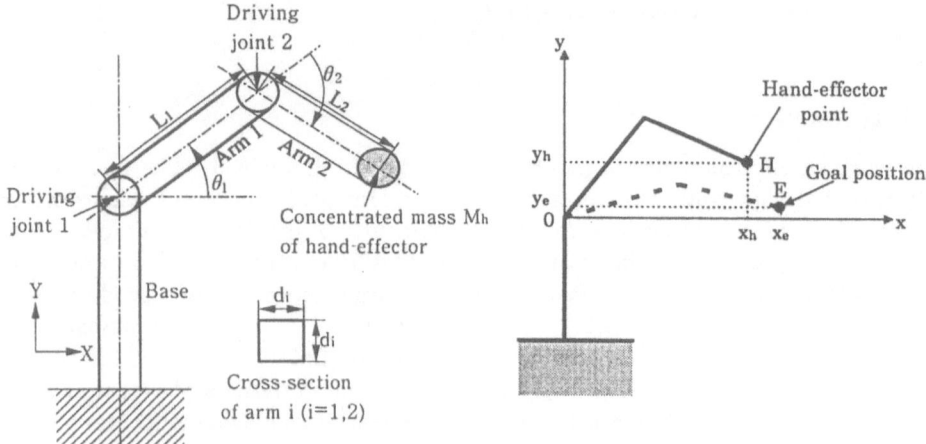

Figure 1. Structural model of an articulated robot having two arms and two driving joints

Figure 2. Coordinate system for evaluating the displacement at the hand-effector point of a robot

2.2. DEFINITION OF OPERATION COST AND THE EVALUATIVE CRITERION

In this paper, the PTP control method using position and velocity feedback by PD type-servoloop (which widely appears in the driving control of robots) is used. Here, the position and velocity feedback controls are used for decreasing the difference between the actual joint angle and the ideal joint angle at an actuator point.

For evaluating the operation cost, the amount of electricity consumed during the operation, that is, the operational energy, is calculated.

The control input E_{ni} at driving joint i is obtained using joint angle $\theta_i(t)$, joint angular velocity $\dot{\theta}_i(t)$ and ideal angle θ_{di} as follows:

$$E_{ni} = -k_{pi}(\theta_i(t) - \theta_{di}) - k_{vi}\dot{\theta}_i(t) \qquad (1)$$

where k_{pi} and k_{vi} are the position feedback gain and the velocity feedback gain, respectively.

In this paper, the response of the joint angle for the driving control input E_{ni} of arms is simulated using eq.(1) and the relation between the control input E_{ni} and the angular acceleration $\ddot{\theta}_i$ described in the following section (2.3).

Then, here, the amount of electricity consumed at each driving joint is calculated by integrating the square of the feedback control input E_{ni} over the operational time t_c. Then, the summation P for the total amount of energy consumed at all the joints is calculated as follows:

$$P = \sum_{i=1}^{2} \int_{0}^{t_c} E_{ni}(t)^2 dt \qquad (2)$$

Here, P is considered as the operation cost. This operation cost is used as a criterion for expressing difficulty of arm control.

2.3. DEFINITIONS OF DYNAMIC ACCURACY AND OPERATIONAL EFFICIENCY AND THE EVALUATIVE CRITERIA

The dynamic accuracies are related with both the dynamic characteristics due to the feedback control at an actuator point and the vibrational responses of elastic arms caused by inertia forces of robot motions. Here, the dynamic characteristics due to the feedback control for decreasing the difference between the actual joint angle and the ideal joint angle at an actuator point and the vibrational responses of elastic arms are integratedly simulated at small intervals of time from the starting time of motion until the time the hand effector point reaches a goal position. At each interval of time, the following three kinds of simulations: 1) the control simulation, 2) the kinematic displacement simulation and 3) the vibrational displacement simulation are conducted successively.

In the control simulation, the driving torques τ_i are calculated based on the difference between the ideal joint angles θ_{di} and the actual joint angles θ_i. The torque τ_i is identical with the control input E_{ni} in eq.(1).

In the kinematic displacement simulation, the equation of motion for the rotational movement of rigid arms is formulated and the relationship between the torque τ_i which is applied to joint i and the angular acceleration $\ddot{\theta}_i$ of joint i is obtained using the Newton-Euler method under the approximation of rigid arms. Then, the changes of the joint angles due to the driving toruques τ_i obtained in the control simulation are obtained.

In the vibrational displacement simulation, the equation of motion for elastic arms is formulated based on the dynamic characteristic values obtained in the kinematic displacement simulation and the vibrational displacement (v_x, v_y) at the hand effector point is obtained.

Finally, the displacement u at the hand effector point is obtained by adding the kinematic displacement and the vibrational displacement (v_x, v_y) as follows:

$$u = \sqrt{\{v_x + \sum_{i=1}^{2} L_i(sin\theta_i - sin\theta_{di})\}^2 + \{v_y + \sum_{i=1}^{2} L_i(cos\theta_i - cos\theta_{di})\}^2} \qquad (3)$$

The operational time t_c and the dynamic accuracy y_d are defined depending on the kinds of assigned jobs and based on the displacement u at the hand effector point as follows:

(1) In the case where dynamic accuracy constraint is given.

As shown in Figure 3, the time interval t_c from the starting time of arm motion until the time when the hand effector point reaches a goal position while satisfying the dynamic accuracy constraint is used as the evaluative factor of the operational time.

(2) In the case where operational time constraint is given.

The maximum amplitude y_d of dynamic displacement u at the hand effector point from the goal position after the planned operational time is used as the evaluative factor of the dynamic accuracy.

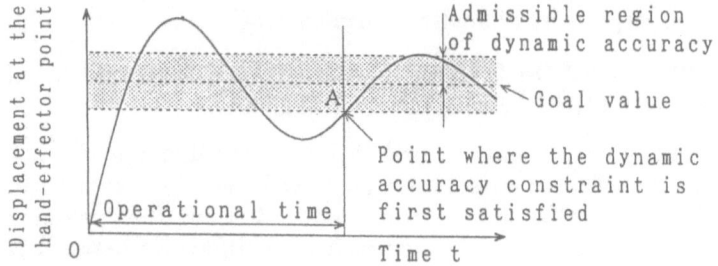

Figure 3. Definition of operational time in the case where dynamic accuracy constraint is given

3. Relationships Among Evaluative Characteristics

The relationships between dynamic operational accuracy or operational efficiency and the operation cost for design variables of the arms in the structural models of industrial robots are clarified. These relationships are basically used in design decision making of robots.

Figure 4 shows the relationship between the operational time and the operation cost under the dynamic accuracy constraint in the structural model of an industrial robot having two arms and two driving joints as shown in Figure 1. For the design optimization of robots, a smaller operational time and a smaller operation cost are preferable. The line was obtained by the ε-constarint method, where the operational time was minimized by successive changes of the upper limit of the operation cost constraint. The solutions on the curve from point A to point B correspond to the Pareto optimum solution set of a multiobjective optimization problem having the two objectives of minimization of the operation cost and minimization of the operational time. On the Pareto optimum solution set, the operation cost and the operational time have a conflicting relationship. So, the optimal solutions should be selected from the solutions on the curve from point A to point B. Solutions on the right region from point B don't include the optimum solution since the operational time also increases for an increase in the operation cost. A relation similar to the one in Figure 4 is obtained between the dynamic accuracy and the operation cost under the operational time constraint.

In design decision making of products of which principal characteristics have conflicting relationships, such as shown in Figure 4, the formulation of the multiobjective design optimization problem [1][2] where the characteristics are simultaneously evaluated on the same stage is effective.

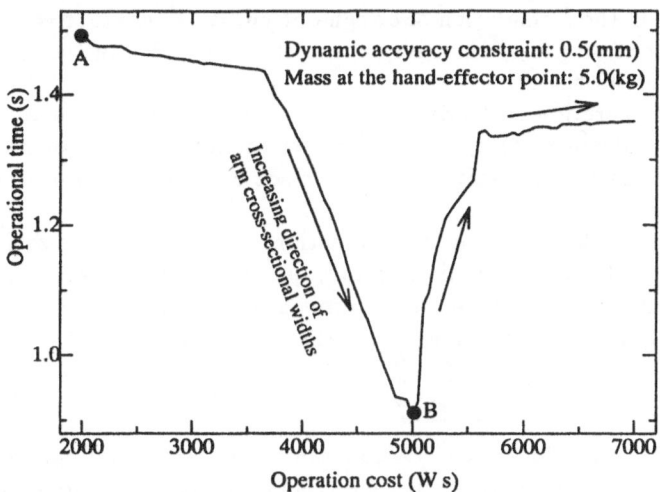

Figure 4. The relationship between the operational time and the operation cost under the dynamic accuracy constraint

4. Design Decision Making Problems Based on Job Categorization

4.1. FORMULATION OF DESIGN OPTIMIZATION PROBLEMS

Industrial robots are usually used for effectively and efficiently conducting specific jobs in manufacturing shops. In such a working environment, a broader application than necessary is not important but enhancement of the product perfomance required for the specific job is truely necessary.

In execution of design optimization, the most important product performance usually has the required level or the goal value. Maximization of one or several product performances under satisfaction of the requirement brings about the profit of the product.

Industrial robots are used for many kinds of jobs. Examples of these jobs would be transporting, welding, spraying, assembling, inspecting, etc. Working environments are diversified in nature, but, jobs in which industrial robots are used in manufacturing shops are roughly categorized into two types of jobs. This division takes place from the standpoints of required accuracy and operational efficiency and is as follows:

(1) Jobs having higher requirement levels for operational efficiency than for operational accuracy (for example, transporting jobs): Requirement levels for operational efficiency are strictly defined and higher operational accuracy brings about profits of simplification of control systems, etc.

(2) Jobs having higher requirement levels for operational accuracy than for operational efficiency (for example, assembling jobs): Requirement levels for operational accuracy are strictly defined and higher operational efficiency brings about

profits by shortening of operational time, etc.

Definite evaluative requirements which must be satisfied without fail are set as the constraints in the formulation of design optimization. Evaluative requirements which bring about profits by their improvement are included in the formulation as the objective functions.

Decrease of operation cost brings about profits of simplification of driving and control systems, decrease of operational power cost,etc. So, the operation cost is always included in the objective functions since a smaller operation cost brings about profits.

For type (1) of the foregoing categorization, the vector optimization problem of minimization of both dynamic accuracy y_d and operation cost J_0 under the constraint of the operational time t_c is formulated as follows:

Minimize [dynamic accuracy y_d, operation cost J_0]
$$\left. \begin{array}{c} \text{subject to } t_c \leq t_c^U \\ y_s \leq y_s^U \end{array} \right\} \qquad (4)$$

For type (2) of the foregoing categorization, the vector optimization problem of minimization of both operational time t_c and operation cost J_0 under the constraint of the dynamic accuracy y_d is formulated as follows:

Minimize [operational time t_c, operation cost J_0]
$$\left. \begin{array}{c} \text{subject to } y_d \leq y_d^U \\ y_s \leq y_s^U \end{array} \right\} \qquad (5)$$

In each formulation, two objective functions exist and the constraint of the static displacement y_s is added for ensuring satisfaction of the static operational accuracy.

Based on consideration for the practical working environment of industrial robots, the foregoing formulations of design decision making are divided into more practical detailed categorization according to the required levels of product perfomances as follows (practical figures in the constraints correspond to those used in the applied examples of Section 5):

Type 1. Jobs having high and definite requirement levels for operational efficiency: for example, transporting jobs.

Minimize [dynamic accuracy y_d, operation cost J_0]
$$\text{subject to } t_c \leq 0.7(s)$$
$$y_s \leq 0.5(mm)$$
$$M_h = 5.0(kg)$$

Type 2. Jobs having definite requirement levels for high speed transportation of heavy objects under the strict static accuracy constraint.

Minimize [dynamic accuracy y_d, operation cost J_0]
$$\text{subject to } t_c \leq 0.8(s)$$
$$y_s \leq 0.5(mm)$$
$$M_h = 30.0(kg)$$

Type 3. Jobs having definite requirement levels for operational accuracy: for example, welding jobs.

Minimize [operational time t_c, operation cost J_0]

$$\text{subject to } y_d \leq 0.5(mm)$$
$$y_s \leq 0.07(mm)$$
$$M_h = 5.0(kg)$$

Type 4. Jobs having extremely high requirement levels for static and dynamic accuracies: for example, assembling jobs of precise elements and products.

Minimize [operational time t_c, operation cost J_0]

$$\text{subject to } y_d \leq 0.1(mm)$$
$$y_s \leq 0.03(mm)$$
$$M_h = 5.0(kg)$$

4.2. DESIGN DECISION MAKING PROCEDURES

The design optimization problems defined in section 4.1 are two-objective optimization problems in which the objective functions are "minimization of the operational time and minimization of the operation cost" or "minimization of the characteristic concerning the operational accuracy and minimization of the operation cost". There are many methods for solving these kinds of optimization problems, but here the min-max multiobjective optimization strategy is used.

g_j and ω_j (j=1, 2) are the objective functions and the weighting factors for the corresponding objective functions, respectively. Here, g_j is normalized by the value of the objective function which is obtained for representative design variable values. The whole feasible space of the design variables is denoted D.

The optimum design variables $\mathbf{d}^*$ $(= \{d_1^*, d_2^*, ..., d_n^*\}^T)$ are formulated as follows:

$$\mathbf{d}^* = minmax\{\omega_1 g_1, \ \omega_2 g_2\} \tag{6}$$

where $\omega_1 \geq 0$, $\omega_2 \geq 0$, and $\omega_1 + \omega_2 = 1$

The initial values of the weighting factors in eq.(6) are given by the decision maker, considering the design circumstance of the product being regarded. If the design solutions obtained are not satisfactory, the weighting factors are altered until satisfactory solutions are obtained.

5. Applied Examples

The design optimization procedures described in section 4 were applied to structural models of the robot having two arms and two joints as shown in Figure 1. The values of the parameters used in the applied examples are shown in Table 1. The design variables for optimization are the cross-sectional widths of robot arms.

In the min-max multiobjective optimization strategy in eq.(6), g_1 corresponds to t_c in the robots of types 3 and 4 given in section 4.1 and corresponds to y_d in the robots of types 1 and 2, and g_2 corresponds to P. Normalization of the evaluative characteristics was conducted for the values of the evaluative characteristics where d_1 and d_2 are 50 mm and 50 mm, respectively.

The optimized results of the design variables (d_1 and d_2), the dynamic accuracy y_d, the operational time t_c and the operation cost P for the assigned jobs categorized in section 4.1 are shown in Table 2, where "$\star$" indicates the coincidence with the upper limit of the constraint. Here, the weighting factors were: $\omega_1=0.8$ and $\omega_2=0.2$.

It can be understood that the optimum design solutions are different depending on the types of assigned jobs.

TABLE 1. Values of parameters used in the applied examples

Parameters	Two-arms,Two-joints robot	Parameters	Two-arms,Two-joints robot
Arm material	Aluminum	Position feedback gain	400(Type1,3,4)
Mass density	2.8×10^3 kg/m^3	of joint 1 (Nm/rad)	900(Type2)
Young's modulus	7.1×10^{10} N/m^2	Velocity feedback gain	100(Type1,3,4)
Poisson's ratio	0.33	of joint 1 (Nm·s/rad)	300(Type2)
Length of arm 1	500mm	Position feedback gain	200(Type1,3,4)
Length of arm 2	400 mm	of joint 2 (Nm/rad)	600(Type2)
Starting positional angle of arm 1	15°	Velocity feedback gain	50(Type1,3,4)
Ending positional angle of arm 1	60°	of joint 2 (Nm·s/rad)	200(Type2)
Starting positional angle of arm 2	15°		
Ending positional angle of arm 2	60°		

TABLE 2. Optimized solutions for each of categorized jobs in the case of a robot having two arms and two driving joints

Type of job	d_1 (mm)	d_2 (mm)	y_d (mm)	t_c (s)	P (W·s)
Type 1	84	54	4.1	0.7*	5400
Type 2	150	60	12.1	0.9*	42000
Type 3	90	45	0.5*	0.93	4950
Type 4	85	44	0.1*	1.09	4800

6. Concluding remarks

A methodology for obtaining optimum designs of industrial robots based on job categorization was proposed in which the structural characteristics and the control characteristics are concurrently evaluated using multiobjective optimization techniques.

References

1. Eschenauer, H., Koski, J. and Osyczka, A. (eds.): *Multicriteria Design Optimization*, Springer-Verlag, 1990.

2. Stadler, W. (ed.): *Multicriteria Optimization in Engineering and in Sciences*, Plenum Press, 1988.

3. Messac, A. and Malek, K.: Control-Structure Integrated Design, *AIAA Journal* **30**, 8 (1992), 2124-2131.

STRUCTURAL SHAPE OPTIMIZATION AND CONVEX PROGRAMMING METHODS

W.H. ZHANG and C. FLEURY
Aerospace Laboratory, LTAS, University of Liège
Rue Ernest-Solvay, 21, B-4000, Belgium

1. Introduction

Nowadays, convex programming methods are recognized as an efficient approach to solve a variety of structural optimization problems. Their integration with F.E. computing codes as well as modern CAD systems provides a powerful tool to fulfil realistic engineering design tasks. In this paper, two topics are addressed. The first one is concerned with shape optimization techniques which include automatic selection of design variables based on parametric CAD model, appropriate implementation of semi-analytical sensitivity analysis method and mesh perturbators for velocity field evaluations. The second one is about convex programming methods. A comparative study of different methods will make sure that using general and high-quality approximation schemes are essential to improve the efficiency of the design procedure when considered problems are characterized by different types of constraints (e.g. static, dynamic) and different types of design variables (e.g. sizing, configurational, topological). In this context, the GMMA (Generalized Method of Moving Asymptotes) and DQA (Diagonal Quadratic Approximation) methods are proposed here. Numerical examples are given to illustrate concerned issues and different methods. A discussion will be made about results.

2. Shape Optimization

Assume that considered problems have the following formulation:

$$
\begin{aligned}
Min \quad & g_0(x) \\
& g_j(x) \leq 0 \qquad j = 1, m \\
& h_k(x) = 0 \qquad k = 1, l \\
& \underline{x_i} \leq x_i \leq \overline{x_i} \qquad i = 1, n
\end{aligned}
\tag{1}
$$

Inequality constraints are imposed here to structural responses (e.g. displacements, stresses) whereas equality constraints represent geometric relations

D. Bestle and W. Schielen (eds.), IUTAM Symposium on Optimization of Mechanical Systems, 357–364.
© *1996 Kluwer Academic Publishers.*

which are required to ensure the regularity and the smoothness of the boundary in design.

2.1. SELECTION OF INDEPENDENT SHAPE DESIGN VARIABLES

Due to the non-linearity of equality constraints, convex programming methods are not directly applicable to (1). Our solution is to transform the above problem into a compact form by retaining only independent design variables. This strategy is, in fact, the same as the parametric design methodology used in mechanical drawings [1].

Mathematically, the selection of independent variables can be carried out by linearing equality constraints at the current design point x^0,

$$A \, \delta x = 0 \qquad \text{with } A_{l \times n} = \{a_{ki}\} = \{\frac{\partial h_k}{\partial x_i}\}, \qquad (n > l) \qquad (2)$$

By splitting the design vector $x = [y, z]$ and the Jacobian matrix $A = [A_1, A_2]$, the system (2) can be consequently written as

$$\delta y = -A_1^{-1} A_2 \delta z = -\tilde{A} \delta z \qquad (\tilde{A} = A_1^{-1} A_2)$$

$$\text{only if } \det(A_1) \neq 0 \qquad (3)$$

It means that z is the desired vector of independent shape design variables. In addition, this relation can be also used to check the validity of a set of design variables when they are intuitively chosen. The necessary condition is that the related submatrix A_1 is non-singular.

2.2. SEMI-ANALYTICAL SENSITIVITY ANALYSIS METHOD AND MESH PERTURBATORS

For static problems, the derivative of finite element system equations Kq=g leads to

$$\frac{\partial q}{\partial z_i} = K^{-1} (\frac{\partial g}{\partial z_i} - \frac{\partial K}{\partial z_i} q) \qquad (4)$$

The semi-analytical method means that derivatives of the stiffness matrix K and the load vector g in (4) are performed by finite difference.

$$\frac{\partial q}{\partial z_i} = K^{-1} (\delta g - \delta K \, q) \frac{1}{\delta z_i} \qquad (5)$$

with $\qquad \delta g \approx g(c^*) - g(c^0)$, $\delta K \approx K(c^*) - K(c^0)$ $\qquad\qquad$ (6)

In these expressions, δz_i is the step size of the ith design variable. c^* and c^0 denote respectively nodal coordinates of the perturbed and initial mesh.

The mesh perturbator aims at evaluating the velocity field v to construct the perturbed mesh c^* through

$$c^* = c^0 + v\delta z_i \qquad (v = \frac{\partial c}{\partial z_i}) \qquad\qquad (7)$$

This relation can be interpreted as the deformation law of the mesh.

General mesh perturbators should be independent of mesh generators. In practice, mesh perturbators can be built up either by geometrical approach or by physical approach. The first one, such as transfinite mapping method limited to structured meshes, boundary node approach and the smoothing method for general unstructured meshes, depends uniquely upon the geometry and/or the topology of the mesh. As to physical approach, it interprets the velocity field as a unknown displacement vector which is obtained by solving the associated F.E. equations. A comparative study in [2] showed that the transfinite mapping method is almost as efficient as the physical approach.

The efficiency of sensitivity analysis can be improved if the semi-analytical method (5) is implemented in accordance with the velocity field v. The expression (5) is in fact only appropriate when perturbed meshes of non-zero velocity field are few. However, when the perturbed meshes cover nearly the entire structural domain, the implementation of (5) should be made as follows:

$$\frac{\partial q}{\partial z_i} = K^{-1}(g(c^*) - g - K(c^*)q + Kq) \frac{1}{\delta z_i} = K^{-1}(g(c^*) - K(c^*)q) \frac{1}{\delta z_i} \qquad (8)$$

in which finite difference calculations completely disappear. This scheme can be exactly extended to dynamic problems.

3. Convex Approximation Methods

Due to the fact that the objective function and constraints are implicit functions of design variables in the original design problem, it is a common practice to use convex approximation methods to construct a sequence of explicit subproblems and then solved by dual approach. These approximation methods can be basically classified into monotonic and non-monotonic approximations. The first family can be typically illustrated by the following GMMA method and the second one by quadratic approximation.

3.1. GENERALIZED METHOD OF MOVING ASYMPTOTES (GMMA)

The proposed GMMA approximation [3] for a given function $g_j(x)$ is expressed as:

$$g_j(x) \approx c_j + \sum_+ p_{ij}/(u_{ij}-x_i) + \sum_- q_{ij}/(x_i-l_{ij}) \tag{9}$$

with coefficients equal to:

$$p_{ij} = \frac{\partial g_j(x^0)}{\partial x_i}(u_{ij}-x_i^0)^2 \text{ , if } \frac{\partial g_j(x^0)}{\partial x_i} > 0$$

$$q_{ij} = -\frac{\partial g_j(x^0)}{\partial x_i}(x_i^0-l_{ij})^2, \text{ if } \frac{\partial g_j(x^0)}{\partial x_i} < 0 \tag{10}$$

$$c_j = g_j(x^0) + \sum_+ \frac{\partial g_j(x^0)}{\partial x_i}(u_{ij}-x_i^0) + \sum_- \frac{\partial g_j(x^0)}{\partial x_i}(x_i^0-l_{ij})$$

The expansion (9) provides exact function value and gradient at the developing point x^0. As compared with the original version of MMA [4], the basic feature of GMMA is that each function $g_j(x)$ has its proper moving asymptote l_{ij} or u_{ij} to each design variable x_i. This property allows to characterize each constraint function independently. GMMA is monotonic because the gradient has a unchanged sign despite of values of design variables. Therefore, this method is especially suitable to static sizing optimization problems because involved objective function and constraints are quasi-monotonic functions. In addition, GMMA is a flexible and general approximation, moving asymptotes can adjust the convexity of the approximation. When moving asymptotes take limiting values, GMMA will be simplified exactly into the CONLIN scheme [5] or linear approximation. To evaluate moving asymptotes, a two-point function value fitting scheme is devised by utilizing the function value at the preceding iteration.

3.2. DIAGONAL QUADRATIC APPROXIMATION (DQA)

In this approximation, only diagonal terms of second order derivatives are retained to ensure its separability.

$$g_j(x) \approx g_j(x^0) + \sum_i \frac{\partial g_j(x^0)}{\partial x_i}(x_i-x_i^0) + \frac{1}{2}\sum_i \frac{\partial^2 g_j(x^0)}{\partial x_i^2}(x_i-x_i^0)^2 \tag{11}$$

This simplification is beneficial to express each primal design variable in terms of dual variables in dual approach.

The non-monotonic property of (11) makes it possible to be well adapted to some class of design problems. For example, when we consider truss geometry design using nodal coordinates as design variables, it is found that the objective function defined by the structural weight and constraints defined by stresses and nodal displacements are non-monotonic functions with respect to nodal coordinates. As a result, neither CONLIN nor MMA is able to give convergent solution although move-limit concepts early used in [6] could reduce, to some extent, the difficulty. This is because the approximation is so poor that the iteration history oscillates. Our recent numerical results prove that the ideal solution is to use the DQA method Moreover, when different types of design variables are simultaneously involved in the design problem, attempts can be also made to construct mixed approximations of the following form

$$g_j(x) \approx c_j + \sum_{i \in A} P_{ij}/(x_i - m_{ij}) + \sum_{i \in B} b_{ij}(x_i - x_i^0) + \frac{1}{2}a_{ij}(x_i - x_i^0)^2 \qquad (12)$$

where monotonic terms in (9) and non-monotonic terms in (11) are combined. This formulation needs to be further investigated.

4. Numerical Examples

4.1. SHAPE DESIGN OF A TORQUE ARM

This example was initially studied by Bennett and Botkin [7]. It is now shown in figure 1. The problem is to find optimal boundary shape for outside and inside contours. The geometric model is described in terms of straight line segments and circular arcs. Eight independent design variables are identified for the moving boundaries. Related data are given below:

$\mu = 0.3$, $t = 0.3$ cm $E = 20.74 \ 10^6$ N/cm^2 $\rho = 7.81 \ 10^{-3}$ kg/cm^3

$F_x = 2789$ N $F_y = 5066$ N $\bar{\sigma} = 8 \ 10^4$ N/cm^2

$d_{min.} = 1$ cm (minimal separation between different boundaries)

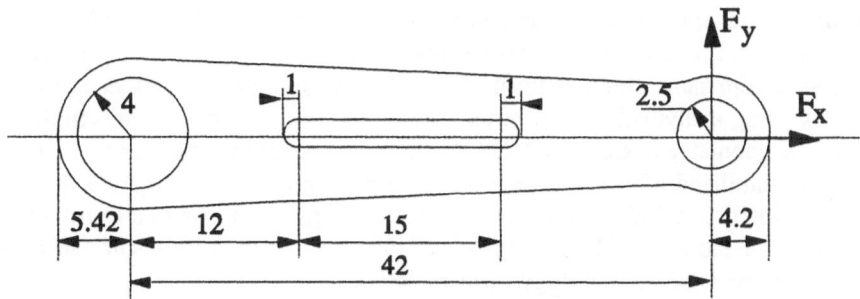

Figure 1. shape design of a torque arm (in cm)

The CONLIN method is used for this example. The solution given below is obtained after eight iterations with a weight reduction of about 50% from 0.8606 kg to 0.4367 kg. As shown in figure 2, the final design is feasible, the maximum von-mises stress increases from 4.21 10^4 N/cm^2 to 7.88 10^4 N/cm^2.

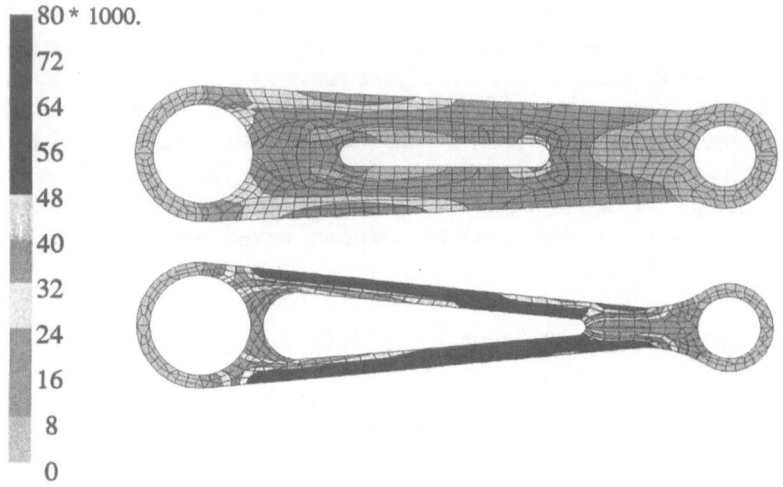

Figure 2. Isovalues of von-mises stress for the initial and final design

4.2. WEIGHT MINIMIZATION OF A 10-BAR TRUSS

As illustrated in figure 3, vertical loads are attached only at two lower nodes. Constraints are defined as the limitation of the stresses ($\bar{\sigma}$ = 2.5*10^4 psi) in all bars and vertical displacements at four nodes ($\bar{u}$ = 2 in). The material has Young's modulus E = 10^7 psi and density ρ = 0.1 lb/in^3. Initial section areas are a^0 =20 in^2.

Two cases are tested: firstly, only section areas are used as design variables; secondly, nodal coordinates are uniquely used as design variables. Results are summarized in table 1. Note that for each pair of values, the first one is the weight and the second one measures the constraint violation when it is greater than one. It can be seen that both CONLIN and GMMA generate monotonic iteration history and converge to the same solution in the first case. Asymptote parameters are evaluated either by using analytical second order derivatives or by fitting scheme. In the second case of the geometry design, the DQA method demonstrates its efficiency. The design process is stabilized after eight iterations. The solution shown in figure 3 has a weight of 6879.4 lb. This solution is reasonable. Its topology is similar to a cantilever beam. Two nodes at the right side are superposed into one point. This result can be in fact easily confirmed by the topology design method.

| Ite. | CONLIN (case 1) | GMMA (case 1) A: analytical asymptotes B: fitting asymptotes | | DQA (case 2) |
		A	B	
1	8392.9 0.9848	8392.9 0.9848	8392.9 0.9848	8392.93 0.98489
2	6135.9 0.9777	5887.5 1.0043	5887.5 1.0043	6681.85 1.81653
3	5860.0 0.9887	5804.4 0.9921	5819.3 0.9926	6658.62 1.25340
4	5738.0 0.9905	5692.8 0.9910	5720.2 0.9924	6800.71 1.04586
5	5624.3 0.9898	5569.2 0.9897	5618.2 0.9917	6891.16 0.99834
6	5501.2 0.9887	5427.5 0.9873	5507.8 0.9907	6881.11 1.00011
7	5363.7 0.9873	5253.9 0.9839	5384.4 0.9893	6880.44 0.99976
8	5206.5 0.9907	5084.6 0.9992	5241.2 0.9890	6879.40 1.00011
9	5086.8 0.9996	<u>5076.7</u> 1.0000	5098.9 0.9987	6879.51 0.99999
10	5077.8 0.9999	5076.7 1.0000	5081.8 0.9999	6879.42 1.00002
11	<u>5076.7</u> 1.0000		5077.2 0.9999	<u>6879.44</u> 1.00000
12	5076.7 1.0000		<u>5076.7</u> 1.0000	

Table 1. Iteration history for 10-bar truss problem

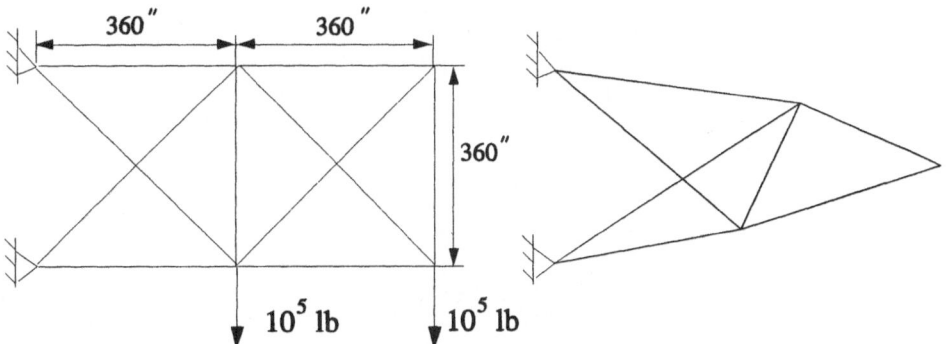

Figure 3. Sizing and geometry design of the 10-bar truss

5. Conclusions

Shape optimization techniques and convex approximation methods are studied in this paper. The selection of independent design variables and the improvement of the sensitivity analysis method aim to, on one hand, enhance the efficiency of the design process, on the other hand, to reduce the difficulty of solution.

The DQA method is highly recommended to truss geometry design problems due to their non-monotonic behaviours. Move-limits are generally not necessary. While the family of GMMA method is more suitable to sizing, shape optimization and other type of problems.

6. References

1. Light, R. and Gossard, D. (1982) Modification of geometric models through variational geometry, *CAD* 14, 209-214.
2. Zhang, W.H. and Beckers, P. (1989) Comparison of different sensitivity analyses approaches for structural shape optimization, in C.A. Brebbia and S. Hernandez (eds.), Recent Advances, Springer-Verlag, Southampton, pp.347-356.
3. Zhang, W.H. and Fleury, C. (1994) Recent advances in convex approximation methods for structural optimization, in B.H.V. Topping and M. Papadrakakis (eds.), *Advances in structural optimization*, Civil-Comp Press, Edinburgh, pp.83-90.
4. Svanberg, K. (1987) Method of moving asymptotes - a new method for structural optimization, *Int. j. numer. methods eng.* 24, 359-373.
5. Fleury, C. (1989) First and second order convex approximation strategies in structural optimization, *structural optimization* 1, 3-10.
6. Kuritz, S.P. and Fleury. C. (1989) Mixed variables structural optimization using convex linearization techniques, *Eng. opt.* 15, 27-41.
7. Bennett, J.A. and Botkin, M.E. (1984) Structural shape optimization with geometric description and adaptive mesh refinement, *AIAA J.* 23, 458-464.

AUTHOR INDEX

365

Mechanics

SOLID MECHANICS AND ITS APPLICATIONS
Series Editor: G.M.L. Gladwell

Aims and Scope of the Series

The fundamental questions arising in mechanics are: *Why?*, *How?*, and *How much?* The aim of this series is to provide lucid accounts written by authoritative researchers giving vision and insight in answering these questions on the subject of mechanics as it relates to solids. The scope of the series covers the entire spectrum of solid mechanics. Thus it includes the foundation of mechanics; variational formulations; computational mechanics; statics, kinematics and dynamics of rigid and elastic bodies; vibrations of solids and structures; dynamical systems and chaos; the theories of elasticity, plasticity and viscoelasticity; composite materials; rods, beams, shells and membranes; structural control and stability; soils, rocks and geomechanics; fracture; tribology; experimental mechanics; biomechanics and machine design.

1. R.T. Haftka, Z. Gürdal and M.P. Kamat: *Elements of Structural Optimization.* 2nd rev.ed., 1990   ISBN 0-7923-0608-2
2. J.J. Kalker: *Three-Dimensional Elastic Bodies in Rolling Contact.* 1990
                                                 ISBN 0-7923-0712-7
3. P. Karasudhi: *Foundations of Solid Mechanics.* 1991   ISBN 0-7923-0772-0
4. *Not published*
5. *Not published.*
6. J.F. Doyle: *Static and Dynamic Analysis of Structures.* With an Emphasis on Mechanics and Computer Matrix Methods. 1991   ISBN 0-7923-1124-8; Pb 0-7923-1208-2
7. O.O. Ochoa and J.N. Reddy: *Finite Element Analysis of Composite Laminates.*
                                                 ISBN 0-7923-1125-6
8. M.H. Aliabadi and D.P. Rooke: *Numerical Fracture Mechanics.*   ISBN 0-7923-1175-2
9. J. Angeles and C.S. López-Cajún: *Optimization of Cam Mechanisms.* 1991
                                                 ISBN 0-7923-1355-0
10. D.E. Grierson, A. Franchi and P. Riva (eds.): *Progress in Structural Engineering.* 1991
                                                 ISBN 0-7923-1396-8
11. R.T. Haftka and Z. Gürdal: *Elements of Structural Optimization.* 3rd rev. and exp. ed. 1992
                                      ISBN 0-7923-1504-9; Pb 0-7923-1505-7
12. J.R. Barber: *Elasticity.* 1992   ISBN 0-7923-1609-6; Pb 0-7923-1610-X
13. H.S. Tzou and G.L. Anderson (eds.): *Intelligent Structural Systems.* 1992
                                                 ISBN 0-7923-1920-6
14. E.E. Gdoutos: *Fracture Mechanics.* An Introduction. 1993   ISBN 0-7923-1932-X
15. J.P. Ward: *Solid Mechanics.* An Introduction. 1992   ISBN 0-7923-1949-4
16. M. Farshad: *Design and Analysis of Shell Structures.* 1992   ISBN 0-7923-1950-8
17. H.S. Tzou and T. Fukuda (eds.): *Precision Sensors, Actuators and Systems.* 1992
                                                 ISBN 0-7923-2015-8
18. J.R. Vinson: *The Behavior of Shells Composed of Isotropic and Composite Materials.* 1993
                                                 ISBN 0-7923-2113-8
19. H.S. Tzou: *Piezoelectric Shells.* Distributed Sensing and Control of Continua. 1993
                                                 ISBN 0-7923-2186-3

Kluwer Academic Publishers – Dordrecht / Boston / London

Mechanics

SOLID MECHANICS AND ITS APPLICATIONS
Series Editor: G.M.L. Gladwell

Kluwer Academic Publishers – Dordrecht / Boston / London

Mechanics

FLUID **MECHANICS AND ITS APPLICATIONS**

Series Editor: R. Moreau

Aims and Scope of the Series

The purpose of this series is to focus on subjects in which fluid mechanics plays a fundamental role. As well as the more traditional applications of aeronautics, hydraulics, heat and mass transfer etc., books will be published dealing with topics which are currently in a state of rapid development, such as turbulence, suspensions and multiphase fluids, super and hypersonic flows and numerical modelling techniques. It is a widely held view that it is the interdisciplinary subjects that will receive intense scientific attention, bringing them to the forefront of technological advancement. Fluids have the ability to transport matter and its properties as well as transmit force, therefore fluid mechanics is a subject that is particularly open to cross fertilisation with other sciences and disciplines of engineering. The subject of fluid mechanics will be highly relevant in domains such as chemical, metallurgical, biological and ecological engineering. This series is particularly open to such new multidisciplinary domains.

1. M. Lesieur: *Turbulence in Fluids.* 2nd rev. ed., 1990   ISBN 0-7923-0645-7
2. O. Métais and M. Lesieur (eds.): *Turbulence and Coherent Structures.* 1991
                                                                      ISBN 0-7923-0646-5
3. R. Moreau: *Magnetohydrodynamics.* 1990   ISBN 0-7923-0937-5
4. E. Coustols (ed.): *Turbulence Control by Passive Means.* 1990   ISBN 0-7923-1020-9
5. A.A. Borissov (ed.): *Dynamic Structure of Detonation in Gaseous and Dispersed Media.* 1991
                                                                      ISBN 0-7923-1340-2
6. K.-S. Choi (ed.): *Recent Developments in Turbulence Management.* 1991
                                                                      ISBN 0-7923-1477-8
7. E.P. Evans and B. Coulbeck (eds.): *Pipeline Systems.* 1992   ISBN 0-7923-1668-1
8. B. Nau (ed.): *Fluid Sealing.* 1992   ISBN 0-7923-1669-X
9. T.K.S. Murthy (ed.): *Computational Methods in Hypersonic Aerodynamics.* 1992
                                                                      ISBN 0-7923-1673-8
10. R. King (ed.): *Fluid Mechanics of Mixing.* Modelling, Operations and Experimental Techniques. 1992   ISBN 0-7923-1720-3
11. Z. Han and X. Yin: *Shock Dynamics.* 1993   ISBN 0-7923-1746-7
12. L. Svarovsky and M.T. Thew (eds.): *Hydroclones.* Analysis and Applications. 1992
                                                                      ISBN 0-7923-1876-5
13. A. Lichtarowicz (ed.): *Jet Cutting Technology.* 1992   ISBN 0-7923-1979-6
14. F.T.M. Nieuwstadt (ed.): *Flow Visualization and Image Analysis.* 1993 ISBN 0-7923-1994-X
15. A.J. Saul (ed.): *Floods and Flood Management.* 1992   ISBN 0-7923-2078-6
16. D.E. Ashpis, T.B. Gatski and R. Hirsh (eds.): *Instabilities and Turbulence in Engineering Flows.* 1993   ISBN 0-7923-2161-8
17. R.S. Azad: *The Atmospheric Boundary Layer for Engineers.* 1993 ISBN 0-7923-2187-1
18. F.T.M. Nieuwstadt (ed.): *Advances in Turbulence IV.* 1993   ISBN 0-7923-2282-7
19. K.K. Prasad (ed.): *Further Developments in Turbulence Management.* 1993
                                                                      ISBN 0-7923-2291-6
20. Y.A. Tatarchenko: *Shaped Crystal Growth.* 1993   ISBN 0-7923-2419-6

Kluwer Academic Publishers – Dordrecht / Boston / London

Mechanics

FLUID MECHANICS AND ITS APPLICATIONS
Series Editor: R. Moreau

Kluwer Academic Publishers – Dordrecht / Boston / London

Mechanics

From 1990, books on the subject of *mechanics* will be published under two series:
FLUID MECHANICS AND ITS APPLICATIONS
Series Editor: R.J. Moreau
SOLID MECHANICS AND ITS APPLICATIONS
Series Editor: G.M.L. Gladwell

Prior to 1990, the books listed below were published in the respective series indicated below.

MECHANICS: DYNAMICAL SYSTEMS
Editors: L. Meirovitch and G.Æ. Oravas

MECHANICS OF STRUCTURAL SYSTEMS
Editors: J.S. Przemieniecki and G.Æ. Oravas

Mechanics

MECHANICS OF ELASTIC AND INELASTIC SOLIDS
Editors: S. Nemat-Nasser and G.Æ. Oravas

MECHANICS OF SURFACE STRUCTURES
Editors: W.A. Nash and G.Æ. Oravas

Mechanics

Mechanics

Mechanics

Kluwer Academic Publishers - Dordrecht / Boston / London